D1084322

The Mechanical Mind in History

The Mechanical Mind in History

edited by Philip Husbands, Owen Holland, and Michael Wheeler

A Bradford Book
The MIT Press
Cambridge, Massachusetts
London, England

For information about special quantity discounts, please email special_sales@mitpress.mit.edu

This book was set in Stone Serif and Stone Sans on 3B2 by Asco Typesetters, Hong Kong. Printed and bound in the United States of America.

Library of Congress Cataloging-in-Publication Data

The mechanical mind in history / edited by Philip Husbands, Owen Holland, and Michael Wheeler.
  p. cm.
Includes bibliographical references and index.
ISBN 978-0-262-08377-5 (hardcover : alk. paper) 1. Artificial intelligence—History.
2. Artificial intelligence—Philosophy. I. Husbands, Phil. II. Holland, Owen.
III. Wheeler, Michael, 1960–
Q335.M3956   2008
006.309—dc22   2007035271

10 9 8 7 6 5 4 3 2 1

Contents

Preface

Time present and time past
Are both perhaps present in time future
And time future contained in time past
—T. S. Eliot, "Four Quartets"

In the overlit arena of modern science, where progress must be relentless, leading to pressure to dismiss last year's ideas as flawed, it is all too easy to lose track of the currents of history. Unless we nurture them, the stories and memories underpinning our subjects slip through our fingers and are lost forever. The roots of our theories and methods are buried, resulting in unhelpful distortions, wrong turns, and dead ends.

The mechanization of mind—the quest to formalize and understand mechanisms underlying the generation of intelligent behavior in natural and artificial systems—has a longer and richer history than many assume. This book is intended to bring some of it back to life. Its scope is deliberately broad, ranging from cybernetic art to Descartes's often underestimated views on the mechanical mind. However, there is some emphasis on what we regard as hitherto underrepresented areas, such as the often overlooked British cybernetic and precybernetic thinkers, and cybernetic influences in politics.

Contributions come from a mix of artists, historians, philosophers, and scientists, all experts in their particular fields. The final section of this book is devoted to interviews with pioneers of machine intelligence, neuroscience, and related disciplines. All those interviewed emerged as major figures during the middle years of the twentieth century, probably the most explosively productive period yet in the search for the key to the mechanical mind. Their memories give fascinating insights into the origins of some of the most important work in the area, as well as adding color to many of the people and places whose names echo through the chapters of this book. The interviews are not presented as verbatim transcripts of the original conversations—such things rarely make for easy reading; instead, they are edited transcripts that have been produced in collaboration with the interviewees. Facts and figures have been thoroughly checked and endnotes have been added to make the pieces as useful as possible as historical testaments.

A substantial introductory chapter sets out the aims of this collection, putting the individual contributions into the wider context of the history of mind as machine while showing how they relate to each other and to the central themes of the book.

We'd like to acknowledge the help of a number of people who lent a hand at various stages of the production of this book. Thanks to Jordan Pollock, whose advocacy of this project when it was at the proposal stage helped to get it off the ground; to Lewis Husbands, for clerical assistance; and to Bob Prior at the MIT Press for his support and encouragement (not to mention patience) throughout. Of course this volume would be nothing without all the hard work and commitment of our contributors—many thanks to all of them.

The Mechanical Mind in History

1 Introduction: The Mechanical Mind

Philip Husbands, Michael Wheeler, and Owen Holland

Through myths, literature, and popular science, the idea of intelligent machines has become part of our public consciousness. But what of the actual science of machine intelligence? How did it start? What were the aims, influences, ideas, and arguments that swirled around the intellectual environment inhabited by the early pioneers? And how did the principles and debates that shaped that founding period persist and evolve in subsequent research? As soon as one delves into these questions, one finds oneself enmeshed in the often obscured roots of ideas currently central to artificial intelligence, artificial life, cognitive science, and neuroscience. Here one confronts a rich network of forgotten historical contributions and shifting cross-disciplinary interactions in which various new questions emerge, questions such as: What intellectual importance should we give to little-known corners of the history of the mechanical mind, such as cybernetic art, the frequently overlooked British cybernetic and pre-cybernetic thinkers, and cybernetic influences in politics? And, more generally, how is our understanding of the science of machine intelligence enriched once we come to appreciate the important reciprocal relationships such work has enjoyed, and continues to enjoy, with a broad range of disciplines? Moreover, issues that we sometimes address from within an essentially ahistorical frame of reference take on a new, historicized form. Thus one wonders not "What is the relationship between the science of intelligent machines and the sciences of neuroscience and biology?" but, rather, "In different phases of its history, how has the science of intelligent machines interacted with the sciences of neuroscience and biology?" Of course, once one has taken proper account of the past, the present inevitably looks different. So, having forged a path through the history of the mechanical mind, one is driven to ask: How far have we really come in the search for the mechanization of mind? What have we actually learned? And where should we go next?

The issues raised in the previous paragraph were what inspired, and subsequently drove the development of, the present volume. Unsurprisingly, given the nature and scope of these issues, the volume is essentially and massively cross-disciplinary in character, bringing together papers by scientists, artists, historians, and philosophers. Moreover, some of the best sources of engaging and illuminating insights into any field of study are the personal memories of those who shaped that field. It is here that the drama of science becomes manifest, along with previously undetected connections and influences. To capture these dimensions of our topic, we have chosen to supplement the usual diet of papers with a number of interviews with highly influential thinkers, most of whom were deeply involved in the birth of the field and have been major contributors to it ever since.

So is the mechanization of mind possible? In a sense this is our question, but that sense needs to be carefully specified. We are not focusing here on something analogous to the now-standard distinction between strong and weak artificial intelligence, so our question is not, "Is it possible to build a machine that *really* instantiates mental states and processes as opposed to 'merely' simulating them?" We are interested in the attempt to explain mind scientifically as a wholly mechanical process—mind as, or perhaps as generated by, an intelligent machine. Given that simulations are established weapons in the scientist's explanatory tool kit—in physics, biology, economics and elsewhere—we take this latter issue to be orthogonal to the "real mind versus simulated mind" debate. Second, we are not focusing, at least not principally, on the attempt to mechanize mind in the sense of building a complete functioning mechanical mind, presumably as an aspect of an integrated mobile robotic platform. The primary issue is not the mechanization of *a* mind. Rather, given science's strategy of abstracting to the key elements of a phenomenon in order to explain it, mechanical models of subsets of mind (for instance, mechanical models of individual psychological capacities such as reasoning or perception) are at the heart of the mechanization of mind, in the specific sense of the attempt to explain mind scientifically as a wholly mechanical process. These are the mechanisms that explain mind as machine.

So far, so good. But what sort of machine do we need for this task? This is where things get most interesting, and where, we believe, the present collection makes a genuine intellectual contribution that goes beyond that of historical scholarship. For what the various papers and memoirs here do is illustrate anew the rich kaleidoscope of diverse and interacting notions of mechanism that historically have figured in the shifting landscape of the mechanical mind. In the pages ahead we shall see mind mechanized as an analogue electrical system of wires, valves, and resistors; as an organized suite of chemical interactions; as a self-organizing electromechanical device, as a team of special-purpose mechanisms; as an automated general-purpose information processor; as an abstract deterministic process specified by state-transition rules (such as a Turing machine); as an integrated collection of symbol-manipulating mechanisms; and as an autonomous network of subsymbolic or nonsymbolic mechanisms. We shall see some of these notions deployed in combination as different aspects of the mental machine, and we shall see some of them pitted against each other in debates over the fundamental character of that machine. In addition, we shall see how some of these different notions have influenced and been influenced by the matrix of cross-disciplinary connections identified earlier.

In the remainder of this chapter, the contributions to this book are put into the wider context of the history of mind as machine. This is not intended to be a comprehensive history, or anything like it, but is merely a sketch that helps to show how the chapters relate to each other and to the central themes of the book. This volume offers a wide range of original material, with some emphasis on underexplored areas, such as British cybernetics, and the relationship between the mechanical mind and the arts. It is intended to complement more specific histories (such as those of the cybernetic period, including Heims 1991; Dupuy 2000) as well as more general surveys of the field (McCorduck 1979; Dyson 1997; Cordeschi 2002; and Boden's recent heroic two-volume history of cognitive science [2006]).

Looking at some discussions of the history of artificial intelligence, one would be forgiven for thinking that the mechanization of mind began, or at least took off properly, with the advent of the digital computer and the pioneering work of thinkers such as Allen Newell and Herbert Simon in the second half of the 1950s. But that is a very narrow and ultimately misleading view of history. There is a prehistory of what we now commonly think of as artificial intelligence in the cybernetic movements of the 1940s and 1950s—movements of which Newell and Simon themselves were deeply aware, incidentally. Moreover, there is a pre-prehistory of artificial intelligence that one might reasonably suggest began with (and this will come as a surprise to some readers) René Descartes (1596–1650). Descartes is often portrayed as the archenemy of mind as machine, but in fact he used clocks (relative rarities in his time) and the complex, animal-like automata that (among other things) moved, growled, spoke, and sang for the entertainment of the wealthy elite of seventeenth-century Europe as models for a range of what we would now think of as psychological capacities. Crucially, however, Descartes thought that some psychological capacities, in particular, reason, remained beyond the reach of a "mere" mechanism (Descartes 1637).

Soon afterward, however, the British philosopher Thomas Hobbes (1588–1679) went further than Descartes to become perhaps the first real champion of the mechanization of mind. He played a crucial role in establishing the intellectual climate that would result in attempts to understand the physical processes underlying intelligent behavior, and would later allow the emergence of the modern science of machine intelligence. Although today he is usually remembered as an ethical and political philosopher, Hobbes was one of the most important natural philosophers of his day. His materialist stance emphasized the machinelike qualities of nature, suggesting the possible creation of artificial animals: artificial intelligences and artificial life. In attacking Descartes's separation of mind and body, Hobbes argued that all of human intelligence is the product of physical mechanisms: that mind is a property of suitably organized matter.

The idea of mind as machine, then, stretches back over several centuries. As hinted at above, Descartes was not as hostile to the idea of mechanistic explanations of intelligent behavior as he is often portrayed today. Michael Wheeler explores this theme in some depth in his chapter, "God's Machines: Descartes on the Mechanization of Mind." He shows that Descartes's position was that machines (in the sense relevant to the mechanization of mind) are essentially collections of special-purpose mechanisms, and that no single machine could incorporate the enormous number of special-purpose mechanisms that would be required for it to reproduce human-like behaviour. By looking at contemporary work in biologically-inspired AI, Wheeler asks to what extent we can yet answer Descartes.

Although Hobbes's *Leviathan* included a combinatorial theory of thinking (Hobbes 1651), details of possible mechanisms for intelligence were very sketchy. It was some time before much progress was made in this direction: the eighteenth century saw the construction of many ingenious mechanical automata, including chess-playing Turks and flatulent ducks, but it wasn't until the nineteenth century that major breakthroughs occurred, including the design of Charles Babbage's programmable Analytical Engine.

The son of a London banker, Babbage (1791–1871) was a brilliant mathematician and engineer who held the same chair at Cambridge University that Newton had occupied. Inspired by Leibniz, whose work was in turn influenced by Hobbes, in 1821 he designed his mechanical Difference Engine for calculating accurate mathematical tables—something of enormous practical importance at the time. However, Babbage's interest in calculating machines ran deeper than the production of mathematical tables. He envisioned such engines as powerful tools for science, hoping that their whirring cogs would shed new light on the workings of nature. In this spirit, in 1834 he began work on his revolutionary Analytical Engine, a general, programmable machine. The engine was to read instructions from sets of punched cards, adapted from those used in Jacquard looms (invented in 1801 to automate textile weaving), and to manipulate partial results in its own internal memory. Rather than being designed to perform just one set of calculations, the machine was intended to be a completely general computing engine; in theory, it could be programmed to perform any calculation. In chapter 2, "Charles Babbage and the Emergence of Automated Reason," Seth Bullock explores the context in which Babbage's work emerged, highlighting the debates on the possibility of automated reason, which covered economic, social, and moral ground. He also shows how Babbage was able to demonstrate the wider applicability of his machines by developing the first computational model intended to help further study of a scientific problem (in this case one in geology).

In 1843 Augusta Ada, Countess of Lovelace (1815–1852) translated into English a paper on the Analytical Engine written by the mathematician Luigi Menabrea (Lovelace 1843). Ada was the daughter of Lord Byron, the great poet. Her parents separated almost immediately after her birth, and Lady Byron raised Ada to appreciate mathematics and science, in part because of her own interest in these areas, but also because she hoped it would drive out any Byronic madness her daughter might have inherited. In collaboration with Babbage, Ada added extensive notes to the manuscript, which make it clear that they both understood the importance of the general nature of the Engine. Ada wrote of its potential to act as a "thinking, reasoning machine." The notes include a detailed description of a method for using the Engine to calculate Bernoulli numbers. This is widely regarded as the first computer program, although there is some controversy over whether the primary author was Lovelace or Babbage. Ada was perhaps the first person to see the possibility of using computational engines in the arts, writing of the Analytic Engine's potential to compose music and generate graphics.

The Analytical Engine was never completed; its construction became mired in manufacturing and bureaucratic difficulties that resulted in the British government's withdrawing funding. In 1991 a team at the Science Museum in London constructed the Difference Engine Number 2 according to Babbage's detailed designs. It worked perfectly. In most respects Babbage's remarkable vision of a universal machine anticipated the modern digital computer age by more than a century.

While Babbage was struggling to construct his engines, the English mathematician George Boole (1815–1864), the self-educated son of a Lincoln cobbler, was building a formal system of logic which went on to serve as a cornerstone of all modern digital technology, but which was also intended to capture the structure of reasoning and thinking (Boole 1854). In Boolean algebra, logical relationships between entities are formalized and manipulated. Variables representing the entities are restricted to two possible values, true or false—1 or 0. By uniting logic with mathematics, in particular binary arithmetic, Boole laid the foundations for the flow of bits and bytes that power our digital age. He died after developing a fever following a soaking in a rainstorm. His demise was unwittingly aided by his wife, who, believing that a cure should mirror the cause, threw buckets of cold water over him as he lay shivering in bed.

Where Babbage and his predecessors developed schemes for describing and automating reasoning at a fairly high, abstract level, one of the first people to try to ground intelligence in brain function was Alfred Smee (1818–1877), a brilliant scientist and engineer who held the somewhat bizarre position of surgeon to the Bank of England. (His father was secretary of the bank and the position was specially created in the hope of tapping into Alfred's inventive flair. It did: he developed electrotype plate printing of banknotes, which greatly reduced problems with forged notes.) Smee pioneered theories of the operation of the nervous system, speculating on how its electrical networks were organized. He also formulated ideas about artificial sense organs and a type of very early artificial neural network.

During the early decades of the twentieth century, advances in electrical engineering and early electronics fed into formal theories of the operation of neurons, as well as greatly improving experimental techniques in the developing field of neurophysiology. This allowed great pioneers such as Lord Adrian (1889–1977) and Charles Sherrington (1857–1952) to lay the foundations for the modern view of the nervous system by greatly advancing knowledge of the electrical properties of nerve cells (Adrian 1928; Sherrington 1940). Communications theory was also emerging in engineering circles; as we shall see, future developments in this area would later have a significant impact on approaches to the mechanization of mind.

At about the same time that Adrian and Sherrington were making great strides in understanding neurons, D'Arcy Thompson was trying to fathom how biological structures develop. In 1917 he published his celebrated book

On Growth and Form (Thompson 1917). As Margaret A. Boden argues in chapter 3, "D'Arcy Thompson: A Grandfather of A-Life," this pioneering work of mathematical biology, in which Thompson sought to develop a quantitative approach to biological forms and processes of growth, not only helped to pave the way for modern theoretical biology but also prefigured the contemporary field of artificial life (or A-Life), the study of life in general, abstract terms. As well as influencing Alan Turing's work on morphogenesis, of which more later, it emphasized the embodied nature of natural intelligence, a theme that has become increasingly central to contemporary cognitive science (Pfeifer and Scheier 1999; Wheeler 2005).

The notion of embodied mechanical intelligence was, quite literally, thrust center stage in the years between the world wars, when Karel Čapek's play *R.U.R.* introduced the world to robots, in the process forging the associated myths and images that now permeate our culture. In "The Robot Story: Why Robots Were Born and How They Grew Up," Jana Horáková and Jozef Kelemen give a detailed account of the origins of Čapek's work, tracing its roots to the dreams and folk tales of old Europe. They show how it was a product of its troubled times and how the idea of robots was interpreted in different ways in Europe and America as it seeped into the collective unconscious. The new dreams and images thus created undoubtedly inspired future generations of machine intelligence researchers.

Smee's early desire to unite the workings of the mind with the underlying neural mechanisms, and to develop machines around the principles uncovered, was a theme that reemerged very strongly in the mid-twentieth century. It was in this period that machine intelligence really took off. At the same time advances in understanding the nervous system continued apace. Kenneth Craik (1914–1945) was an influential, if now often forgotten, figure in the flurry of progress that occurred. Craik was a brilliant Scottish psychologist, based at Cambridge University, who pioneered the study of human-machine interfaces, and was a founder of cognitive psychology and also of cybernetic thinking. He died tragically young, in a road accident on the last day of the war in Europe, his potential surely not fully realized. His classic 1943 book, *The Nature of Explanation* (Craik 1943), introduced the radical and influential thesis that the brain is a kind of machine that constructs small-scale models of reality that allow anticipation of external events. Disgruntled with mainstream philosophy of mind and much of psychology, and inspired by the strides Adrian and his colleagues were making, he maintained that explanations of intelligence should incorporate an understanding of the underlying neural processes. Craik's influence on the development of cybernetics, on both sides of the Atlantic, is discussed in Philip Husbands and Owen Holland's chapter on the Ratio Club.

At the same time as Craik was starting to develop his ideas, in another part of Cambridge the mathematician Alan Turing (1912–1954) was about to publish a startling paper on one of David Hilbert's open problems in mathematics, the *Entscheidungsproblem* ("decision problem"), namely: Is it possible to define a formal procedure that could be used to decide whether any given mathematical assertion was provable. Turing's highly original approach to the problem was to define a kind of simple abstract machine (Turing 1936). By using such a machine as a very general way of constructing a formal procedure in mathematics, he was able to show that it followed that the answer to the problem was no. The concept of the Turing machine, as it became known, now serves as the foundation of modern theories of computation and computability. In the paper Turing explicitly drew a parallel between the operation of such a machine and human thought processes. Turing also introduced a more general concept that was to have an immense practical impact: the Universal Turing Machine. This machine could interpret and then execute the set of instructions defining *any* given standard Turing machine (each of which corresponded to a particular formal procedure or algorithm). Thus, the Universal Turing Machine embodies the central principle of the computer as we know it today: a single machine that can perform any well-defined task as long as it is given the appropriate set of instructions, or program. A hundred years after Babbage, and by a very different route, Turing envisaged a completely general supermachine. This time the vision was to come to fruition.

Donald Michie's chapter, "Alan Turing's Mind Machines," draws on his experience as one of Turing's close colleagues in wartime code-cracking work at Bletchley Park, the headquarters of Britain's cryptography efforts, to give insights into the development of Turing's ideas and the early computers that flowed from them. He argues that Turing's unfashionable and often resisted obsession with tackling combinatorial problems with brute-force computation, partly born of his wartime experience with cryptanalytical problems, helped to shape the way computers came to be used. He shows that computer analyses of combinatorial domains such as chess, inspired by Turing's work, are still of great importance today in yielding new approaches to the difficult problem of transparency in complex computer-based decision systems.

In a complementary chapter, Andrew Hodges asks "What did Alan Turing Mean by 'Machine'?" He focuses on the title of Turing's unpublished 1948 report "Intelligent Machinery" (Turing 1948) to explore what Turing intended by an "intelligent machine." Turing saw central roles for the new digital computers in the development of machine intelligence and in the exploration of brain mechanisms through simulations, both of which came to pass. Hodges argues that although the central thrust of Turing's thought was that the action of brains, like that of any machine, could be captured by classical computation, he was aware that there were potential problems in connecting computability with physical reality.

The Second World War was to prove a major catalyst for further advances in mechanistic conceptions of intelligence as well as in the development of practical computers. In Britain there was little explicitly biological research carried out as part of the war effort, so most biologists were drafted into the main thrust of scientific research on communications and radar. As explained in chapter 6, this was to have the extremely important effect of exposing these biologists to some electronics and communication theory as well as to engineers and mathematicians who were experts in these areas. This mixing of people and disciplines led to an important two-way flow of ideas that was to prove highly significant in advancing the formal understanding of the nervous system as well as developments in machine intelligence. There was much discussion of electronic brains, and the intense interest in the subject carried over into peacetime.

In the early 1940s a circle of scientists intent on understanding general principles underlying behavior in animals and machines began to gather around the MIT mathematician Norbert Wiener (1894–1964). Inspired by Wiener's classified work on automatic gun aiming, Arturo Rosenblueth, Wiener, and Julian Bigelow (1943) published a paper on the role of feedback mechanisms in controlling behavior. This work triggered great interest among other American scientists in new approaches to the mechanization of mind. Influenced by Wiener's ideas, but also aware of Craik's and Turing's work, the group was initially composed of a small number of mathematicians and engineers (Wiener, John von Neumann, Bigelow, Claude Shannon, Walter Pitts) and brain scientists (Rafael Lorente de Nó, Rosenblueth, Warren McCulloch). A series of meetings sponsored by the Macy Foundation saw the group expand to incorporate the social sciences. Wiener named the enterprise cybernetics; the publication of his book *Cybernetics, or Control and Communication in the Animal and the Machine* (Wiener 1948), along with the proceedings of the Macy meetings (von Foerster 1950–55), did much to spread its influence and popularity. As well as Wiener's book, notable developments that came under the cybernetic umbrella included McCulloch and Pitts's seminal work on mathematical descriptions of neuronal networks (McCulloch and Pitts 1943; Pitts and McCulloch 1947), providing the first examples of artificial neural networks, and Shannon's information theory (Shannon and Weaver 1949). McCulloch and Pitts modeled neuronal networks in terms of connected logic units and showed that their nets were equivalent to Universal Turing Machines, implicitly suggesting a close link between the nervous system and the digital computer. Information theory, which provided a mathematical framework for designing and understanding communication channels, is another foundation stone of the digital age. It also provided new ideas about the operating principles of biological senses and what kinds of processing might be going on in the nervous system.

In Britain, where war work had also familiarized many scientists with feedback mechanisms and early information theory, a parallel group formed, the Ratio Club. The club was founded and organized by John Bates, a neurologist at the National Hospital for Nervous Diseases in London. The other twenty carefully selected members were a mixed group of mainly young neurophysiologists, engineers, and mathematicians, with the center of gravity firmly toward the brain sciences. This illustrious group included W. Ross Ashby, Horace Barlow, Thomas Gold, Jack Good, Donald MacKay, Alan Turing, W. Grey Walter, and Albert Uttley. Most members had a strong interest in developing "brainlike" devices, either as a way of formalizing and exploring theories about biological brains, or as a pioneering effort in creating machine intelligence, or both. Most meetings of the club occurred between September 1949 and July 1953. During this extremely productive period various members made highly significant contributions to cybernetics and related fields. Husbands and Holland's chapter, "The Ratio Club: A Hub of British Cybernetics," for the first time tells the story of this remarkable group. Horace Barlow's very significant contributions to neuroscience, including his introduction into it of important information-theoretic concepts (Barlow 1959), were heavily influenced by the club. Members pioneered a wide range of techniques and ideas that are proving to be ever more influential. For instance, Grey Walter (1910–1977), a leader in electroencephalographic (EEG) research, built the first autonomous mobile robots, controlled by simple electronic nervous systems (Walter 1953). W. Ross Ashby (1903–1972), who had actually published on the role of feedback in adaptive systems several years before Rosenblueth, Wiener, and Bigelow (Ashby 1940), further developed such notions, culminating in their demonstration in his adaptive Homeostat machine (Ashby 1952); and Turing, whose seminal paper on machine intelligence (Turing 1950) was published during the club's lifetime, pioneered the use of computational models in biology in his groundbreaking work on morphogenesis, which showed how regular patterns could be formed by appropriately parameterized reaction-diffusion systems—work that called up the spirit of D'Arcy Thompson (Turing 1952).

Ashby, who is now widely acknowledged as the most important theorist of cybernetics after Wiener—partly through the influence of his books (Ashby 1952, 1956)—had a singular vision that he had developed in isolation for many years before becoming part of the scientific establishment in the late 1940s. His unique philosophy, which stressed the dynamic nature of brain mechanisms and the interactions between organism and environment, is explored by Peter Asaro in chapter 7, "From Mechanisms of Adaptation to Intelligence Amplifiers: The Philosophy of W. Ross Ashby." Asaro sheds light on what kind of machine Ashby thought the brain was and how its principles might be captured in an artificial device.

Parallel developments in the United States also focused on biologically inspired brainlike devices, including work by researchers such as Frank Rosenblatt and Marvin Minsky on the construction of electronic artificial neural networks that were able to perform simple learning tasks. Oliver Selfridge, a grandson of the founder of London's famous Selfridge's department store, had left Britain at the age of fourteen to study with Wiener at MIT. In the mid-1950s he developed his breakthrough Pandemonium system, which learned to recognize visual patterns, including alphanumeric characters (Selfridge 1959). The system employed a layered network of processing units that operated in parallel and made use of explicit feature detectors that only responded to certain visual stimuli—a more general mechanism than the specific detectors that had recently been shown to exist in biological vision systems by Horace Barlow in the form of "fly detectors" in the frog's retina (Barlow 1953). Neural mechanisms that are selectively responsive to certain general features (for instance, edge and convexity detectors) were subsequently shown to exist in natural vision systems by Jerry Lettvin, Humberto Maturana, Warren McCulloch, and Walter Pitts (1959).

Most prominent among the second wave of British cyberneticists were Stafford Beer (1926–2002) and Gordon Pask (1928–1996), who were both particularly influenced by Ashby. Beer took cybernetic ideas into the world of industrial management and became a highly successful consultant to corporations and governments alike. In "Santiago Dreaming," Andy Beckett tells the story of how in the early 1970s the Allende administration in Chile engaged Beer to design and develop a revolutionary electronic communication system in which voters, workplaces, and the government were to be linked together by a kind of "socialist internet." Pask was an eccentric figure who strode around in an Edwardian cape while pursuing radical ideas far from the mainstream. In "Gordon Pask and His Maverick Machines," Jon Bird and Ezequiel Di Paolo highlight Pask's willingness to explore novel forms of machine, often in collaboration with Beer, in his quest to better understand principles of self-organization that would illuminate the mechanisms of intelligence. These included a "growing" electrochemical device intended to act as an artificial ear. They show how Pask's work is relevant to current research in AI and A-life, and how key questions he posed have not yet been answered.

Pask, like other machine intelligence researchers before and since, was interested in applying his ideas in the visual arts. As Paul Brown shows in chapter 11, "The Mechanization of Art," Wiener's and Ashby's ideas were quickly appreciated by a number of artists, such as Nicolas Schöffer, who in the mid-1950s pioneered a kind of autonomous kinetic art, cybernetic sculptures. Brown traces the cultural, as well as scientific, antecedents of this work in an account of how the mechanization of art developed over the centuries. He focuses on its growth during part of the second half of the twentieth century, a period that saw the influential 1968 Institute of Contemporary Arts (London) exhibition *Cybernetic Serendipity*, which featured Pask's installation *Colloquy of Mobiles*. He reminds us that a number of artists working in this field, such as Edward Ihnatowicz (1926–1988), pioneered approaches to autonomous systems, prefiguring today's growing dialogue between artists and scientists in this area.

In 1956 two young American academics, John McCarthy and Marvin Minsky, organized a long workshop at Dartmouth College to develop new directions in what they termed *artificial intelligence*. McCarthy in particular proposed using newly available digital computers to explore Craik's conception of intelligent machines as using internal models of external reality, emphasizing the power of symbolic manipulation of such models. At the workshop, Allen Newell (1927–1992) and Herbert Simon (1916–2001), influenced by aspects of Selfridge's work, demonstrated a symbolic reasoning program that was able to solve problems in mathematics. This was the beginning of the rise of logic-based, symbol-manipulating computer programs in the study of machine intelligence. This more abstract, software-bound paradigm came to dominate the field and pulled it away from its biologically inspired origins. For a while the term "artificial intelligence," or AI, was exclusively associated with this style of work. This paradigm, which to some extent harked back to the older ideas of Boole and Leibniz, also served as a new kind of abstract model of human reasoning, becoming very influential in psychology and, later, in cognitive science.

Roberto Cordeschi illustrates some of the tension between cybernetic and early AI theories in his chapter, "Steps Toward the Synthetic Method: Symbolic Information Processing and Self-Organizing Systems in Early Artificial Intelligence Modeling." He compares two theories of human cognitive processes, one by the Ratio Club member and cyberneticist Donald Mackay (1922–1987), the other by Newell and Simon. MacKay's model is constructed around his notion of self-organizing systems, whereas Newell and Simon's is based on high-level symbol manipulation. Cordeschi explores epistemological issues raised by each.

The new AI movement in the United States gained significant financial and industrial support in the 1960s, as it began to dominate the arena while the influence and impetus of cybernetics fell away. However, work in neural nets, adaptive and self-organizing systems, and other outgrowths of cybernetics did not disappear altogether. As the weaknesses of the mainstream AI approaches became apparent and the adaptive-systems methods improved, with a number of crucial advances in artificial neural networks and machine learning, the tide turned (see Anderson and Rosenfeld 1998 for an excellent oral history of the rise and fall and rise of artificial neural networks). Since the late 1980s, biologically inspired and subsymbolic approaches have swept back to take center stage. These include an emphasis on whole embodied artificial "creatures" that must adapt to real unforgiving environments. Their brains run on onboard digital computers, as Turing foresaw more than fifty years ago. Work in machine intelligence has again become much more closely aligned with research in the biological sciences. Many of the ideas and methods developed by the great pioneers of the mid-twentieth century have once more come to the fore— the mechanization-of-mind project, although still very far from completion, appears to be back on track. Which is not to say that there is agreement on the best way forward.

One of the most prominent critics of classical AI, or good old-fashioned AI—GOFAI—was Hubert Dreyfus. In "Why Heideggerian AI Failed and How Fixing It Would Require Making It More Heideggerian," he turns the spotlight on one of GOFAI's replacements. Informed by personal experiences and encounters at MIT (the high temple of AI, new and old), Dreyfus tells of how he watched the symbol-processing approach degenerate, and of how it was replaced by what he terms "Heideggerian AI," a movement that began with the work of Rodney Brooks and colleagues (Brooks 1999).

This work puts central emphasis on acting in the world and thus concentrates on the development of mobile autonomous robots. Dreyfus explains why, in his view, this style of AI has also failed and suggests how it should be fixed, calling on Walter Freeman's neurodynamics and stressing the importance of the specifics of how particular bodies interact with their environments.

The final section of the book offers a series of interviews, conducted by one of the editors, with major figures whose careers were firing into life in the middle of the last century, an astonishingly fertile period in the search for the secrets of mechanical intelligence. We are given vivid accounts of how these great scientists' ideas developed and of who influenced them. Certain themes and characters echo through these interviews, giving fresh perspective on material earlier in the book.

John Maynard Smith, one of the great evolutionary biologists of the twentieth century, who originally trained as an engineer, gives us an insight into the spirit of science immediately after the Second World War as well as into the early influence of cybernetics on developmental and evolutionary biology. John Holland, the originator of genetic algorithms, recounts how his theories of adaptive systems were in turn influenced by biology, then reflects on recent developments and considers why, in the late 1980s, there was a great resurgence of interest in complex adaptive systems. Oliver Selfridge, one of the pioneers of machine learning, tells us what it was like to be at the heart of the MIT cybernetics enterprise in the 1940s and 1950s, and how he helped Minsky and McCarthy to establish the field of AI. Regretting GOFAI's lack of interest in learning and adaptation during its heyday, he gives his views on where the field should go now. The great neuroscientist Horace Barlow paints a picture of life in Lord Adrian's department at Cambridge University during the late 1940s and tells how the Ratio Club profoundly influenced his subsequent career. Toward the end of his interview he makes the highly pertinent point that as neuroscience has developed over the past fifty years, it has fragmented into specialized subareas. So although knowledge has increased to an enormous extent, there is now a greater need than ever for an overarching theory. The theorists, experimentalists, and modelers must all combine in a coherent way if we are ever to understand the nervous system in sufficient detail to formulate its principles. Jack Cowan, a pioneer of neural networks and computational neuroscience, gives a unique perspective on activity in machine intelligence in the UK and the United States in the late 1950s and early 1960s. He recounts how his ideas developed under the influence of some of the great pioneers of cybernetics, and how those ideas flourished throughout his subsequent career.

From positions of authority, with access to extraordinarily wide perspectives, these pioneers look back at what has been achieved, and comment on how far we still have to go, in the mechanization of mind. All are optimistic for the long term, but stress the enormous complexity of the task. In short, although much has been achieved and great progress has been made in understanding the details of specific mechanisms and competences, in terms of the overall picture, we have not yet come very far at all. This message serves as a useful antidote to the wild ravings of those who claim that we will soon be downloading our minds into silicon (although it is not clear whether this will be before or after our doors are kicked in by the superintelligent robots that these same people claim will take over the world and enslave us).

References

Adrian, Edgar Douglas (Lord Adrian). 1928. *The Basis of Sensation*. London: Christophers.

Anderson, J. A., and E. Rosenfeld. 1998. *Talking Nets: An Oral History of Neural Networks*. Cambridge, Mass.: MIT Press.

Ashby, W. Ross. 1940. "Adaptiveness and Equilibrium." *Journal of Mental Science* 86: 478.

———. 1952. *Design for a Brain*. London: Chapman & Hall.

———. 1956. *An Introduction to Cybernetics*. London: Chapman & Hall.

Barlow, Horace B. 1953. "Summation and Inhibition in the Frog's Retina." *Journal of Physiology* 119: 69–88.

———. 1959. "Sensory Mechanism, the Reduction of Redundancy, and Intelligence." In *Mechanisation of Thought Processes: Proceedings of a Symposium held at the National Physical Laboratory on 24–27 November 1958*, edited by D. Blake and Albert Uttley. London: Her Majesty's Stationery Office.

Boden, M. A. 2006. *Mind as Machine: A History of Cognitive Science*. Oxford: Oxford University Press.

Boole, G. 1854. *An Investigation of the Laws of Thought, on Which Are Founded the Mathematical Theories of Logic and Probabilities*. London: Macmillan.

Brooks, R. A. 1999. *Cambrian Intelligence: The Early History of the New AI*. Cambridge, Mass.: MIT Press.

Cordeschi, Roberto. 2002. *The Discovery of the Artificial: Behavior, Mind and Machines Before and Beyond Cybernetics*. Dordrecht: Kluwer Academic Publishers.

Craik, Kenneth J. W. 1943. *The Nature of Explanation*. Cambridge: Cambridge University Press.

Descartes, René. 1637/1985. "Discourse on the Method of Rightly Conducting One's Reason and Seeking the Truth in the Sciences." In *The Philosophical Writings of Descartes*, edited by J. Cottingham, R. Stoothoff, and D. Murdoch. Volume 1. Cambridge: Cambridge University Press.

Dupuy, J.-P. 2000. *The Mechanization of the Mind*. Translated from French by M. B. DeBevoise. Princeton: Princeton University Press.

Dyson, George. 1997. *Darwin Among the Machines*. Reading, Mass.: Addison-Wesley.

Foerster, H. von, ed. 1950–55. *Cybernetics, Circular Causal and Feedback Mechanisms in Biological and Social Systems: Published Transactions of the 6th, 7th, 8th, 9th and 10th Conferences*. 5 volumes. New York: Josiah Macy Jr. Foundation.

Heims, S. 1991. *Constructing a Social Science for Postwar America: The Cybernetics Group, 1946–1953*. Cambridge, Mass.: MIT Press.

Hobbes, Thomas. 1651. *Leviathan*, London: Andrew Crooke.

Lettvin, Jerry Y., H. R. Maturana, Warren S. McCulloch, and Walter H. Pitts. 1959. "What the Frog's Eye Tells the Frog's Brain." *Proceedings of the IRE* 47: 1940–59.

Lovelace, Ada. 1843. "Notes on L. Menabrea's 'Sketch of the Analytical Engine Invented by Charles Babbage, Esq.'" In *Taylor's Scientific Memoirs*. Volume 3. London: J. E. & R. Taylor.

McCorduck, P. 1979. *Machines Who Think: A Personal Inquiry into the History and Prospect of Artificial Intelligence*. San Francisco: Freeman.

McCulloch, Warren S., and Walter Pitts. 1943. "A Logical Calculus of the Ideas Immanent in Nervous Activity." *Bulletin of Mathematical Biophysics* 5: 115–33.

Pfeifer, R., and C. Scheier. 1999. *Understanding Intelligence*. Cambridge, Mass.: MIT Press.

Pitts, Walter, and Warren S. McCulloch. 1947. "How We Know Universals: The perception of Auditory and Visual Forms." *Bulletin of Mathematical Biophysics* 9: 127–47.

Rosenblueth, A., Norbert Wiener, and J. Bigelow. 1943. "Behavior, Purpose and Teleology." *Philosophy of Science* 10, no. 1: 18–24.

Selfridge, Oliver G. 1959. "Pandemonium: A Paradigm for Learning." In *The Mechanisation of Thought Processes*, edited by D. Blake and A. Uttley. Volume 10, *National Physical Laboratory Symposia*. London: Her Majesty's Stationery Office.

Shannon, Claude, and W. Weaver. 1949. *The Mathematical Theory of Communication.* Chicago: University of Illinois Press.

Sherrington, Charles. 1940. *Man on His Nature.* Oxford: Oxford University Press.

Thompson, D. W. 1917. *On Growth and Form.* Cambridge: Cambridge University Press.

Turing, Alan M. 1936. "On Computable Numbers, with an Application to the Entscheidungsproblem." *Proceedings of the London Mathematical Society* 42, no. 2: 230–65.

———. 1948/2004. "Intelligent Machinery." Report for the National Physical Laboratory. In *The Essential Turing,* edited by A. M. Turing and B. J. Copeland. Oxford: Oxford University Press. Available at www.turingarchive.org.

———. 1950. "Computing Machinery and Intelligence." *Mind* 59: 433–60.

———. 1952. "The Chemical Basis of Morphogenesis." *Philosophical Transactions of the Royal Society of London* (series B) 237: 37–72.

Walter, W. Grey. 1953. *The Living Brain.* London: Duckworth.

Wheeler, Michael. 2005. *Reconstructing the Cognitive World.* Cambridge, Mass.: MIT Press.

Wiener, Norbert. 1948. *Cybernetics, or Control and Communication in the Animal and the Machine.* Cambridge, Mass.: MIT Press.

2   Charles Babbage and the Emergence of Automated Reason

Seth Bullock

Charles Babbage (1791–1871) (figure 2.1) is known for his invention of the first automatic computing machinery, the Difference Engine and later the Analytical Engine, thereby prompting some of the first discussions of machine intelligence (Hyman 1982). Babbage's efforts were driven by the need to efficiently generate tables of logarithms—the very word "computer" having originally referred to people employed to calculate the values for such tables laboriously by hand. Recently, however, historians have started to describe the wider historical context within which Babbage was operating, revealing how he, his contemporaries, and their students were influential in altering our conception of the workforce, the workplace, and the economics of industrial production in a Britain increasingly concerned with the automation of labor (Schaffer 1994).

While it was clear that all manner of unskilled manual labour could be achieved by cleverly designed mechanical devices, the potential for the same kind of machinery to replicate mental labor was far more controversial. Were reasoning machines possible? Would they be useful? Even if they were, was their use perhaps less than moral? Babbage's contribution to this debate was typically robust. In demonstrating how computing machinery could take part in (and thereby partially automate) academic debate, he challenged the limits of what could be achieved with mere automata, and stimulated the next generation of "machine analysts" to conceive and design devices capable of moving beyond mere mechanical calculation in an attempt to achieve full-fledged automated reason.

In this chapter, some of the historical research that has focused on Babbage's early machine intelligence and its ramifications will be brought together and summarized. First, Babbage's use of computing within academic research will be presented. The implications of this activity on the wider question of machine intelligence will then be discussed, and the relationship between automation and intelligibility will be explored.

Figure 2.1
Charles Babbage in 1847. Source: http://www.kevryr.net/pioneers/gallery/ns_
babbage2.htm (in public domain).

Intermittently throughout these considerations, connections between the
concerns of Babbage and his contemporaries and those of modern artificial
intelligence (AI) will be noted. However, examining historical activity
through modern lenses risks doing violence to the attitudes and significances of the agents involved and the complex causal relationships between
them and their works. In order to guard against the overinterpretation of
what is presented here as a "history" of machine intelligence, the paper
concludes with some caveats and cautions.

The Ninth Bridgewater Treatise

In 1837, twenty-two years before the publication of Darwin's *On the Origin
of Species* and over a century before the advent of the first modern computer, Babbage published a piece of speculative work as an uninvited Ninth

Bridgewater Treatise (Babbage 1837; see also Babbage 1864, chapter 29, "Miracles," for a rather whimsical account of the model's development). The previous eight works in the series had been sponsored by the will of Francis Henry Egerton, the Earl of Bridgewater and a member of the English clergy. The will's instructions were to make money available to commission and publish an encyclopedia of natural theology describing "the Power, Wisdom, and Goodness of God, as manifested in the Creation" (Brock 1966; Robson 1990; Topham 1992).

In attempting such a description, natural theologists tended to draw attention to states of affairs that were highly unlikely to have come about by chance and could therefore be argued to be the work of a divine hand. For instance, the length of the terrestrial day and seasons seem miraculously suited to the needs and habits of plants, man, and other animals. Natural theologists also sought to reconcile scientific findings with a literal reading of the Old Testament, disputing evidence that suggested an alarmingly ancient earth, or accounting for the existence of dinosaur bones, or promoting evidence for the occurrence of the great flood. However, as Simon Schaffer (1994) points out, natural theology was also "the indispensable medium through which early Victorian savants broadcast their messages" (p. 224).

Babbage's contribution to the Bridgewater series was prompted by what he took to be a personal slight that appeared in the first published and perhaps most popular Bridgewater Treatise. In it, the author, Reverend William Whewell, denied "the mechanical philosophers and mathematicians of recent times any authority with regard to their views of the administration of the universe" (Whewell 1834, p. 334, cited in Schaffer 1994, p. 225). In reply, Babbage demonstrated a role for computing machinery in the attempt to understand the universe and our relationship to it, presenting the first published example of a *simulation model.*

In 1837, Babbage was one of perhaps a handful of scientists capable of carrying out research involving computational modeling. In bringing his computational resources to bear on a live scientific and theological question, he not only rebutted Whewell and advanced claims for his machines as academic as well as industrial tools, but also sparked interest in the extent to which more sophisticated machines might be further involved in full-blown reasoning and argument.

The question that Babbage's model addressed was situated within what was then a controversial debate between what Whewell had dubbed catastrophists and uniformitarians. Prima facie, this dispute was internal to geology, since it concerned the geological record's potential to show evidence of divine intervention. According to the best field geologists of the day, geological change "seemed to have taken place in giant steps: one geological environment contained a fossil world adapted to it, yet the next stratum showed a different fossil world, adapted to its own environment but not obviously derivable from the previous fossil world" (Cannon 1960, p. 7). Catastrophists argued for an interventionist interpretation of this evidence, taking discontinuities in the record to be indicators of the occurrence of miracles—violations of laws of nature. In contrast, uniformitarians argued that allowing a role for sporadic divine miracles interrupting the action of natural processes was to cast various sorts of aspersions on the Deity, suggesting that His original work was less than perfect, and that He was constantly required to tinker with his Creation in a manner that seemed less than glorious. Moreover, they insisted that a precondition of scientific inquiry was the assumption that the entire geological record must be assumed to be the result of unchanging processes. Miracles would render competing explanations of nature equally valid. No theory could be claimed to be more parsimonious or coherent than a competing theory that invoked necessarily inexplicable exogenous influences. As such, the debate was central to understanding whether and how science and religion might legitimately coexist.

Walter Cannon (1960) argues that it is important to recognize that this debate was not a simple confrontation between secular scientists and religious reactionaries that was ultimately "won" by the uniformitarians. Rather, it was an arena within which genuine scientific argument and progress took place. For example, in identifying and articulating the degree to which the natural and physical world fitted each other, both currently and historically, and the startling improbability that brute processes of contingent chance could have brought this about, authors such as Whewell laid a foundation upon which Darwin's evolutionary theory sat naturally.

Babbage's response to the catastrophist position that apparent discontinuities were evidence of divine intervention was to construct what can now be recognized as a simple simulation model (see figure 2.2). He proposed that his suitably programmed Difference Engine could be made to output a series of numbers according to some law (for example, the integers, in order, from 0 onward), but then at some predefined point (say 100,000) begin to output a series of numbers according to some different law such as the integers, in order, from 200,000 onward. Although the output of such a Difference Engine (an analogue of the geological record) would feature a discontinuity (in our example the jump from 100,000 to 200,000), the underlying process responsible for this output would have

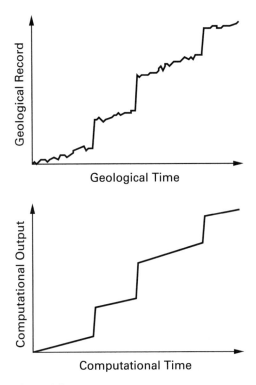

Figure 2.2
Babbage's (1836) evolutionary simulation model represented the empirically observed history of geological change as evidenced by the geological record (upper panel) as the output of a computing machine following a program (lower panel). A suitably programmed computing machine could generate sequences of output that exhibited surprising discontinuities without requiring external influence. Hence discontinuities in the actual geological record did not require "catastrophic" divine intervention, but could be the result of "gradualist" processes.

remained constant—the general law, or program, that the machine was obeying would not have changed. The discontinuity would have been the result of the naturally unfolding mechanical and computational process. No external tinkering analogous to the intervention of a providential deity would have taken place.

Babbage not only described such a program in print but demonstrated a working portion of his Difference Engine carrying out the calculations described (see figure 2.3). At his Marylebone residence, he surprised a stream of guests drawn from society and academia with machine behavior that suggested a new way of thinking about both automata and miracles.

Figure 2.3
Difference Engine. Source: http://www.kevryr.net/pioneers/gallery/ns_babbage5.htm
(in public domain).

Doran Swade (1996) describes how Darwin, recently returned from his voyages on the *Beagle*, was urged by Charles Lyell, the leading geologist, to attend one of Babbage's "soirées where he would meet fashionable intelligentsia and, moreover, 'pretty women'" (p. 44). Schaffer (1994) casts Babbage's surprising machine as providing Darwin with "an analogue for the origin of species by natural law without divine intervention" (pp. 225–26).

In trying to show that discontinuities were not necessarily the result of meddling, but could be the natural result of unchanging processes, Babbage cultivated the image of God as a programmer, engineer, or industrialist, capable of setting a process in motion that would accomplish His intentions without His intervening repeatedly. In Victorian Britain, the notion of God as draughtsman of an "automatic" universe, one that would run unassisted, without individual acts of creation, destruction, and so forth, proved attractive. This conception was subsequently reiterated by several other natural philosophers, including Darwin, Lyell, and Robert Chambers, who argued that it implied "a grander view of the Creator—One who operated by general laws" (Young 1985, p. 148). However, here we are less interested in the theological implications of Babbage's work, and more concerned with the manner in which he exploited his computational machinery in order to achieve an academic goal.

Babbage clearly does not attempt to capture the full complexity of natural geology in his machine's behavior. Indeed, the analogy between the Difference Engine's program and the relevant geological processes is a crude one. However, the formal resemblance between the two was sufficient to enable Babbage's point to be made. His computing machine is thus clearly being employed as a model, and a model of a particular kind—an idealized conceptual tool rather than a realistic facsimile intended to "stand in" for the real thing.

Moreover, the model's goal is not to shed light directly on geological discontinuity per se. Its primary function is to force an audience to reflect on their own reasoning processes (and on those of the authors of the preceding eight legitimate Bridgewater Treatises). More specifically, the experiment encourages viewers to (re)consider the grounds upon which one might legitimately identify a miracle, suggesting that a mere inability to understand some phenomenon as resulting from the continuous action of natural law is not sufficient, for the continuous action of some "higher law," one discernible only from a more systemic perspective, could always be responsible. Thus, Babbage's is an "experiment" that brings no new data to light, it generates no geological facts for its audience, but seeks to rearrange their theoretical commitments.[1]

Babbage approached the task of challenging his audiences' assumptions as a stage magician might have done (Babbage 1837, p. 35):

Now, reader, let me ask how long you will have counted before you are firmly convinced that the engine, supposing its adjustments to remain unaltered, will continue whilst its motion is maintained, to produce the same series of natural numbers? Some minds perhaps are so constituted, that after passing the first hundred terms they will be satisfied that they are acquainted with the law. After seeing five hundred terms, few will doubt; and after the fifty-thousandth term the propensity to believe that the succeeding term will be fifty thousand and one, will be almost irresistible.

Key to his argument was the surprise generated by mechanical discontinuity. That a process unfolding "like clockwork" could nevertheless confound expectation simultaneously challenged the assumed nature of both mechanical and natural processes and the power of rational scientific induction. In this respect, Babbage's argument resonates with some modern treatments of "emergent behavior." Here, nonlinearities in the interactions between a system's components give rise to unexpected (and possibly irreducible, that is, quasi-miraculous) global phenomena, as when, for instance, the presumably simple rules followed by insects generate complex self-regulating nest architectures (Ladley and Bullock 2005), or novel forms emerge from shape grammars (March 1996a, 1996b). For Babbage, however, any current inability on our part to reconcile some aggregate property with the constitution and organization of the system that gives rise to it is no reason to award the phenomenon special status. His presumption is that for some more sophisticated observer, reconciling the levels of description will be both possible and straightforward, nonlinearity or no nonlinearity.

Additionally, there is a superficial resemblance between the catastrophist debate of the nineteenth century and the more recent dispute over the theory of punctuated equilibria introduced by Niles Eldredge and Stephen Jay Gould (1973). Both arguments revolve around the significance of what appear to be abrupt changes on geological time scales. However, where Babbage's dispute centered on whether change could be explained by one continuously operating process or must involve two different mechanisms—the first being geological processes, the second Divine intervention—Gould and Eldredge did not dispute that a single evolutionary process was at work. They take pains to point out that their theory does not supersede phylogenetic gradualism, but augments it. They wish to account for the two apparent modes of action evidenced by the fossil record—long periods of stasis, short bursts of change—not by invoking two processes but by explaining the unevenness of evolutionary change. In this respect, the theory that Eldredge and Gould supply attempts to meet a modern challenge: that of explaining nonlinearity, rather than merely accommodating it. Whereas Babbage's aim was merely to demonstrate that a certain kind of nonlinearity was logically possible in the absence of exogenous interference, Gould and Eldredge exemplify the attempt to discover how and why nonlinearities arise from the homogeneous action of low-level entities.

Babbage, too, spent some time developing theories with which he sought to explain how specific examples of geological discontinuity could have arisen as the result of unchanging and continuously acting physical geological processes. One example of apparently rapid geological change that had figured prominently in geological debate since being depicted on the frontispiece of Lyell's *Principles of Geology* (1830) was the appearance of the Temple of Serapis on the edge of the Bay of Baiae in Pozzuoli, Italy (see figure 2.4). The surfaces of the forty-two-foot pillars of the temple are characterized by three regimes. The lower portions of the pillars are smooth, their central portions have been attacked by marine creatures, and above this region the pillars are weathered but otherwise undamaged. These abrupt changes in the character of the surfaces of the pillars were taken by geologists to be evidence that the temple had been partially submerged for a considerable period of time.

For Lyell (1830), an explanation could be found in the considerable seismic activity that had characterized the area historically. It was well known that eruptions could cover land in considerable amounts of volcanic material and that earthquakes could suddenly raise or lower tracts of land. Lyell reasoned that a volcanic eruption could have buried the lower portion of the pillars before an earthquake lowered the land upon which the temple stood into the sea. Thus the lower portion would have been preserved from erosion, while a middle portion would have been subjected to marine perforations and an upper section to the weathering associated with wind and rain.

Recent work by Brian Dolan (1998) has uncovered the impact that Babbage's own thoughts on the puzzle of the pillars had on this debate. Babbage, while visiting the temple, noted an aspect of the pillars that had hitherto gone undetected: a patch of calciated stone located between the central perforated section and the lower smooth portion. He inferred that this calciation had been caused, over considerable time, by calcium-bearing spring waters that had gradually flooded the temple, as the land upon which it stood sank lower and lower. Eventually this subsidence caused

Figure 2.4
The Temple of Serapis. The frontispiece for the first six volumes of Lyell's Principles
of Geology. By permission of the Syndics of Cambridge University.

the temple pillars to sink below sea level and resulted in the marine erosion
evident on the middle portion of the columns.

Thus Babbage's explanation invoked gradual processes of cumulative
change, rather than abrupt episodes of discontinuous change, despite the
fact that the evidence presented by the pillars is that of sharply separated
regimes. Babbage's account of this gradual change relied on the notion
that a central, variable source of heat, below the earth's crust, caused expansion and contraction of the land masses above it. This expansion or
contraction would lead to subsidence or elevation of the land masses
involved. Babbage exploited the power of his new calculating machine in
attempting to prove his theory, but not in the form of a simulation model.
Instead, he used the engine to calculate tables of values that represented
the expansion of granite under various temperature regimes, extrapolated
from empirical measurements carried out with the use of furnaces. With

these tables, Babbage could estimate the temperature changes that would have been necessary to cause the effects manifested by the Temple of Serapis (see Dolan 1998, for an extensive account of Babbage's work on this subject).

Here, Babbage is using a computer, and is moving beyond a gradualist account that merely tolerates discontinuities, such as that in his Bridgewater Treatise, to one that attempts to explain them. In this case his engine is not being employed as a simulation model but as a prosthetic calculating device. The complex, repetitive computations involved in producing and compiling his tables of thermal expansion figures might normally have been carried out by "computers," people hired to make calculations manually. Babbage was able to replace these error-prone, slow, and costly manual calculations with the action of his mechanical reckoning device.

Like simulation modeling, this use of computers has become widespread across modern academia. Numerical and iterative techniques for calculating, or at least approximating, the results of what would be extremely taxing or tedious problems have become scientific mainstays. However, this kind of automated extrapolation differs significantly from the simulation described above. Just as the word "intelligence" itself can signify, first, the possession or exercise of superior cognitive faculties and, second, the obtainment or delivery of useful information, such as military intelligence, for Babbage, machine intelligence could either refer to some degree of automated reasoning or (less impressively) the "manufacture" of information (Schaffer 1994). While Babbage's model of miracles and his automatic generation of thermal expansion tables were both examples of "mechanized intelligence," they differed significantly in that the first was intended to take part in and thereby partially automate thought processes directed at understanding, whereas the second exemplified his ability to "manufacture numbers" (Babbage 1837, p. 208). This subtle but important difference was not lost upon Babbage's contemporaries, and was central to unfolding discussions and categorizations of mental labor.

Automating Reason

For his contemporaries and their students, the reality of Babbage's machine intelligence and the prospect of further advances brought to the foreground questions concerning the extent to which mental activity could and should be automated. The position that no such activity could be achieved "mechanically" had already been somewhat undermined by the success of unskilled human calculators and computers, who were able to efficiently generate correct mathematical results while lacking an understanding of the routines that they were executing.

National programs to generate navigational and astronomical tables of logarithmic and trigonometric values (calculated up to twenty-nine decimal places!) would not have been possible in practice without this redistribution of mental effort. Babbage himself was strongly influenced by Baron Gaspard De Prony's work on massive decimal tables in France from 1792, where he had employed a division of mathematical labor apparently inspired by his reading of Adam Smith's *Wealth of Nations* (see Maas 1999, pp. 591–92).

[De Prony] immediately realised the importance of the principle of the division of labour and split up the work into three different levels of task. In the first, "five or six" eminent mathematicians were asked to simplify the mathematical formulae. In the second, a similar group of persons "of considerable acquaintance with mathematics" adapted these formulae so that one could calculate outcomes by simply adding and subtracting numbers. This last task was then executed by some eighty predominantly unskilled individuals. These individuals were referred to as the computers or calculators.

Babbage's Difference Engine was named after this "method of differences," reducing formulae to combinations of addition and subtraction. However, there was a clear gulf separating true thinking from the mindless rote activity of computers, whether human or mechanical. For commentators such as the Italian mathematician and engineer Luigi Federico Menebrea, whose account of a lecture Babbage gave in Turin was translated into English by Ada Lovelace (Lovelace 1843), there appeared little chance that machinery would ever achieve more than the automation of this lowest level of mental activity. In making this judgment, Menebrea "pinpointed the frontiers of the engine's capacities. The machine was able to calculate, but the mechanization of our 'reasoning faculties' was beyond its reach, unless, Menebrea implicitly qualified, the rules of reasoning themselves could be algebraised" (Maas 1999, p. 594–95).

For Menebrea it was apparently clear that such a mental calculus would never be achieved. But within half a century, just such algebras were being successfully constructed by George Boole and John Venn. For some, the potential for mechanizing such schemes seemed to put reasoning machines within reach, but for others, including Venn himself, the objections raised by Menebrea still applied.

Simon Cook (2005) describes how Venn, in his "On the Diagrammatic and Mechanical Representation of Propositions and Reasonings" of 1880, clearly recognized considerable potential for the automation of his logical formalisms but went on to identify a strictly limited role for such machinery. The nature of the labor involved in logical work, Venn stated (p. 340),

involves four "tolerably distinct steps": the statement of the data in accurate logical language, the putting of these statements into a form fit for an "engine to work with," thirdly the combination or further treatment of our premises after such a reduction, and finally interpretation of the results. In Venn's view only the third of these steps could be aided by an engine.

For Venn, then, computing machinery would only ever be useful for automating the routine process of thoughtlessly combining and processing logical terms that had to be carefully prepared beforehand and the resulting products analyzed afterward.

This account not only echoes De Prony's division of labor, but, to modern computer scientists, also bears a striking similarity to the theory developed by David Marr (1982) to describe the levels of description involved in cognitive science and artificial intelligence. For Marr, any attempt to build a cognitive system within an information-processing paradigm involves first a statement of the cognitive task in information-processing terms, then the development of an algorithmic representation of the task, before an implementation couched in an appropriate computational language is finally formulated. Venn's steps also capture this march from formal conception to computational implementation. Rather than stressing the representational form employed at each stage, Venn concentrates on the associated activity, and, perhaps as a result, considers a fourth step not included by Marr: the *interpretation* of the resulting behavior, or output, of the computational process. We will return to the importance of this final step.

Although Venn's line on automated thought was perhaps the dominant position at that time, for some scholars Babbage's partially automated argument against miracles had begun to undermine it. Here a computer took part in scientific work not by automating calculation, but in a wholly different way. The engine was not used to compute a result. Rather, the substantive element of Babbage's model was the manner in which it changed over time. In the scenario that Babbage presented to his audience, his suitably programmed Difference Engine will, in principle, run forever. Its calculation is not intended to produce some end product; rather, the ongoing calculation is itself the object of interest. In employing a machine in this way, as a model and an aid to reasoning, Babbage "dealt a severe blow to the traditional categories of mental philosophy, without positively proving that our higher reasoning faculties could be mechanized" (Maas 1999, p. 593).

Recent historical papers have revealed how the promise of Babbage's simulation model, coupled with the new logics of Boole and Venn, inspired two of the fathers of economic science to design and build automated reasoning machines (Maas 1999; Cook 2005). Unlike Babbage and Lovelace, the names Stanley Jevons (1835–1882) and Alfred Marshall (1842–1924) are not well known to students of computing or artificial intelligence. However, from the 1860s onward, first Jevons and then Marshall brought about a revolution in the way that economies were studied, effectively establishing modern economics. It was economic rather than biological or cognitive drivers that pushed both men to consider the role that machinery might play in automating logical thought processes.

Jevons pursued a mathematical approach to economics, exploring questions of production, currency, supply and demand, and so forth and developing his own system of logic (the "substitution of similars") after studying and extending Boole's logic. His conviction that his system could be automated such that the logical consequences of known states of affairs could be generated efficiently led him to the design of a "logical piano... capable of replacing for the most part the action of thought required in the performance of logical deduction" (Jevons 1870, p. 517). But problems persisted, again limiting the extent to which thought could be automated. Jevons's logical extrapolations relied upon the substitution of like terms, such as "London" and "capital of England." The capacity to decide which terms could be validly substituted appeared to resist automation, becoming for Jevons "a dark and inexplicable gift which was starkly to be contrasted with calculative, mechanical rationality" (Maas 1999, p. 613). Jevons's piano, then, would not have inclined Venn to alter his opinion on the limitations of machine logic.

Cook (2005) has recently revealed that Marshall (who, upon Jevons's early death by drowning in 1882, would eventually come to head the marginalist revolution within economics) also considered the question of machine intelligence. In "Ye Machine," the third of four manuscripts thought to have been written in the late 1860s to be presented to the Cambridge Grote Club, he described his own version of a machine capable of automatically following the rules of logic. However, in his paper he moves beyond previous proponents of machine intelligence in identifying a mechanism capable of elevating his engine above mere calculation, to the realm of creative reason. Menebrea himself had identified the relevant respect in which these calculating machines were significantly lacking in his original discussion of Babbage's engines. "[They] could not come to any correct results by 'trial and guess-work', but only by fully written-out procedures" (Maas 1999, p. 593). It was introducing this kind of exploratory behavior that Marshall imagined. What was required were the kinds of surprising mechanical jumps staged by Babbage in his drawing room. Marshall (Cook 2005, p. 343) describes a machine with the ability to process logical rules that,

"like Paley's watch," might make others like itself, thus giving rise to "hereditary and accumulated instincts." Due to accidental circumstances the "descendents," however, would vary slightly, and those most suited to their environment would survive longer: "The principle of natural selection, which involves only purely mechanical agencies, would thus be in full operation."

As such, Marshall had imagined the first example of an explicitly evolutionary algorithm, a machine that would surprise its user by generating and testing new "mutant" algorithmic tendencies. In terms of De Prony's tripartite division of labor, such a machine would transcend the role of mere calculator, taking part in the "adapting of formulae" function heretofore carried out by only a handful of persons "of considerable acquaintance with mathematics." Likewise, Marshall's machine broke free of Venn's restrictions on machine intelligence. In addition to the task of mechanically combining premises according to explicitly stated logics, Marshall's machine takes on the more elevated task of generating new, superior logics and their potentially unexpected results.

Andy Clark (1990) has described the explanatory complications introduced by this move from artificial intelligences that employ explicit, manually derived logic to those reliant on some automatic process of design or adaptation. Although the descent through Marr's "classical cascade" involved in the manual design of intelligent computational systems delivers, as a welcome side effect, an understanding of how the system's behavior derives from its algorithmic properties, no such understanding is guaranteed where this design process is partially automated. For instance, Marr's computational algorithms for machine vision, once constructed, were understood by their designer largely as a result of his gradual progression from computational to algorithmic and implementational representations. The manual design process left him with a grasp of the manner in which his algorithms achieved their performance. By contrast, when one employs artificial neural networks that learn how to behave or evolutionary algorithms that evolve their behavior, a completed working system demands further interpretation—Venn's fourth step—before the way it works can be understood.

The involvement of automatic adaptive processes thus demands a partial inversion of Marr's cascade. In order to understand an adaptive machine intelligence, effort must be expended recovering a higher, algorithmic-level representation of how the system achieves its performance from a working implementation-level representation. The scale and connectivity of the elements making up these kinds of adaptive computational system can make achieving this algorithmic understanding extremely challenging.

For at least one commentator on machine intelligence, it was exactly the suspect intelligibility of automatic machine intelligence that was objectionable. The Rev. William Whewell was a significant Victorian figure, having carved out a role for himself as historian, philosopher, and critic (see figure 2.5). His principal interest was in the scientific method and the role of induction within it. For Whewell, the means with which scientific questions were addressed had a moral dimension. We have already heard how Whewell's dismissal of atheist mathematicians in his Bridgewater Treatise seems to have stimulated Babbage's work on simulating miracles (though Whewell was likely to have been targeting the mathematician Pierre-Simon Laplace rather than Babbage). He subsequently made much more explicit

Figure 2.5
The Rev. William Whewell in 1835.

attacks on the use of machinery by scientists—a term he had coined in 1833.

Whewell brutally denied that mechanised analytical calculation was proper to the formation of the academic and clerical elite. In classical geometry "we tread the ground ourselves at every step feeling ourselves firm," but in machine analysis "we are carried along as in a rail-road carriage, entering it at one station, and coming our of it at another.... It is plain that the latter is not a mode of exercising our own loco-motive powers.... It may be the best way for men of business to travel, but it cannot fitly be made a part of the gymnastics of education. (Schaffer 1994, pp. 224–25)

The first point to note is that Whewell's objection sidesteps the issues of performance that have occupied us so far. Here, it was irrelevant to Whe-well that machine intelligence might generate commercial gain through accurate and efficient calculation or reasoning. A legitimate role within science would be predicated not only on the ability of computing machines to replicate human mental labor but also on their capacity to aid in the revelation of nature's workings. Such revelation could only be achieved via diligent work. Shortcuts would simply not do. For Whewell it was the journey, not the destination, that was revelatory. Whewell's objec-tion is mirrored by the assertion sometimes made within artificial intelli-gence that if complex but inscrutable adaptive algorithms are required in order to obtain excellent performance, it may be necessary to sacrifice a complete understanding of how exactly this performance is achieved—"We are engineers, we just need it to work." Presumably, Whewell would have considered such an attitude alien to academia.

More prosaically, the manner in which academics increasingly rely upon automatic "smart" algorithms to aid them in their work would have wor-ried Whewell. Machine intelligence as typically imagined within modern AI (for example, the smart robot) may yet be a distant dream, but for Whewell and Babbage, it is already upon us in the automatically executed statistical test, the facts, figures, opinions, and arguments instantaneously harvested from the Internet by search engines, and so forth. Where these shortcuts are employed without understanding, Whewell would argue, aca-demic integrity is compromised.

There are also clear echoes of Whewell's opinions in the widespread ten-dency of modern theoreticians to put more faith in manually constructed mathematical models than automated simulation models of the same phe-nomena. While the use of computers to solve mathematical equations nu-merically (compare Babbage's thermal expansion calculations) is typically regarded as unproblematic, there is a sense that the complexity—the impenetrability—of simulation models can undermine their utility as scientific tools (Grimm 1999; Di Paolo et al. 2000).

However, it is in Marshall's imagined evolving machine intelligence that the apotheosis of Whewell's concerns can be found. In the terms of Whewell's metaphor, not only would Marshall be artificially transported from problem to solution by such a machine, but he would be ferried through deep, dark, unmapped tunnels in the process. At least the rail tracks leading from one station to another along which Whewell's imagined locomotive must move had been laid by hand in a process involving much planning and toil. By contrast, Marshall's machine was free to travel where it pleased, arriving at a solution via any route possible. While the astonishing jumps in the behavior of Babbage's machine were not surprising to Babbage himself, even the programmer of Marshall's machine would be faced with a significant task in attempting to complete Venn's "interpretation" of its behavior.

Conclusion

This chapter has sought to highlight activities relevant to the prehistory of artificial intelligence that have otherwise been somewhat neglected within computer science. In gathering together and presenting the examples of early machine intelligence created by Babbage, Jevons, and Marshall, along with contemporaneous reflections on these machines and their potential, the chapter relies heavily on secondary sources from within a history of science literature that should be of growing importance to computer science. Although this paper attempts to identify a small number of issues that link contemporary AI with the work of Babbage and his contemporaries, it is by no means a piece of historical research and the author is no historian. Despite this, in arranging this material here on the page, there is a risk that it could be taken as such.

Babbage's life and work have already been the repeated subject of Whiggish reinterpretation—the tendency to see history as a steady linear progression (see Hyman 1990 for a discussion). In simplifying or ignoring the motivations of our protagonists and the relationships between them, there is scope here, too, for conveying the impression of an artificially neat causal chain of action and reaction linking Babbage, Whewell, Jevons, Marshall, and others in a consensual march toward machine intelligence driven by the same questions and attitudes that drive modern artificial intelligence. Such an impression would, of course, be far from the truth. The degree to which each of these thinkers engaged with questions of machine intelligence varied wildly: for one it was the life's work; for another, a brief interest. And even with respect to the output of each individual, the elements highlighted here range from significant signature works to obscure footnotes or passing comments. It will be left to historians of science to provide an accurate account of the significances of the activities presented here. This chapter merely seeks to draw some attention to them.

Given the sophistication already evident in the philosophies associated with machine intelligence in the nineteenth century, it is perhaps surprising that a full-fledged philosophy of technology, rather than science, has only recently begun to emerge (Ihde 2004). In the absence of such a discipline, artificial intelligence and cognitive philosophy, especially that influenced by Heidegerrian themes, have played a key role in extending our understanding of the role that technology has in influencing the way we think (see, for example, Dreyfus 2001). If we are to cope with the rapidly expanding societal role of computers in, for instance, complex systems modeling, adaptive technologies, and the Internet, we must gain a firmer grasp of the epistemic properties of the engines that occupied Babbage and his contemporaries.

Unlike an instrument, that might simply be a pencil, engines embody highly differentiated engineering knowledge and skill. They may be described as "epistemic" because they are crucially generative in the practice of making scientific knowledge.... Their epistemic quality lies in the way they focus activities, channel research, pose and help solve questions, and generate both objects of knowledge and strategies for knowing them. (Carroll-Burke 2001, p. 602)

Acknowledgments

This chapter owes a significant debt to the painstaking historical research of Simon Schaffer, Brian Dolan, Simon Cook, Harro Maas, and, less recently, Walter Cannon.

Note

1. See Bullock (2000) and Di Paolo, Noble, and Bullock (2000) for more discussion of Babbage's simulation model and simulation models in general.

References

Babbage, Charles. 1837. *Ninth Bridgewater Treatise: A Fragment.* 2nd edition. London: John Murray.

Brock, W. H. 1966. "The Selection of the Authors of the Bridgewater Treatises." *Notes and Records of the Royal Society of London* 21: 162–79.

Bullock, Seth. 2000. "What Can We Learn from the First Evolutionary Simulation Model?" In *Artificial Life VII: Proceedings of the Seventh International Conference On Artificial Life*, edited by M. A. Bedau, J. S. McCaskill, N. H. Packard, and S. Rasmussen, pp. 477–86. Cambridge, Mass.: MIT Press.

Cannon, W. 1960. "The Problem of Miracles in the 1830s." *Victorian Studies* 4: 4–32.

Carroll-Burke, P. 2001. "Tools, Instruments and Engines: Getting a Handle on the Specificity of Engine Science." *Social Studies of Science* 31, no. 4: 593–625.

Clark, A. 1990. "Connectionism, Competence and Explanation." In *The Philosophy of Artificial Intelligence*, edited by Margaret A. Boden, pp. 281–308. Oxford: Oxford University Press.

Cook, S. 2005. "Minds, Machines and Economic Agents: Cambridge Receptions of Boole and Babbage." *Studies in the History and Philosophy of Science* 36: 331–50.

Darwin, Charles. 1859. *On the Origin of Species*. London: John Murray.

Di Paolo, E. A., J. Noble, and Seth Bullock. 2000. "Simulation Models as Opaque Thought Experiments." In *Artificial Life VII: Proceedings of the Seventh International Conference On Artificial Life*, edited by M. A. Bedau, J. S. McCaskill, N. Packard, and S. Rasmussen, pp. 497–506. Cambridge, Mass.: MIT Press.

Dolan, B. P. 1998. "Representing Novelty: Charles Babbage, Charles Lyell, and Experiments in Early Victorian Geology." *History of Science* 113, no. 3: 299–327.

Dreyfus, H. 2001. *On the Internet*. London: Routledge.

Eldredge, N., and Steven Jay Gould. 1973. "Punctuated Equilibria: An Alternative to Phyletic Gradualism." In *Models in Paleobiology*, edited by T. J. M. Schopf, pp. 82–115. San Francisco: Freeman, Cooper.

Grimm, V. 1999. "Ten Years of Individual-Based Modelling in Ecology: What We Have Learned and What Could We Learn in the Future?" *Ecological Modelling* 115: 129–48.

Hyman, R. A. 1982. *Charles Babbage: Pioneer of the Computer*. Princeton: Princeton University Press.

———. 1990. "Whiggism in the History of Science and the Study of the Life and Work of Charles Babbage." *IEEE Annals of the History of Computing* 12, no. 1: 62–67.

Ihde, D. 2004. "Has the Philosophy of Technology Arrived?" *Philosophy of Science* 71: 117–31.

Jevons, W. S. 1870. "On the Mechanical Performance of Logical Inference." *Philosophical Transactions of the Royal Society* 160: 497–518.

Ladley, D., and Seth Bullock. 2005. "The Role of Logistic Constraints on Termite Construction of Chambers and Tunnels." *Journal of Theoretical Biology* 234: 551–64.

Lovelace, Ada. 1843. "Notes on L. Menabrea's 'Sketch of the Analytical Engine invented by Charles Babbage, Esq.'" *Taylor's Scientific Memoirs.* Volume 3. London: J. E. & R. Taylor.

Lyell, Charles. 1830/1970. *Principles of Geology.* London: John Murray; reprint, London: Lubrecht & Cramer.

Maas, H. 1999. "Mechanical Rationality: Jevons and the Making of Economic Man." *Studies in the History and Philosophy of Science* 30, no. 4: 587–619.

March, L. 1996a. "Babbage's Miraculous Computation Revisited." *Environment and Planning B: Planning & Design* 23(3): 369–76.

———. 1996b. "Rulebound Unruliness." *Environment and Planning B: Planning & Design* 23: 391–99.

Marr, D. 1982. *Vision.* San Francisco: Freeman.

Robson, J. M. 1990. "The Fiat and the Finger of God: The Bridgewater Treatises." In *Victorian Crisis in Faith: Essays on Continuity and Change in 19th Century Religious Belief,* edited by R. J. Helmstadter and B. Lightman. Basingstoke, U.K.: Macmillan.

Schaffer, S. 1994. "Babbage's Intelligence: Calculating Engines and the Factory System." *Critical Inquiry* 21(1): 203–27.

Swade, D. 1996. "'It Will Not Slice a Pineapple': Babbage, Miracles and Machines." In *Cultural Babbage: Technology, Time and Invention,* edited by F. Spufford and J. Uglow, pp. 34–52. London: Faber & Faber.

Topham, J. 1992. "Science and Popular Education in the 1830s: The Role of the Bridgewater Treatises." *British Journal for the History of Science* 25: 397–430.

Whewell, W. 1834. *Astronomy and General Physics Considered with Reference to Natural Theology.* London: Pickering.

Young, R. M. 1985. *Darwin's Metaphor: Nature's Place in Victorian Culture.* Cambridge: Cambridge University Press.

3 D'Arcy Thompson: A Grandfather of A-Life[1]

Margaret A. Boden

It's well known that three core ideas of A-life were originated many years ago, but couldn't be appreciated—still less, explored—until vastly increased computer power and computer graphics became available. Alan Turing's diffusion equations and John von Neumann's cellular automata were introduced with a fair degree of theoretical detail in the early 1950s. As for genetic algorithms, these were glimpsed at the same time by von Neumann, and defined by John Holland in the early 1960s. But it wasn't until the late 1980s that any of these could be fruitfully implemented.

What's not so well known is that various issues that are prominent in current A-life were being thought about earlier still, even before the First World War. In 1917, Sir D'Arcy Wentworth Thompson (1860–1948), professor of zoology at the University of St. Andrews, published *On Growth and Form*. He was asking biological questions, and offering biological answers, very much in the spirit of A-life today.

The book was immediately recognized as a masterpiece, mostly because of the hugely exciting ideas and the many fascinating examples, but also because of the superb, and highly civilized, prose in which it was written. Countless readers were bewitched by it, and begged for a second edition. That appeared during the next World War, in 1942, six years before Thompson's death. It had grown from just under 800 to 1,116 pages—there was plenty to chew on there.

So why isn't it more famous now? The reason is much the same as the reason why Turing's (1952) paper on reaction-diffusion–based morphogenesis became widely known only fairly recently. Biologists, and especially embryologists, in the 1950s could see that Turing's work might be highly relevant, indeed fundamental, to their concerns. But lacking both specific biochemical knowledge and computational power to handle the sums, they couldn't *do* anything with it.

The same was true of John Holland's work (1962, 1975). I remember being hugely impressed by the paper he gave at a twenty-person weekend meeting held in Devon in 1981 (Selfridge, Rissland, and Arbib 1984). Not only had he tackled evolutionary programming, but he'd solved the credit-assignment problem, a recurring, and seemingly intractable, problem in contemporary AI. I'd never heard of him, and when I got home from the Devonshire countryside I asked my AI colleagues why they weren't shouting his name to the rooftops. Some replied that his work wasn't usable (he'd done the mathematics, but not the programming)—and some had never heard of him, either.

Similarly, D'Arcy Thompson's wartime readers were intrigued, even persuaded, by his book. But putting it into biological practice wasn't intellectually—or, rather, technologically—feasible. Today, we're in a better position to appreciate what he was trying to do, and even to carry on where he left off.

In sum, if Turing and von Neumann (with Ross Ashby and W. Grey Walter) were the fathers of A-life, D'Arcy Thompson was its grandfather. I don't just mean that he could have been, if anyone had still been listening. For at least one person *was* listening: *On Growth and Form* was one of only six references cited by Turing at the end of his morphogenesis paper. For that reason alone D'Arcy Thompson is worthy of respect. But in the post–World War II period, his name was still one to conjure with. I came across *On Growth and Form* as a medical student in the mid-1950s, and was entranced. Many others were, too, which is presumably why an abridged (though still weighty) version was published some years later (Thompson 1992). In short, D'Arcy Thompson inspired not only Turing, but others as well.

Who Was D'Arcy Thompson?

D'Arcy Thompson—he's hardly ever referred to merely as Thompson—was born in 1860, just a year after the publication of *The Origin of Species*, and was already middle-aged when Queen Victoria died in 1901. He survived both world wars, dying at the age of almost ninety in 1948. That was the year in which the Manchester Mark I computer (sometimes known as the Manchester Automatic Digital Machine, or MADM), for which Turing was the first programmer, became operational.

If D'Arcy Thompson had an exceptional span in years, he also had an extraordinary span in intellectual skills. He was a highly honored classical scholar, who translated the authoritative edition of Aristotle's *Historia Ani-*

malium (Thompson 1910). In addition, he was a biologist and mathematician. Indeed, he was offered chairs in classics and mathematics as well as in zoology.

While still a teenager (if "teenagers" existed in Victorian England), he edited a small book of essays based on studies from the Museum of Zoology in Dundee (Thompson 1880), but he soon graduated to larger tomes. In his early twenties, he prepared a bibliography nearly three hundred pages long of the work on invertebrates that had been published since his birth (Thompson 1885). At that young age he also edited and translated a German biologist's scattered writings on how flowers of different types are pollinated by insects. (In broom, for instance, the stamens "explode" when the bee lands on the keel of the flower, and the style curls upwards so that the stigma strikes the bee's back.) The result was a 670-page volume for which Charles Darwin (1809–1882) wrote the preface (Thompson 1883).

Forty years later, he was commenting on ancient Egyptian mathematics in *Nature* (Thompson 1925), and analyzing thirty years' worth of data on the size of the catches made by fishermen trawling off Aberdeen (Thompson 1931). And just before the appearance of the second edition of *On Growth and Form*, he put together a collection of some of his essays (Thompson 1940) whose subjects ran from classical biology and astronomy through poetry and medicine to "Games and Playthings" from Greece and Rome. The collection included popular pieces originally written for *Country Life*, *Strand Magazine*, and *Blackwood's Magazine* (Thompson 1940). His last book, which appeared a few months before he died, was *Glossary of Greek Fishes*: a "sequel" to his volume on all the birds mentioned in ancient Greek texts (Thompson 1895/1947). Clearly, then, D'Arcy Thompson was a man of parts.

Some of the titles mentioned might suggest that he was a list maker. On the contrary, he was a great intellect and a superb wordsmith. His major book has been described by the biologist Peter Medawar as "beyond comparison the finest work of literature in all the annals of science that have been recorded in the English tongue" (Medawar 1958, p. 232). And his intoxicating literary prose was matched by his imaginative scientific vision.

Biomimetics: Artefacts, but Not A-Life

For all his diverse skills, D'Arcy Thompson was no Charles Babbage. So he wasn't playing around with computers, electronic or not. Nor was he playing around with any other gizmos. In short, he wasn't doing biomimetics.

Biomimetics involves making material analogues of the physical stuff of living things, in order to investigate its physico-chemical properties. Vaulted roofs modeled on leaf-structure count, since they are testing and exemplifying the tensile properties of such physical structures. But automata don't. Even if the movements of specific bodily organs are being modeled, as in Jacques de Vaucanson's flute player, which moved its tongue, lips, and fingers (1738/1742/1979)—the physical *stuff* is not.

Perhaps the first example of biomimetics, and certainly one of the most startling, was due to the British scientist Henry Cavendish (1731–1810). In 1776, Cavendish nominated Captain James Cook for election to the Royal Society. Having just completed his second great voyage of discovery, Cook had exciting tales to tell of exotic fish and alien seas. But so did Cavendish. For, in the very same year, he'd built an artificial electric fish and laid it in an artificial sea (Wu 1984; Hackman 1989).

Its body was made of wood and sheepskin, and its electric organ was two pewter discs, connected by a brass chain to a large Leyden battery; its habitat was a trough of salt water. Cavendish's aim was to prove that "animal electricity" is the same as the physicist's electricity, not an essentially different, vital, phenomenon. His immobile "fish" wouldn't have fooled anyone into thinking it was a real fish, despite its fish-shaped leather body, but—and this was the point—it did deliver a real electric shock, indistinguishable from that sent out by a real torpedo fish.

Cavendish intended his artificial fish to deliver an intellectual shock, as well as a real one. His aim was to demystify a vital phenomenon, to show the continuity between the physical and the organic, and, of course, to display the physical principle underlying the living behavior.

He thought this shocking hypothesis to be so important that he invited some colleagues into his laboratory to observe the experiment—so far as we know, the only occasion on which he did so (Wu 1984: 602). Certainly, such an invitation from the shy, taciturn Cavendish was a remarkable event: an acquaintance said that he "probably uttered fewer words in the course of his life than any man who ever lived to fourscore years, not at all excepting the monks of la Trappe."[2]

(Oliver Sacks [2001] has suggested that Cavendish's unsociability was due to Asperger's syndrome. If so, he was perhaps in good company: the same posthumous "diagnosis" has been made of Einstein and Newton [Baron-Cohen and James 2003].)

But if Cavendish's doubly shocking demonstration was an exercise in biology, and simultaneously in physics, it wasn't an exercise in mathematics. That is to say, it wasn't an early example of A-Life.

A-life is abstract in nature. On the one hand, it's concerned with "life as it could be," not only "life as it is" (Langton 1989). On the other hand, it studies life-as-it-is not by putting it under the microscope, or twirling it around in a test-tube, but by seeking its logical-computational principles. Even A-life work on biochemistry is looking for abstract principles, not—or not only—for specific molecules (see, for example, Drexler 1989; Szostak, Bartel, and Luisi 2001; Kauffman 2003).

Cavendish's experiment couldn't have been done without the artificial fish in its bath of conducting fluid, because his aim was to reproduce the same physical phenomenon, electrical conductivity, that occurs in some living things. Biomimetics requires physical mimesis. But A-life doesn't.

Someone might even say that A-life doesn't need *any* artefacts: not fish-in-fluid, nor computers, either. If artefacts are needed at all, then in principle, just three will suffice: pencil, paper, and armchair. Some hugely important early A-life work was done either without the aid of computers, or, in Turing's case, with the aid only of very primitive machines. In practice, however, computers are almost always needed.

It's possible, in other words, for someone to do mathematical biology without being able to do computational biology. They may be able to define the mathematical principles, and even to intuit their general implications, without being able to calculate their consequences in any detail. That's precisely the position that D'Arcy Thompson was in. After all, computers weren't a feature of the Edwardian age.

First Steps in Mathematical Biology

Isolated examples of mathematically expressed biological research were scattered in the pre-twentieth-century literature. But mathematical biology as an all-encompassing and systematic approach was attempted only after the turn of the century—by D'Arcy Thompson.

Although Darwin had written the preface for Thompson's first "real" book, Thompson had become increasingly critical of Darwinian theory. An early intimation of this was in his paper "Some Difficulties of Darwinism," given in 1894 to an Oxford meeting of the British Association for the Advancement of Science (one of Babbage's many brainchildren, in 1831). His book, over twenty years later, explained at length why he felt Darwinism to be inadequate as an explanation of the living creatures we see around us.

Like some maverick modern biologists (Webster and Goodwin 1996; Goodwin 1994; Kauffman 1993), he regarded natural selection as strictly secondary to the origin of biological form. The origin of form, he said, must be explained in a different way.

He integrated a host of individual biological facts within a systematic vision of the order implicit in living organisms. That is, he used various ideas from mathematics not only to describe, but also to explain, fundamental features of biological form. He wasn't content, for example, to note that patterns of leaf-sprouting on plants may often be described by a Fibonacci number series, such as 0,1,1,2,3,5,8,13,21 ... He converted this finding from a mathematical curiosity into a biologically intelligible fact, by pointing out that this is the most efficient way of using the space available.

Significantly, he often combined "pure" mathematical analysis with the equations of theoretical physics. In this way he tried to explain not only specific anatomical facts, such as the width and branching patterns of arteries relative to the amount of blood to be transported, but also why certain forms appear repeatedly in the living world.

D'Arcy Thompson referred to countless examples of actual organisms, but he had in mind also *all possible* life forms. As he put it (Thompson 1942, p. 1026):

[I] have tried in comparatively simple cases to use mathematical methods and mathematical terminology to describe and define the forms of organisms. . . . [My] study of organic form, which [I] call by Goethe's name of Morphology, is but a portion of that wider Science of Form which deals with the forms assumed by matter under all aspects and conditions, and, in a still wider sense, with *forms which are theoretically imaginable* [emphasis added].

For D'Arcy Thompson, then, the shapes of animals and plants aren't purely random: we can't say, "Anything goes." To the contrary, developmental and evolutionary changes in morphology are constrained by underlying general principles of physical and mathematical order.

Goethe's Morphology

As he clearly acknowledged, D'Arcy Thompson's work was closely related to Johann von Goethe's (1749–1832) rational morphology. Goethe had coined the word "morphology," meaning the study of organized things. It refers not just to their external shape but also to their internal structure and development and, crucially, *their structural relations to each other*. Goethe intended morphology to cover both living and inorganic nature, even including crystals, landscape, language, and art, but D'Arcy Thompson's interest was in its application to biology.

In his "An Attempt to Interpret the Metamorphosis of Plants," Goethe (1790/1946) had argued that superficially different parts of a flowering plant—such as sepals, petals, and stamens—are derived by transformations from the basic, or archetypal, form: the leaf. Later, he posited an equivalence (homology) between the arms, front legs, wings, and fins of different animals. All these, he said, are different transformations of the forelimb of the basic vertebrate type. And all bones, he claimed, are transformations of vertebrae. In other words, he combined meticulous naturalistic observation with a commitment to the fundamental unity of nature.

For instance, Goethe is widely credited with a significant discovery in comparative anatomy, namely, that the intermaxillary bone, which bears the incisors in a rabbit's jaw, exists in a reduced form in the human skeleton, as it does in other vertebrates. (Strictly speaking, he rediscovered this fact [Sherrington 1942, 21f], and restated the claim that sepals are a type of leaf [Goethe 1790, 73].) The issue was "significant" because some people had used the bone's seeming absence to argue that God created a special design for human beings, marking them off from the animals. Goethe, by contrast, related human skulls to the archetypal vertebrate skull, much as he related sepals to the archetypal leaf.

Goethe didn't think of morphological transformations as temporal changes, still less as changes due to Darwinian evolution, which was yet to be discovered. Rather, he saw them as abstract, quasi-mathematical derivations from some Neoplatonic ideal in the mind of God. But these abstractions could be temporally instantiated.

So in discussing the development of plants, for instance, he referred to actual changes happening in time as the plant grows. He suggested that sepals or petals would develop under the influence of different kinds of sap, and that external circumstances could lead to distinct shapes, as of leaves developing in water or in air—a suggestion that D'Arcy Thompson took very seriously, as we'll see.

The point of interest here is that Goethe focused attention on the restricted range of basic forms ("primal phenomena") in the organic world. He encouraged systematic comparison of them, and of the transformations they could support. He also suggested that only certain forms are possible: we can imagine other living things, but not just *any* life forms. In a letter of 1787 (see Nisbet 1972, p. 45), he wrote:

With such a model (of the archetypal plant [*Urpflanz*] and its transformations)...one will be able to contrive an infinite variety of plants. They will be *strictly logical* plants—in other words, even though they may not actually exist, they could exist.

They will not be mere picturesque and imaginative projects. They will be imbued with inner truth and necessity. And the same will be applicable to all that lives [emphasis added].

Similarly, in his essay on plant metamorphosis (1790), he wrote, "Hypothesis: All is leaf. This simplicity makes possible the greatest diversity."

Critics soon pointed out that he overdid the simplicity. He ignored the roots of plants, for instance. His excuse was telling (Nisbet 1972, p. 65):

[The root] did not really concern me, for what have I to do with a formation which, while it can certainly take on such shapes as fibres, strands, bulbs and tubers, remains confined within these limits to a dull variation, in which endless varieties come to light, but without any intensification (of archetypal form); and it is this alone which, in the course marked out for me by my vocation, could attract me, hold my attention, and carry me forward.

To ignore apparent falsifications of one's hypothesis so shamelessly seems utterly unscientific in our Popperian age. And some of Goethe's contemporaries complained about it, too. But his attitude stemmed from his idealist belief in the essential unity of science and aesthetics. He even compared the plant to a superb piece of architecture, whose foundations, the roots, are of no interest to the viewer. More generally, "Beauty is the manifestation of secret laws of nature which, were it not for their being revealed through beauty, would have remained unknown for ever" (Nisbet 1972, p. 35). For Goethe, and perhaps for D'Arcy Thompson, too, this language had an import much richer than the familiar appeals to theoretical "simplicity," "symmetry," or "elegance."

Questions about such abstract matters as the archetypal plant were very unlike those being asked by most physiologists at the time. If a body is not just a flesh-and-blood mechanism but a transformation of an ideal type, how it happens to work—its mechanism of cords and pulleys—is of less interest than its homology.

Indeed, for the holist Goethe the mechanism may even depend on the homology. Perhaps it's true that a certain kind of sap, a certain chemical mechanism, will induce a primordial plant part to develop into a sepal rather than a petal. But what's more interesting in this view is that sepals and petals are the structural possibilities on offer. How one describes the plant or body part in the first place will be affected by the type, and the transformations, supposedly expressed by it.

It's not surprising, then, that Goethe was out of sympathy with the analytic, decompositional methods of empiricist experimentalism. By the same token, anyone following in his footsteps, as D'Arcy Thompson did, would be swimming against that scientific tide.

Initially, Goethe's morphology attracted scepticism even from descriptive (nonexperimental) biologists. But shortly before his death, his ideas were publicly applauded by Etienne Geoffroy Saint-Hilaire (Merz 1904, vol. 2, p. 244). Geoffroy Saint-Hilaire agreed with Goethe that comparative anatomy should be an exercise in "rational morphology," a study of the successive transformations—rational, not temporal—of basic body plans.

After his death, Goethe's work was cited approvingly even by Thomas Huxley and the self-proclaimed mechanist Hermann von Helmholtz (1821–1894). Indeed, Helmholtz credited Goethe with "the guiding ideas [of] the sciences of botany and anatomy...by which their present form is determined," and praised his work on homology and transformation as "ideas of infinite fruitfulness" (Helmholtz 1853/1884, pp. 34, 30).

"Infinite fruitfulness" isn't on offer every day. So why were Goethe's ideas largely forgotten by the scientific community? Surely, such an encomium from such a high-profile scientist, and committed mechanist, as Helmholtz would be enough to guarantee close, and prolonged, attention?

Normally, yes. However, only six years after Helmholtz spoke of Goethe's "immortal renown" in biology, Darwin published *On the Origin of Species by Means of Natural Selection* (1859/1964). This radically changed the sorts of inquiry that biologists found relevant. One might even say that it changed the sorts of enquiry that biologists found *intelligible* (see Jardine 1991). Biological questions were now posed in ways that sought answers in terms of either mechanistic physiology or Darwinian evolution.

Soon, genetics became an additional source of inquiry. The neo-Darwinian mix of physiology, evolution, and genetics was a heady brew. It quickly became the biological orthodoxy, eclipsing *Naturphilosophie* in all its forms. Darwin, like Goethe, encouraged systematic comparisons between different organs and organisms, but he posited no ideal types. He explained morphological similarity in terms of contingency-ridden variation and selective descent, or coincidental likeness between environmental constraints. In short, morphological self-organization largely disappeared as a scientific problem, surviving only in embryology.

Charles Sherrington even said that "were it not for Goethe's poetry, surely it is true to say we should not trouble about his science," and that metamorphosis is "no part of botany today" (Sherrington 1942, pp. 23, 21).

From Morphology to Mathematics

Ironically, Sherrington's remark was published in the very same year as the long-awaited new edition of *On Growth and Form*. Although Goethe himself is now largely ignored by biologists (but see Webster and Goodwin 1991, especially chapters 1 and 5), his questions have survived—thanks, largely, to D'Arcy Thompson.

Like Goethe, whom he quoted with approval several times in his book, D'Arcy Thompson sought an abstract description of the anatomical structures and transformations found in living things—indeed, in all possible things. So, for instance, he discussed the reasons for the spherical shape of soap bubbles. His reference to "forms which are theoretically imaginable" recalls Goethe's reference to "strictly logical plants"—in other words, "life as it could be." And like Goethe, he believed that certain forms were more natural, more likely, than others. In some sense, he thought, there are "primal phenomena."

Also like Goethe—though here, the comparison becomes more strained—he asked questions about the physical mechanisms involved in bodily growth. But his philosophical motivation for those questions was different in an important respect. Although D'Arcy Thompson was sympathetic to some of the claims of the *Naturphilosophen*, he wasn't a fully paid-up member of their club. Indeed, he opened his book by criticizing Kant and Goethe, complaining that they had ruled mathematics out of natural history (Thompson 1942, p. 2).

In part, he was here expressing his conviction that "the harmony of the world is made manifest in Form and Number, and the heart and soul and all the poetry of Natural Philosophy are embodied in the concept of mathematical beauty" (p. 1096ff.). This conviction wasn't shared by his professional colleagues: "Even now, the zoologist has scarce begun to dream of defining in mathematical language even the simplest organic forms" (p. 2). But in part, he was saying that physics—real physics—is crucially relevant for understanding "form."

The idealist Goethe had seen different kinds of sap as effecting the growth of sepal or petal, but for him those abstract possibilities had been generated by the divine intelligence self-creatively immanent in nature. D'Arcy Thompson, by contrast, argued that it is real physical processes, instantiating strictly physical laws, which generate the range of morphological possibilities. Certainly, those laws conform to abstract mathematical relationships—to projective geometry, for example. But biological forms are made possible by underlying material-energetic relations.

Accordingly, D'Arcy Thompson tried to relate morphology to physics, and to the dynamical processes involved in bodily growth. He suggested that very general physical (as opposed to specific chemical or genetic) constraints could interact to make some biological forms possible, or even necessary, while others are impossible.

Had he lived today, D'Arcy Thompson would doubtless have relished the work of Ralph Linsker (1986, 1988, 1990) and Christoph von der Malsburg (1973, 1979) on the self-organization of feature detectors in the sensory cortex, for it explains why we should expect to find systematic neuroanatomical structure in the brain, as opposed to a random ragbag of individually effective detector cells. Moreover, the "why" isn't a matter of selection pressures, but of spontaneous self-organization. But this recent research required computational concepts and computing power (not to mention anatomical data) that Thompson simply didn't have. He could use only the mathematics and physics available in the early years of the century.

Although D'Arcy Thompson wasn't the first biologist to study bodies, he might be described as the first biologist who took *embodiment* seriously. The physical phenomena he discussed included diffusion, surface forces, elasticity, hydrodynamics, gravity, and many others. And he related these to specific aspects of bodily form.

His chapter "On Magnitude," for example, argued both that size can be limited by physical forces and that the size of the organism determines which forces will be the most important. Gravity is crucial for mice, men, and mammoths, but the form and behavior of a water beetle may be conditioned more by surface tension than by gravity. A bacillus can in effect ignore both, being subject rather to Brownian motion and fluid viscosity. Similarly, the fixed ratio between volume and surface area is reflected, in a single cell or a multicellular animal, in respiratory surfaces such as the cell membrane, feathery gills, or alveolar lungs. Again, his fascinating discussion of "The Forms of Cells" suggested, among many other things, that the shape and function of cilia follow naturally from the physics of their molecular constitution.

Perhaps the best-known chapter of *On Growth and Form*, the one that had the clearest direct influence, was "On the Theory of Transformations, or the Comparison of Related Forms." This employed a set of two-dimensional Cartesian grids to show how differently shaped skulls, limb bones, leaves, and body forms are mathematically related. One form could generate many others, by enlargement, skewing, and rotation.

So, instead of a host of detailed comparisons of individual body parts bearing no theoretical relation with each other, anatomists were now being offered descriptions having some analytical unity.

To be sure, these purely topological transformations couldn't answer questions about more radical alterations in form. The gastrulation of an embryo, for example, couldn't be explained in this way (see Turing 1952). And only very few zoologists, of whom Medawar was one, tried to use D'Arcy Thompson's specific method of analysis. But his discussion inspired modern-day allometrics: the study of the ratios of growth rates of different structures, in embryology and taxonomy.

More Admiration than Influence

One didn't need to be doing allometrics to admire D'Arcy Thompson. By midcentury, he was widely revered as a scientist of exceptional vision (Hutchinson 1948; Le Gros Clark and Medawar 1945). The second edition of *On Growth and Form* was received with excitement in 1942, the first (only five hundred copies) having sold out twenty years before. Reprints had been forbidden by D'Arcy Thompson himself, while he worked on the revisions, and second-hand copies had been fetching ten times their original price.

However, only a decade after the second edition, which people had awaited so eagerly for years, the advent of molecular biology turned him virtually overnight into a minority taste. As we've seen, much the same had happened to his muse, Goethe, whose still-unanswered biological questions simply stopped being asked when Darwin's theory of evolution came off the press in 1859. By the end of the 1960s, only a few biologists regarded D'Arcy Thompson as more than a historical curiosity.

One of these was Conrad Waddington (1905–1975), a developmental biologist at the University of Edinburgh (his theory of "epigenesis" influenced Jean Piaget, the prominent developmental psychologist; see Boden 1994, 98–101). Waddington continually questioned the reductionist assumption that molecular biology can—or, rather, will—explain the many-leveled self-organization of living creatures. It's hardly surprising, then, that D'Arcy Thompson was often mentioned in his "by invitation only" seminars on theoretical biology, held in the late 1960s at the Rockefeller Foundation's Villa Serbelloni on Lake Como (Waddington 1966–1972).

But Waddington, too, was a maverick, more admired than believed. His theory of epigenesis couldn't be backed up by convincing empirical evidence, whether in the developing brain or in the embryo as a whole. Only after his death did his ideas gain ground. Significantly, the proceedings of the first A-life conference were dedicated to him (Langton 1989, p. xiii).

D'Arcy Thompson's most devoted admirers, however, had to concede that it was difficult to turn his vision into robust theoretical reality. Despite his seeding of allometrics, his direct influence on biology was less strong than one might expect, given the excitement one still experiences on reading his book.

Even the subsequent attempts to outline a mathematical biology eschewed his methods. Joseph Woodger's (1929, 1937) axiomatic biology, for instance, owed more to mathematical logic and the positivists' goal of unifying science (Neurath 1939) than to D'Arcy Thompson. And Turing's mathematical morphology employed numerically precise differential equations, not geometrical transformations. In short, D'Arcy Thompson figured more as inspirational muse than as purveyor of specific biological theory or fact.

The reason why his influence on other biologists, although "very great," was only "intangible and indirect" (Medawar 1958, p. 232) is implied by his own summary comment. At the close of his final chapter, he recalled the intriguing work of a naval engineer who, in 1888, described the contours and proportions of fish "from the shipbuilder's point of view." He suggested that hydrodynamics must limit the form and structure of swimming creatures. But he admitted that he could give no more than a hint of what this means, in practice. In general, he said (Thompson 1942, p. 1090):

Our simple, or simplified, illustrations carry us but a little way, and only half prepare us for much harder things. . . . *If the difficulties of description and representation could be overcome*, it is by means of such co-ordinates in space that we should at last obtain an adequate and satisfying picture of the processes of deformation and the directions of growth. (emphasis added)

Echoes in A-Life

This early exercise in mathematical biology resembled current work in A-life in various ways. So much so that one would expect D'Arcy Thompson, were he to return today, to recognize the theoretical point of most work in A-life, even though he'd be bemused by its high-tech methodology.

For instance, he'd be fascinated by Dimitri Terzopoulos's lifelike computer animation of fish, with its detailed interplay of hydrodynamics and bodily form (Terzopoulos, Tu, and Gzeszczuk 1994). These "fish" weren't robots, but software creatures existing in a computer-generated virtual world. Whereas Cavendish's "fish" was a solitary object lying inert in a dish of water, these were constantly in motion, sometimes forming hunter-hunted pairs or co-moving schools. Each one was an autonomous system, with simple perceptual abilities that enabled it to respond to the world and to its fellows. The major bodily movements, with their associated changes in body shape, resulted from twelve internal muscles (conceptualized as springs). The computerized fish learned to control these in order to ride the (simulated) hydrodynamics of the surrounding seawater. A host of minor movements arose from the definitions of seventy-nine other springs and twenty-three nodal point masses, whose (virtual) physics resulted in subtly lifelike locomotion.

He'd be intrigued, also, by Karl Sims's (1994) A-life evolution of decidedly *unlifelike* behavior, as a result of a specific mistake in the simulated physics. He'd be the first to realize that in a physical world such as that defined (mistakenly) by Sims, these strange "animals" would be better adapted to their environment than those that actually exist. For sure, he'd be interested in programs of research that systematically varied physical parameters to see what sorts of creatures would result. And he'd be fascinated by Randall Beer's studies of locomotion in robot cockroaches (Beer 1990, 1995; Beer and Gallagher 1992). For, unlike Terzopoulos and Sims, Beer subjected his computer creatures to the unforgiving discipline of the real physical world.

He'd applaud Greg Turk's (1991) models of diffusion gradients and would delight in Turk's demonstration of how to generate leopard spots, cheetah spots, lionfish stripes, and giraffe reticulations. And he'd doubtless be pleased to learn that Turk's equations were based on Turing's, which in turn were inspired by D'Arcy Thompson himself.

He'd sympathize with biologists such as Brian Goodwin and Stuart Kauffman, who see evolution as grounded in general principles of physical order (Webster and Goodwin 1996; Goodwin 1994; Kauffman 1993). He'd agree with A-lifers who stress the dynamic dialectic between environmental forces and bodily form and behavior. He might well have embarked on a *virtual* biomimetics: a systematic exploration of the effects of simulated physical principles on simulated anatomies. And he'd certainly share A-life's concern with *life as it could be*—his "theoretically imaginable forms"—rather than life as we know it.

Difficulties of Description

The "difficulties of description and representation" bemoaned by D'Arcy Thompson remained insuperable for more than half a century after publication of those first five hundred copies of his book. Glimpses of how they might be overcome arose in the early 1950s, a few years after his death.

Actually overcoming them took even longer. Or perhaps one should rather say it *is taking* even longer, for we haven't answered all of his questions yet.

Despite the deep affinity of spirit between D'Arcy Thompson's work and A-life research, there are three important, and closely related, differences. Each of these reflects his historical situation—specifically, the fact that his work was done before the invention of computers.

One difference concerns the practical usefulness of computer technology and shows why (contrary to the suggestion noted above) A-life's artefacts are not, in fact, dispensable. The other two concern limitations on the mathematical concepts available when D'Arcy Thompson was writing: in his words, the difficulties of description and representation that needed to be overcome.

First, D'Arcy Thompson was able to consider only broad outlines, largely because he had to calculate the implications of his theories using hand and brain alone. Today, theories with richly detailed implications can be stated and tested with the help of superhuman computational power. The relevant theories concern, for instance, the hydrodynamics of fish; the interactions between various combinations of diffusion gradients; and processes of evolution and coevolution occurring over many thousands of generations.

In addition, we can now study chaotic phenomena, which include many aspects of living organisms, where tiny alterations to the initial conditions of a fully deterministic system may have results utterly different from those in the nonaltered case. These results can't be predicted by approximation, or by mathematical analysis. The only way to find out what they are is to watch the system—or some computer specification of it—run, and see what happens. In all these cases, the "help" A-life gets from computers isn't an optional extra, but a practical necessity.

Second, D'Arcy Thompson's theory, though relatively wide in scope, didn't encompass the most general feature of life: self-organization as such. Instead, it considered many specific examples of self-organization. This isn't surprising. Prior to computer science and information theory, no precise language was available in which this could be discussed.

And third, although he did consider deformations produced by physical forces, D'Arcy Thompson focused more on structure than on process. This is characteristic of precomputational theories in general. In anthropology, for example, Claude Levi-Strauss in the early-1950s posited cognitive *structures*, based on binary opposition, to explain cultural phenomena, leaving his successors—notably Daniel Sperber—to consider the *processes* involved in communication and cultural evolution (see Boden 2006, chapter 8.vi). Prior to computer science, with its emphasis on the exact results of precisely specified procedures, scientists lacked ways of expressing—still less, of accurately modeling and tracking—the details of change.

Uniform physical changes could be described by linear differential equations, to be sure. And Babbage (1838/1991) could even lay down rules, or programs, for his Difference Engine determining indefinitely many "miraculous" discontinuities. But much as Babbage, as he admitted, couldn't program the transformation of caterpillar into butterfly, so D'Arcy Thompson's mathematics couldn't describe the morphological changes and dynamical bifurcations that occur in biological development.

And What Came Next?

One might have expected that cybernetics would provide some of the necessary advances in descriptive ability. The scope of cyberneticians' interests, especially on D'Arcy Thompson's home ground, the UK, was very wide (Boden 2006, chapter 4). Among other things, it included various exercises in mathematical biology, and it used robots and analogue computer modeling as a research technique. The study of "circular causal systems" drew on mainstream ideas about metabolism and reflexology, not on the morphological questions that interested D'Arcy Thompson. But the cybernetic movement considered some central biological concerns now at the core of A-life: adaptive self-organization, the close coupling of action and perception, and the autonomy of embodied agents.

It even made some progress. For instance, Ashby's (1952) "design for a brain," and his Homeostat machine, depicted brain and body as dynamical physical systems. And Grey Walter's (1950) tortoises, explicitly intended as "an imitation of life," showed that lifelike behavioral control can be generated by a very simple system.

However, the cybernetics of the 1950s was hampered both by lack of computational power and by the diversionary rise of symbolic AI. Only much later, and partly because of lessons learned by symbolic AI, could cybernetic ideas be implemented more convincingly. (Even so, recent dynamical approaches suffer a limitation shared by cybernetics: unlike classical AI, they can't easily represent hierarchical structure, or detailed structural change.)

As it turned out, it was physics and computer science, not cybernetics, which very soon after D'Arcy Thompson's death, in 1948, produced mathematical concepts describing the generation of biological form. Indeed, two of the founding fathers of computer science and AI, Turing and von

Neumann, were also the two founding fathers of A-life. (Von Neumann's intellectual range was even greater than Turing's, including chemical engineering for example [Ulam 1958].)

Around midcentury, they each developed accounts of self-organization, showing how simple processes could generate complex systems involving emergent order. They might have done this during D'Arcy Thompson's lifetime, had they not been preoccupied with defense research. While Turing was code-breaking at Bletchley Park, von Neumann was in Los Alamos, cooperating in the Manhattan Project to design the atom bomb.

The end of the war freed some of their time for more speculative activities. Both turned to abstract studies of self-organization. Their new theoretical ideas eventually led to a wide-ranging mathematical biology, which could benefit from the increasingly powerful technology that their earlier work had made possible.

In sum, D'Arcy Thompson didn't get there first. He didn't really get there at all. But he did pave the way.

Notes

1. This chapter draws on chapters 2.vi.d–f and 15.ii–iii of my book *Mind as Machine: A History of Cognitive Science* (Oxford: Oxford University Press, 2006).

2. *Encyclopedia Britannica*, 15th ed., s.v. "Henry Cavendish."

References

Ashby, W. Ross. 1952. *Design for a Brain: The Origin of Adaptive Behaviour*. London: Wiley.

Babbage, C. B. 1838/1991. *The Ninth Bridgwater Treatise: A Fragment*. In *The Works of Charles Babbage*, edited by M. Campbell-Kelly. volume 9. London: Pickering & Chatto.

Baron-Cohen, S., and I. James. 2003. Einstein and Newton Showed Signs of Asperger's Syndrome. *New Scientist*, 3rd May, 10.

Beer, R. D. 1990. *Intelligence as Adaptive Behavior: An Experiment in Computational Neuroethology*. Boston: Academic Press.

———. 1995. "A Dynamical Systems Perspective on Agent-Environment Interaction." *Artificial Intelligence* 72: 173–215.

Beer, R. D., and J. C. Gallagher. 1992. "Evolving Dynamical Neural Networks for Adaptive Behavior." *Adaptive Behavior* 1: 91–122.

Boden, Margaret A. 1994. *Piaget.* 2nd ed. London: HarperCollins.

———. 2006. *Mind as Machine: A History of Cognitive Science.* Oxford: Oxford University Press.

Darwin, Charles R. 1859/1964. *On the Origin of Species by Means of Natural Selection, or, The Preservation of Favoured Races in the Struggle for Life.* London: John Murray; facsimile ed., Cambridge, Mass.: Harvard University Press.

Drexler, K. E. 1989. "Biological and Nanomechanical Systems: Contrasts in Evolutionary Complexity." In *Artificial Life,* edited by C. G. Langton. Redwood City, Calif.: Addison-Wesley.

Goethe, Johann Wolfgang von. 1790/1946. "An Attempt to Interpret the Metamorphosis of Plants (1790)," and "Tobler's Ode to Nature (1782)." *Chronica Botanica* 10, no. 2(1946): 63–126.

Goodwin, B. C. 1994. *How the Leopard Changed Its Spots: The Evolution of Complexity.* London: Weidenfeld & Nicolson.

Hackman, W. D. 1989. "Scientific Instruments: Models of Brass and Aids to Discovery." In *The Uses of Experiment,* edited by D. Gooding, T. Pinch, and S. Schaffer. Cambridge: Cambridge University Press.

Helmholtz, Hermann von. 1853/1884. "On Goethe's Scientific Researches." Translated by H. W. Eve. In *Popular Lectures on Scientific Subjects.* London: Longmans Green, 1884.

Holland, John H. 1962. "Outline for a Logical Theory of Adaptive Systems." *Journal of the Association for Computing Machinery* 9: 297–314.

———. 1975. *Adaptation in Natural and Artificial Systems: An Introductory Analysis with Applications to Biology, Control, and Artificial Intelligence.* Ann Arbor: University of Michigan Press.

Hutchinson, G. E. 1948. "In Memoriam, D'Arcy Wentworth Thompson." *American Scientist* 36: 577–606.

Jardine, N. 1991. *The Scenes of Inquiry: On the Reality of Questions in the Sciences.* Oxford: Clarendon Press.

Kauffman, S. A. 1993. *The Origins of Order: Self-Organization and Selection in Evolution.* Oxford: Oxford University Press.

———. 2003. "Understanding Genetic Regulatory Networks." *International Journal of Astrobiology* (special issue: *Fine-Tuning in Living Systems*) 2: 131–39.

Langton, C. G. 1989. "Artificial Life." In *Artificial Life: The Proceedings of an Interdisciplinary Workshop on the Synthesis and Simulation of Living Systems,* September 1987, edited by C. G. Langton. Redwood City, Calif.: Addison-Wesley.

Le Gros Clark, W. E., and P. B. Medawar, eds. 1945. *Essays on Growth and Form Presented to D'Arcy Wentworth Thompson.* Oxford: Oxford University Press.

Linsker, R. 1986. "From Basic Network Principles to Neural Architecture." *Proceedings of the National Academy of Sciences* 83: 7508–12, 8390–94, 8779–83.

———. 1988. "Self-Organization in a Perceptual Network." *Computer Magazine* 21: 105–17.

———. 1990. "Perceptual Neural Organization: Some Approaches Based on Network Models and Information Theory." *Annual Review of Neuroscience* 13: 257–81.

Malsburg, Christoph von der. 1973. "Self-Organization of Orientation Sensitive Cells in the Striate Cortex." *Kybernetik* 14: 85–100.

———. 1979. "Development of Ocularity Domains and Growth Behavior of Axon Terminals." *Biological Cybernetics* 32: 49–62.

Medawar, Peter B. 1958. "Postscript: D'Arcy Thompson and *Growth and Form.*" *D'Arcy Wentworth Thompson: The Scholar-Naturalist, 1860–1948.* London: Oxford University Press.

Merz, J. T. 1904/1912. *A History of European Thought in the Nineteenth Century.* 4 vols. London: Blackwood.

Neurath, O., ed. 1939. *International Encyclopedia of Unified Science.* 4 volumes. Chicago: University of Chicago Press.

Nisbet, H. B. 1972. *Goethe and the Scientific Tradition.* London: University of London, Institute of Germanic Studies.

Sacks, Oliver. 2001. "Henry Cavendish: An Early Case of Asperger's Syndrome?" *Neurology* 57: 1347.

Selfridge, Oliver G., E. L. Rissland, and M. A. Arbib, eds. 1984. *Adaptive Control in Ill-Defined Systems.* New York: Plenum.

Sherrington, C. S. 1942. *Goethe on Nature and on Science.* Cambridge: Cambridge University Press.

Sims, K. 1994. "Evolving 3D-Morphology and Behavior by Competition." *Artificial Life* 1: 353–72.

Szostak, J. W., D. P. Bartel, and P. L. Luisi. 2001. "Synthesizing Life." *Nature* 409: 387–90.

Terzopoulos, D., X. Tu, and R. Gzeszczuk. 1994. "Artificial Fishes with Autonomous Locomotion, Perception, Behavior, and Learning in a Simulated Physical World." In *Artificial Life IV (Proceedings of the Fourth International Workshop on the Synthesis and Simulation of Living Systems)*, edited by R. A. Brooks and P. Maes. Cambridge, Mass.: MIT Press.

Thompson, D'Arcy Wentworth, ed. 1880. *Studies from the Museum of Zoology in University College, Dundee*. Dundee: Museum of Zoology.

———, trans. and ed. 1883. *The Fertilization of Flowers by Insects*. London: Macmillan.

———. 1885. *A Bibliography of Protozoa, Sponges, Coelenterata and Worms, Including Also the Polozoa, Brachiopoda and Tunicata, for the years 1861–1883*. Cambridge: University Press.

———. 1895/1947. *A Glossary of Greek Birds*. Oxford: Oxford University Press.

———, trans. 1910. *Historia Animalium*. Volume 4, *The Works of Aristotle*, edited by J. A. Smith and W. D. Ross. Oxford: Clarendon Press.

———. 1917/1942. *On Growth and Form*. Cambridge: Cambridge University Press.

———. 1925. "Egyptian Mathematics: The Rhind Mathematical Papyrus—British Museum 10057 and 10058." *Nature* 115: 935–37.

———. 1931. *On Saithe, Ling and Cod, in the Statistics of the Aberdeen Trawl-Fishery, 1901–1929*. Edinburgh: Great Britain Fishery Board for Scotland.

———. 1940. *Science and the Classics*. London: Oxford University Press.

———. 1992. *On Growth and Form: The Complete Revised Edition*. New York: Dover.

Turing, Alan M. 1952. "The Chemical Basis of Morphogenesis." *Philosophical Transactions of the Royal Society* (series B) 237: 37–72.

Turk, G. 1991. "Generating Textures on Arbitrary Surfaces Using Reaction-Diffusion." *Computer Graphics* 25: 289–98.

Ulam, S. M. 1958. "John von Neumann, 1903–1957." *Bulletin of the American Mathematical Society* 64, no. 3(May): 1–49.

Vaucanson, Jacques de. 1738/1742/1979. *An Account of the Mechanism of an Automaton or Image Playing on the German Flute*. Translated by J. T. Desaguliers. Paris: Guerin; London: Parker; facsimile reprints of both, Buren, Berlin: Frits Knuf, 1979.

Waddington, C. H., ed. 1966–1972. *Toward a Theoretical Biology*. 4 volumes. Edinburgh: Edinburgh University Press.

Walter, W. Grey. 1950. "An Imitation of Life." *Scientific American* 182(5): 42–45.

Webster, G., and B. C. Goodwin. 1996. *Form and Transformation: Generative and Relational Principles in Biology*. Cambridge: Cambridge University Press.

Woodger, J. H. 1929. *Biological Principles: A Critical Study*. London: Routledge.

———. 1937. The *Axiomatic Method in Biology*. Cambridge: Cambridge University Press.

Wu, C. H. 1984. "Electric Fish and the Discovery of Animal Electricity." *American Scientist* 72: 598–607.

4 Alan Turing's Mind Machines

Donald Michie

Everyone who knew him agreed that Alan Turing had a very strange turn of mind. To cycle to work at Bletchley Park in a gas mask as protection against pollen, or to chain a tin mug to the coffee-room radiator to ensure against theft, struck those around him as odd. Yet the longer one knew him the less odd he seemed after all. This was because all the quirks and eccentricities were united by a single cause, the last that one would have expected, namely, a simplicity of character so marked as to be by turns embarrassing and delightful, a schoolboy's simplicity, but extreme and more intensely expressed.

When a solution is obvious, most of us flinch away. On reflection we perceive some secondary complication, often a social drawback of some kind, and we work out something more elaborate, less effective, but acceptable. Turing's explanation of his gas mask, of the mug chaining, or of other startling short cuts was "Why not?", said in genuine surprise. He had a deep-running streak of self-sufficiency, which led him to tackle every problem, intellectual or practical, as if he were Robinson Crusoe. He was elected to a fellowship of King's College, Cambridge, on the basis of a dissertation titled "The Central Limit Theorem of Probability," which he had rediscovered and worked out from scratch. It seemed wrong to belittle so heroic an achievement just on the grounds that it had already been done!

Alan Turing's great contribution was published in 1936, when he was twenty-four. While wrestling Crusoe-like with a monumental problem of logic, he constructed an abstract mechanism which had in one particular embodiment been designed and partly built a century earlier by Charles Babbage, the Analytical Engine. As a purely mathematical engine with which to settle an open question, the decidability problem (*Entscheidungsproblem*), Turing created a formalism that expressed all the essential properties of what we now call the digital computer. This abstract mechanism is the Turing machine. Whether or not any given mathematical function can in principle be evaluated was shown by Turing to be reducible to the question of whether a Turing machine, set going with data and an appropriate program of computation on its tape, will ever halt. For a long time I thought that he did not know about Babbage's earlier engineering endeavour. In all the talk at Bletchley about computing and its mathematical models, I never heard the topic of Babbage raised. At that time I was quite ignorant of the subject myself. But according to Professor Brian Randell's paper "The Colossus," delivered to the 1976 Los Alamos Conference on the History of Computing (see Randell 1976), Thomas H. Flowers "recalls lunch-time conversations with Newman and Turing about Babbage and his work." However that may be, the isolation and formal expression of the precise respect in which a machine could be described as "universal" was Turing's.

The universal Turing machine is the startling, even bizarre, centerpiece of the 1936 paper "On Computable Numbers with an Application to the Entscheidungsproblem" (Turing 1936). Despite its title, the paper is not about numbers in the restricted sense, but about whether and how it is possible to compute functions. A function is just a (possibly infinite) list of questions paired with their answers. Questions and answers can, of course, both be encoded numerically if we please, but this is part of the formalities rather than of the essential meaning.

For any function we wish to compute, imagine a special machine to be invented, as shown in figure 4.1. It consists of a read-write head, and a facility for moving from one field ("square," in Turing's original terminology) of an unbounded tape to the next. Each time it does this it reads the symbol contained in the corresponding field of the tape, a 1 or a 0 or a blank. This simple automaton carries with it, in its back pocket as it were, a table of numbered instructions ("states," in Turing's terminology). A typical instruction, say number 23 in the table, might be: "If you see a 1 then write 0 and move left; next instruction will be number 30; otherwise write a blank and move right; next instruction will be number 18."

To compute f(x)—say, the square root of—enter the value of x in binary notations as a string of 1's and 0's on the tape, in this case "110001," which is 49 in binary. We need to put a table of instructions into the machine's back pocket such that once it is set going the machine will halt only when the string of digits on the tape has been replaced by a new one corresponding precisely to the value of f(x). So if the tape starts with 110001, and the table of instructions has been correctly prepared by someone who wishes to compute square roots to the nearest whole number, then when the machine has finished picking its way backward and forward it will leave on the tape the marks "111," the binary code for 7.

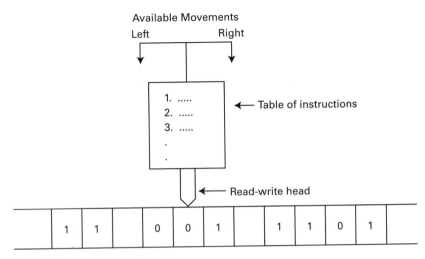

Figure 4.1
Constituents of a Turing machine. If a new "table of instructions" is supplied for each computation, then each use creates a new, special-purpose machine. If a once-and-for-all table ("language") is supplied, so that the specification of any given special machine which it is to simulate is placed on the input tape, then we have a universal Turing machine.

General Computations

When f = square root, we can well imagine that a table of instructions can be prepared to do the job. But here is an interesting question: How do we know this? Could this be knowable in general? Could a systematic procedure be specified to discover for every given function whether it is or is not Turing-computable, in the sense that a table of instructions could or could not be prepared?

In the process of showing that the answer is no, Turing generalized the foregoing scheme. He imagined an automaton of the same kind as that already described, except that it is a general-purpose machine. If we want it to compute the squareroot we do not have to change its instruction table. Instead we merely add to the tape, alongside the encoding of the number whose square root we want, a description of the square-root machine— essentially just its table of instructions. Now what is to stop the general-purpose machine from obeying the symbols of this encoding of the square-root machine's instruction table? "Plenty!" the astute reader at once replies. This new automaton, as so far described, consists again just of a read-write head. It has no "brain," or even elementary understanding of what it reads from the tape. To enable it to interpret the symbols which it encounters, another table of instructions must again be put into its back pocket—this time a master-table the effect of which is to specify a language in the form of rules of interpretation. When it encounters a description, in that language, of any special-purpose Turing machine whatsoever, it is able, by interpreting that description, faithfully to simulate the operations of the given special-purpose machine. Such a general-purpose automaton is a universal Turing machine. With a language in its back pocket, the machine is able to read the instructions "how to compute square roots," then the number, and after that to compute the square root.

Using this construction, Alan Turing was able to prove a number of far-reaching results. There is no space here to pursue these. Suffice it to say that when mathematicians today wish to decide fundamental questions concerned with the effectiveness or equivalence of procedures for function evaluation, or with the existence of effective procedures for given functions, they still have recourse to the simple-minded but powerful formal construction sketched above.

In practical terms the insights derivable from the universal Turing machine (UTM) are as follows: The value of x inscribed on the tape at the start corresponds to the data tape of the modern computing setup. Almost as obvious, the machine description added alongside corresponds to a program for applying f to this particular x to obtain the answer. What, then, is the table of instructions that confers on the UTM the ability to interpret the program? If the computer is a "naked machine" supplied by a manufacturer who provides only what is minimally necessary to make it run, then the table of instructions corresponds to the "order code" of that machine.[1] Accordingly the "machine description" appropriate to square root is a program written in the given order code specifying a valid procedure for extracting the square root. If, however, we ask the same question after we have already loaded a compiler program for, say, the early high-level programming language ALGOL-60, then we have in effect a new universal Turing machine, the "ALGOL-60 machine." In order to be interpretable when the machine runs under this new table of instructions, the square-root program must now be written, not in machine code, but in the ALGOL-60 language. We can see, incidentally, that indefinitely many languages, and hence different UTMs, are constructible.

There are various loose ends and quibbles. To head off misunderstanding I should add that the trivial example "square root" has been selected only for ease of exposition: the arguments hold for arbitrarily complicated problems. Second, what has been stated only applies, strictly, to computers with unbounded memory. Third, the first thing that a modern machine ordinarily does is to "read in" both data and program, putting the contents of the Turing "tape" into memory. The Turing machine formalism does not bother with this step since it is logically immaterial whether the linear store ("tape") is to be conceived as being inside or outside: because it is notionally unbounded, it was doubtless easier originally to picture it as "outside"!

From the standpoint of a mathematician this sketch completes the story of Turing's main contribution. From the point of view of an information engineer such as me, it was only the beginning. In February 1947 Alan Turing delivered a public lecture at the London Mathematical Society. In it he uttered the following (Turing 1947, pp. 122–123):

It has been said that computing machines can only carry out the purposes that they are instructed to do.... But is it necessary that they should always be used in such a manner? Let us suppose that we have set up a machine with certain initial instruction tables, so constructed that these tables might on occasion, if good reason arose, modify these tables. One can imagine that after the machine had been operating for some time, the instructions would have been altered out of recognition, but nevertheless still be such that one would have to admit that the machine was still doing very worthwhile calculations. Possibly it might still be getting results of the type desired when the machine was first set up, but in a much more efficient manner. In such a case one could have to admit that the progress of the machine had not been foreseen when its original instructions were put in. It would be like a pupil who had learnt much from his master, but had added much more by his own work. When this happens I feel that one is obliged to regard the machine as showing intelligence. As soon as one can provide a reasonably large memory capacity it should be possible to begin to experiment on these lines.

Ten years were to pass before the first experiments in machine learning were undertaken, by Arthur Samuels at IBM (Samuels 1959), and thirty-five years before conceptual and programming tools made possible the experimental assault that is gathering force today along the Turing line. For consider modification not only of the data symbols on the UTM tape but also of the machine-description symbols—modification of the program by the program! My own laboratory constituted one of the resources dedicated to this "inductive learning" approach.

In a particular sense, Alan Turing was anti-intellectual. The intellectual life binds its practitioners collectively to an intensely developed skill, just as does the life of fighter aces, of opera stars, of brain surgeons, of yachtsmen, or of master chefs. Strands of convention, strands of good taste, strands of sheer snobbery intertwine in a tapestry of myth and fable to which practitioners meeting for the first time can at once refer for common ground. Somewhere, somehow, in early life, at the stage when children first acquire ritual responsiveness, Turing must have been busy with something else.

Brute-Force Computation

The Robinson Crusoe quality was only one part of it. Not only independence of received knowledge but avoidance of received styles (whether implanted by fashion or by long tradition) gave him a form of pleasure not unmixed with glee. There was much of this in his recurrent obsession with attacking deep combinatorial problems by brute-force computation. This was at the heart of some of his cryptanalytical successes—notably his crucial inroad into the German Enigma cipher while working at Bletchley Park. It is difficult now to remember how startling, and to persons of mathematical taste how grating and offensive, was the notion of near-exhaustive enumeration of cases as an approach to a serious problem. Yet negative reactions to Ken Appel and Wolfgang Haken's computer-aided proof of the four-color theorem (Appel, Haken, and Koch 1977) gives a base from which to extrapolate back to the year 1943, the year my personal acquaintance with Alan Turing was formed at Bletchley Park. At that instant I was on the verge of becoming a founding member of a team led by Turing's prewar mentor, Max Newman, in a mechanized attack on a class of German ciphers collectively known as "Fish." Our machines were special-purpose. But they showed what could be done by vacuum-tube technology in place of electromechanical switching, inspiring both Newman and Turing in their seminal postwar roles in developing the first-ever high-speed general-purpose computing. A digression on this earlier phase is in order.

During the war the Department of Communications of the British Foreign Office was housed at Bletchley Park, Buckinghamshire, where secret work on cryptanalysis was carried out. As part of this work various special machines were designed and commissioned, the early ones being mainly electromechanical, the later ones electronic and much closer to being classifiable as program-controlled computers.

The Bletchley Machines

The first of the electromechanical machines, the "Heath Robinson," was designed by Charles Wynn-Williams at the Telecommunications Research Establishment at Malvern. At Bletchley one of the people with influence on design was Alan Turing. The machine incorporated two synchronized

photoelectric paper tape readers, capable of reading three thousand characters per second. Two loops of five-hole tape, typically more than one thousand characters in length, would be mounted on these readers. One tape
would be classed as data, and would be stepped systematically relative
to the other tape, which carried some fixed pattern. Counts were made of
any desired Boolean function of the two inputs. Fast counting was performed electronically, and slow operations, such as control of peripheral
equipment, by relays. The machine, and all its successors, were entirely automatic in operation, once started, and incorporated an on-line output tele-
printer or typewriter.

Afterward, various improved "Robinsons" were installed, including the
"Peter Robinson," the "Robinson and Cleaver," and the "Super Robinson."
This last one was designed by T. H. Flowers in 1944, and involved four
tapes being driven in parallel. Flowers, like many of the other engineers
involved in the work, was a telephone engineer from the Post Office Research Station.

The electronic machines, known as the Colossi because of their size, were
developed by a team led by Professor Max H. A. Newman, who started the
computer project at Manchester University after the war. Other people directly involved included Tommy Flowers, Allen W. M. Coombs, Sidney W.
Broadhurst, William Chandler, I. J. "Jack" Good, and me. During the later
stages of the project several members of the U.S. armed services were
seconded at various times to work with the project for periods of a year or
more.

Flowers was in charge of the hardware, and in later years designed an
electronic telephone exchange. On his promotion, his place was taken by
Coombs, who in postwar years designed the time-shared transatlantic multichannel voice-communication cable system. After the war, Good was for a
time associated with the Manchester University computer project, and
Coombs and Chandler were involved in the initial stages of the design of
the ACE (automatic computing engine) computer at the National Physical
Laboratory, before building the MOSAIC computer at the Post Office Research Station. Alan Turing was not directly involved in the design of the
Colossus machine, but with others he specified some of the requirements
that the machines were to satisfy. It has also been claimed by Good that
Newman, in supervising the design of the Colossi, was inspired by his
knowledge of Turing's 1936 paper.

In the Colossus series almost all switching functions were performed by
hard valves, which totaled about two thousand. There was only one tape,
the data tape. Any preset patterns which were to be stepped through these

data were generated internally from stored component patterns. These components were stored in ring registers made of thyratrons and could be set manually by plug-in pins. The data tape was driven at 5,000 characters per second. In the Mark 2 version of the machine an effective speed of 25,000 characters per second was obtained by a combination of parallel operations and short-term memory. Boolean functions of all five channels of pairs of successive characters could be set up by plug-board, and counts accumulated in five bi-quinary counters.

The first Colossus was installed by December 1943, and was so successful that three Mark 2 Colossi were ordered. By great exertions the first of these was installed before D-day (June 6, 1944). By the end of the war about ten Colossi had been installed, and several more were on order.

My point of departure for this digression was Alan Turing's readiness to tackle large combinatorial problems by means that entailed brute-force enumeration of cases. His design of the "Bombe" machine for cracking Enigma codes was a success of this character. The Colossus story was also one of exhaustive searches, increasingly with the aid of man-machine cooperation in the search for give-away statistical clues. Some kinds of snobbery conceive "pure thought" as flashes of insight—a kind of mystical ideal. The humdrum truth of the matter is then allowed to escape, namely, that for sufficiently tough problems the winning formula prescribes one part insight to many parts systematic slog. Nowhere can this truth have been more deeply embedded in daily reality than in the gradual delegation at Bletchley of ever more of the intellectual slog to the proliferating new varieties of machines.

Of course the abstract notion of combinational exhaustion was already deeply entrenched in mathematics. But what about the use of a physical device to do it? To make such proposals in earnest seemed to some people equivalent to bedaubing the mathematical subculture's precious tapestry with squirtings from an engineer's oilcan. Writing of an earlier juncture of intellectual history, Plutarch in "The Life of Marcellus," has left an unforgettable account (Plutarch 1917, 473):

Eudoxus and Archylas had been the first originators of this far-famed and highly prized art of mechanics, which they employed as an elegant illustration of geometrical truths, and as a means of sustaining experimentally, to the satisfaction of the senses, conclusions too intricate for proof by words and diagrams.... But what with Plato's indignation at it, and his invectives against it as the mere corruption and annihilation of the one good of geometry—which was thus shamefully turning its back on the unembodied objects of pure intelligence to recur to sensation, and to ask for help...from matter; so it was that mechanics came to be separated from geometry, and, repudiated and neglected by philosophers, took its place as a military art.

It was indeed in a military art, cryptography, that Turing's first practical mechanizations made their debut. It is also of interest that in a paper submitted as early as 1939 (not published until 1943, owing to wartime delays) a mechanizable method is given for the calculation of Georg Riemann's zeta-function suitable for values in a range not well covered by previous work. Why was Turing so interested in this? The answer would undoubtedly serve as another red rag to Plato's ghost, for the point at issue was a famous conjecture in classical pure mathematics: Do all the zeros of the Riemann function lie on the real line? In a postwar paper the oilcan reappears in an attempt to calculate a sufficiency of cases on a computing machine to have a good chance either of finding a counterexample and thus refuting the Riemann hypothesis or, alternatively, of providing nontrivial inductive support. The attempt, which was reported in the 1953 *Proceedings of the London Mathematical Society* (Turing 1953), failed owing to machine trouble.

Machine trouble! Alan's robust mechanical ineptness coupled with insistence that anything needed could be done from first principles was to pip many a practical project at the post. He loved the struggle to do the engineering and extemporization himself. Whether it all worked in the end sometimes seemed secondary. I was recruited at one point to help in recovering after the war some silver he had buried as a precaution against liquidation of bank accounts in the event of a successful German invasion. After the first dig, which ended in a fiasco, we decided that a metal detector was needed. Naturally Alan insisted on designing one, and then building it himself. I remember the sinking of my spirits when I saw the contraption, and then our hilarity when it actually seemed to be working. Alas its range was too restricted for the depth at which the silver lay, so that positive discovery was limited to the extraordinary abundance of metal refuse which lies, so we found, superficially buried in English woodlands.

The game of chess offered a case of some piquancy for challenging with irreverent shows of force the mastery that rests on traditional knowledge. At Bletchley Park, Turing was surrounded by chess masters who did not scruple to inflict their skill upon him. The former British champion Harry Golombek recalls an occasion when instead of accepting Turing's resignation he suggested that they turn the board round and let him see what he could do with Turing's shattered position. He had no difficulty in winning. Programming a machine for chess played a central part in the structure of Turing's thinking about broader problems of artificial intelligence. In this he showed uncanny insight. As a laboratory system for experimental work chess remains unsurpassed. But there was present also, I can personally vouch, a Turing streak of iconoclasm: What would people say if a machine beat a master? How excited he would be today when computer programs based on his essential design are regularly beating masters at lightning chess, and producing occasional upsets at tournament tempo!

Naturally Turing also had to build a chess program (a "paper machine" as he called it). At one stage he and I were responsible for hand-simulating and recording the respective operations of a Turing-Champernowne and a Michie-Wylie paper machine pitted against each other. Fiasco again! We both proved too inefficient and forgetful. Once more Alan decided to go it alone, this time by programming the Ferranti Mark 1 computer to simulate both. His problems, though, were now compounded by "people problems," in that he was not at all sure whether Tom Kilburn and others in the Manchester laboratory, where he was working by that time, really approved of this use for their newly hatched prototype. It was characteristic of Turing, who was in principle anarchistically opposed to the concept of authority or even of seniority, that its flesh-and-blood realizations tended to perplex him greatly. Rather than confront the matter directly, he preferred tacitly to confine himself to nocturnal use of the machine. One way and another, the program was not completed.

It is fashionable (perhaps traditional, so deep are subcultural roots) to pooh-pooh the search-oriented nature of Turing's thoughts about chess. In his Royal Society obituary memoir, Max Newman observes in words of some restraint that "it is possible that Turing under-estimated the gap that separates combinatory from position play." Few yet appreciate that, by setting the ability of the computer program to search deeply along one line of attack on a problem in concert with the human ability to conceptualize the problem as a whole, programmers have already begun to generate results of deep interest. I have not space to follow the point here, but will simply exhibit, in figure 4.2, a paradigm case. Here a program cast in the Turing-Shannon mould, playing another computer in 1977, apparently blundered. The chess masters present, including the former world champion Mikhail Botvinnik, unanimously thought so. But retrospective analysis showed that in an impeccably pure sense the move was not a blunder but a brilliancy, because an otherwise inescapable mate in five (opaque to the watching masters) could by this sacrifice be fended off for another fifteen or more moves.

The equivocal move by Black, who has just been placed in check by the White Queen in the position shown, was 34 ... R–K1, making a free gift of the Rook. The program, Kaissa, had spotted that the "obvious" 34 ... K–N2 could be punished by the following sequence:

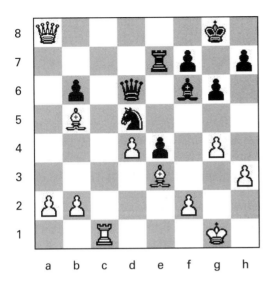

Figure 4.2
The paradigm, derived by Turing and Claude Shannon for game playing, implemented on an IBM three-million-instructions-per-second computer, probes beyond the tactical horizons of even a grand master. In this match from Toronto in 1977, the chess-playing software Kaissa, playing black, continued R–Kl. It looks like a blunder—but was it?

35.	Q–B8 ch.	K × Q (forced)
36.	B–R6 ch.	B–N2 (or K–N1)
37.	R–B8 ch.	Q–Q1
38.	R × Q ch.	R–K1
39.	R × R mate	

Suppose now that we interpret the situations-and-actions world of chess as an analogy of computer-aided air-traffic control, or regulation of oil platforms or of nuclear power stations. If assigned to monitoring duty, Grand Master Botvinnik would undoubtedly have presumed a system malfunction and would have intervened with manual override! Kaissa's deep delaying move (in the parable, affording respite in which to summon ambulances, fire engines, and so forth) would have been nullified.

These examples, taken from the computer chess world of twenty-five years ago, no more than touch the surface of the human mind's predicament, faced by ever more impenetrable complexity. With the likes of Kaissa there was, after all, the thought that it was still within the realm of technical feasibility to equip such a brute-force device with some sort of "self-explanation harness." In the twenty-first century the matter now stands rather differently, at least in the case of chess, if not in the regulation of oil platforms and power stations.

Enter a Mega Monster

A brute-force calculating chess monster, Hydra, has now emerged. Developed in the United Arab Emirates by a four-man team led by Dr. Chrilly Donninger, it runs on thirty-two processors, each enhanced by special FPGA chess hardware (C. Donninger and U. Lorenz 2004). FPGA stands for "field-programmable gate array," a type of logic chip that can be directly programmed, almost as though it were software but running at modern hardware speeds. Hydra can assess potential positions in look-ahead at a rate of 200 million per second. For each of the possible five-piece endgames, Hydra's databases allow the machine to look up the best or equal-best move and theoretical worst-case outcome in every possible situation. Hydra searches in the middle-game typically to depth 18 to 19, and in the endgame to depth 25. At the nominal depth 18 to 19, the longest variations are searched to about depth 40 (the theoretical limit is 127), the shortest one to depth 8. The search tree is strongly nonuniform.

A six-game match between the Hydra chess machine and Britain's number one grand Master, Michael Adams, took place at the Wembley Centre in London from on June 21 to 27, 2005. One of the most lop-sided chess matches in recent memory ended with the nearest thing to a whitewash. In six games at regular time controls Adams succeeded in achieving a single draw, in game 2 with a clever save in an essentially lost position. In the other five games he was crushed by the machine.

Meanwhile comparable trends characterize the technologies that are increasing our dependence while also adding to planetary perils. Human incomprehension of increasingly intricate systems is part of the problem. What chance of "self-explanation harnesses"?

Suppose that a Hydra look-alike, call it the Autocontroller, were converted to act as a nuclear power station control computer. There could come a moment at which, having searched out possible "engineer-intervention/system response" sequences to a depth of, say, 20, the Autocontroller displays a message: "Only 67,348 stability-restoring paths available. Partial explanations of key subpaths can be displayed on request. WARNING: At normal reading speeds total human time to scan explanations is estimated at 57 mins 22 secs; time available before next cluster of control decisions is 3 mins 17 secs. RECOMMENDATION: Press 'Trust Autocontroller' button." What are the harassed control engineers to do?

Broader Horizons

Increasing numbers of industrial and military installations are controlled by problem-solving computing systems. The cloak cast by combinatorial complexity over the transparency of machine functions has thus acquired topical urgency. Computer analyses of chess and other combinatorial domains, originally inspired by Alan Turing, are today yielding new approaches to problems of seemingly irreducible opacity.

Note

1. "Order code," a term used in the early days of computing, is synonymous with "operation code"—the portion of a machine-language instruction that specifies the operation to be performed.

References

Appel, Ken, Wolfgang Haken, and J. Koch. 1977. "Every Planar Map Is Four Colorable." *Journal of Mathematics* 21: 439–567.

Donninger, Chrilly, and U. Lorenz. 2004. "The Chess Monster Hydra." In *Proceedings of the Fourteenth International Conference on Field-Programmable Logic and Applications (FPL)*, edited by J. Becker, M. Platzner, and S. Vernalde. Lecture Notes in Computer Science, volume 3203. New York: Springer.

Plutarch. 1917. *Parallel Lives*. Translated by Bernardotte Perrin. Volume 5. Cambridge, Mass.: Harvard University Press, Loeb Classical Library.

Randell, Brian. 1976. "The Colossus." Technical Report Series No. 90. Newcastle, UK: University of Newcastle, Computing Laboratory.

Samuels, Arthur L. (1959). "Some Studies in Machine Learning Using the Game of Checkers." *IBM Journal of Research & Development* 3, no. 3: 210–29.

Turing, Alan M. 1936. "On Computable Numbers, with an Application to the Entscheidungsproblem." *Proceedings of the London Mathematical Society* 2, no. 42: 230–65.

———. 1947. "Lecture to the London Mathematical Society on 20 February 1947." In *A. M. Turing's ACE Report of 1946 and Other Papers*, edited by B. E. Carpenter and R. W. Doran. Cambridge, Mass.: MIT Press.

———. 1953. "Some Calculations of the Riemann Zeta-Function." *Proceedings of the London Mathematical Society* 3, no. 3: 99–117.

Editors' Note

Donald Michie (1923–2007) Sadly, Donald died in a car accident just as this book was going to press, so his chapter will be one of his last publications. He and his former wife, Dame Anne McClaren, a highly distinguished biologist, died together in the accident. He was educated at Rugby school and Balliol College, Oxford, where he was awarded an open scholarship to study classics in 1942. However, he decided to defer entry and in 1943 enrolled for training in cryptography and was soon recruited to Bletchley Park in Buckinghamshire, Britain's wartime code-cracking headquarters. There he worked with Alan Turing, Jack Good, Max Newman, and others in a highly successful team that made many invaluable contributions to the war effort. During this period Donald made a number of important advances in the use of early computing techniques in cryptology. After the war he took up his place at Oxford but his experiences at Bletchley Park had given him a passion for science, so he switched from classics and received his MA in human anatomy and physiology in 1949. This was followed by a DPhil in genetics, a field in which he made several important contributions, some with Anne McClaren, whom he married in 1952. From about 1960 he decided to concentrate his efforts on machine intelligence—a field he had first become interested in through his work with Turing—and dedicated the rest of his career to it. He did much to galvanize the area in Britain, founding the department of machine intelligence and perception at Edinburgh University in 1966. He made a number of important contributions to machine learning and edited a classic series of books on machine intelligence. In 1984 he founded the Turing Institute in Glasgow, which conducted industrially oriented machine intelligence research for several years. He received numerous honorary degrees and achievement awards of learned societies in computing and artificial intelligence. He had a lifelong commitment to socialism, integrating scientific inquiry with the struggle for social justice.

5   What Did Alan Turing Mean by "Machine"?

Andrew Hodges

Machines and Intelligence

Alan Turing died in June 1954, before the term "artificial intelligence" was established. He might have preferred the term "machine intelligence" or "mechanical intelligence," following the phrase "Intelligent Machinery" in the (then still unpublished) report he wrote in 1948 (Turing 1948/ 2004). This provocative oxymoron captured what he described as a "heretical theory." This article is centered on that 1948 report, and the much more famous philosophical paper that followed it in 1950 (Turing 1950), but it is not intended to add to the detailed attention that has been lavished on Turing's ideas about "intelligence." Turing's 1950 paper is one of the most cited and discussed in modern philosophical literature—and the 1948 work, originally unpublished, has also come to prominence, for instance in the elaborate trial of Turing's networks by Teuscher (2002). Instead, it will examine the other half of Turing's deliberately paradoxical expression: the question of what he meant by "machine" or "mechanical." This is equally important to the theory and practice of artificial intelligence. Whereas previous thinkers had conceived of homunculi, automata, and robots with human powers, the new setting of the digital computer gave a far more definite shape to the conception of the "mechanical."

To examine the meaning of Turing's references to machinery in 1948 we first need to go back to the Turing machine of 1936 (Turing 1936). At first sight it might seem that Turing had mastered the whole area with his definitions and discoveries at that time, leaving little room for comment, but the situation is in fact not so clear.

The Turing Machine and Church's Thesis

We should first look back further, to about 1932. This is when, in a private essay (Turing 1932), Turing showed his youthful fascination with the physics of the brain. It rested on an idea, made popular by Arthur Eddington, that the indeterminacy of quantum mechanics might explain the nature of consciousness and free will. It is important to remember that the conflict between the appearance of free will and the deterministic explanation of physical phenomena has always been a central puzzle in science, vital to the whole materialist standpoint. Turing was aware of it from an early age. It is this question of the *physical content* of mechanistic explanation—focusing on the physical properties of the brain—that underlies the discussion that follows.

When in 1936 Turing (1936) gave an analysis of mental operations appropriate to his discussion of the Entscheidungsproblem, he did not address himself to this general question of free will. He confined himself to considering a human being following some definite computational rule, so as to give a precise account of what was meant by "effective calculability." His assumption of a finite memory and finite number of states of mind is, therefore, only stated in this context. It does not consider what a human mind might achieve when not confined to rule following, and so exercising wilful choice. (In retrospect, these bold assumptions seem to set the stage for Turing's later thesis about how a computer could simulate *all* kinds of mental operations, but we have no way of knowing what Turing's views were in this early period.) Another question that is not addressed in his 1936 work is what could be achieved by a physical machine, as opposed to the model human rule follower.

The reason for emphasizing this negative is that when Church (1937/1997) reviewed Turing's paper in 1937, endorsing its definition of effective calculability, he attributed to Turing a definition of computability expressed in terms of machines of finite spatial dimension:

[Turing] proposes as a criterion that if an infinite sequence of digits 0 and 1 be "computable" that it shall be possible to devise a computing machine, occupying a finite space and with working parts of finite size, which will write down the sequence to any desired number of terms if allowed to run for a sufficiently long time. As a matter of convenience, certain further restrictions are imposed in the character of the machine, but these are of such a nature as obviously to cause no loss of generality—in particular, a human calculator, provided with pencil and paper and explicit instructions, can be regarded as a kind of Turing machine.

What Church wrote was incorrect, for Turing had *not* proposed this criterion. Turing gave a careful model of the human calculator, with an analysis of mental states and memory, which Church's summary ignored, and he said nothing about "working parts" or "finite size." Yet Turing recorded no objection to this description of his work. In his 1938 Ph.D. thesis

(Turing 1938b) he gave a brief statement of the Church-Turing thesis, using the words "purely mechanical process," equated to "what could be carried out by a machine." Turing's expression is less sweeping than Church's, since the words "a machine" could be read as meaning "a Turing machine." But he made no effort whatever to put Church right and insist on his human calculator model. Church (1940) repeated his definition in 1940, and Turing himself, as we shall see, moved seamlessly between humanly applied methods and "machines," even though he had given no analysis or even explanation of what was meant by "machine" comparable with the careful discussion of what he meant by and assumed to be true of a human calculator.

It is puzzling why Church so freely adopted this language of machines in the absence of such an analysis and why Turing apparently went along with it. One possible factor is that the action of the human mathematician carrying out the rules of formal proof "like a machine" was in those pre-computer days so much more complex than any other imaginable machine. Human work was naturally the logicians' focus of attention, and as Turing (1948/1986) put it in his 1948 report, engineered machines were "necessarily limited to extremely straightforward" tasks until "recent times (e.g. up to 1940)." This was a coded reference to his own Enigma-breaking Bombe machines (by no means straightforward) of that year, and confirms that in the 1936 period he saw nothing to learn from extant machinery.[1]

But it is still surprising that Turing did not insert a caveat raising the question of whether there might in principle be machines exploiting physical phenomena (in quantum mechanics and general relativity, say) that would challenge the validity of Church's assumptions based on naive classical ideas of parts, space, and time. Turing had a very good background in twentieth-century physics and as we have noted had already suggested that quantum mechanics might play a crucial role in the functioning of the brain. This question is particularly fascinating because his (1938b) work discussed *uncomputable* functions in relation to the human "intuition" involved in seeing the truth of a formally unprovable Gödel sentence, an apparently nonmechanical action of mind. What role did he think the physical brain was playing in such "seeing"? Unfortunately, it is impossible to know what he thought in this prewar period; his statements avoided the word "brain."

A quite different interpretation has been given however, by the philosopher B. J. Copeland, who has now edited a selection of Turing's papers (Copeland 2004). In this and numerous other publications, Copeland makes much of the idea that by discussing effective calculation of the human calculator, Turing expressly excluded the question of what machines might be able to do. Copeland (2002) has further asserted that Church also endorsed only Turing's formulation of the human rule follower. This is simply not true, as can be seen from Church's review as previously quoted. Copeland has also made the more dramatic claim that Turing expressly allowed for the possibility of machines more powerful than Turing machines. Thus Copeland (2002) suggests that the reason for Turing's restriction to a human calculator was that "among a machine's repertoire of atomic operations there may be those that no human being unaided by machinery can perform." This argument is not, however, to be found in Turing's writing.

Specifically, the extraordinary claim is made by Copeland and Proudfoot (1999) that Turing's "oracle-machine" is to be regarded as a machine that might be physically constructed. Now, Turing's "oracle" is a postulated element in the advanced logical theory of his PhD thesis that "by unspecified means" can return values of an uncomputable function (e.g., say, of any Turing machine, whether it halts or not). Mathematical logicians have taken it as a purely mathematical definition, giving rise to the concept of relative computability (for a review see Feferman 1988). This is not quite the whole story, because Turing was certainly concerned with the extra-mathematical question of how mental "intuition" seems to go beyond the computable. However, it is essentially something postulated for the sake of argument, not something supposed to be an effective means of calculation. An "oracle-machine" is a Turing machine whose definition is augmented so that it can "call the oracle."

Although Turing emphasized that the oracle "cannot be a machine," Copeland asserts that the oracle-machine which calls it *is* a machine. He rests this argument on the observation that Turing introduced the oracle-machine concept as "a new kind of machine." Yet to consider an oracle-machine a machine would obviously contradict Turing's basic statement in his thesis that effectively calculable functions are those that "could be carried out by a machine," and that these are the Turing-computable functions. How could Turing have equated effective calculation with the action of Turing machines, if he was introducing a more powerful "kind of machine" in that same 1938 work? This makes no sense, and to interpret what Turing meant by "new kind of machine," we need only note what "kinds of machines" he had defined in 1936. These were the "automatic" machines and "choice" machines, the former being what we call Turing machines and the latter being a generalized "kind of machine" calling for the intervention of an operator. The oracle-machines follow this model: they are like the choice machines in being only partially mechanical. The steps that call the oracle are, indeed, described by Turing as "non-mechanical."

Copeland and Proudfoot (1999), however, insist that the words "new kind of machine" mean that Turing imagined the oracle-machine as something that might be technologically built to compute uncomputable functions in practice; they announce that the search is now under way for a physical oracle that would usher in a new computer revolution. They draw a picture of the oracle as a finite black box. They further argue (Copeland and Proudfoot 2004) that the oracle can be a nonmechanical part of a machine in the same sense that "ink" can be. A machine prints with ink (which is not a machine); likewise a machine can call on an oracle (which is not a machine). The analogy is untenable: there is nothing inherent in ink (or, more properly, the physical implementation of a logical state) that introduces a function infinitely more complex than that of the machine itself. In contrast, the whole point of an oracle is that it does just this. Later we shall see further evidence that Turing never saw an oracle-machine as a purely mechanical process.

To summarize, Turing's loose use of the expression "kind of machine" to introduce a class of partially mechanical concepts should not be allowed to confuse the issue. Rather, what we learn from the classic texts is that Church and Turing seem to have supposed, without detailed analysis, that the "purely mechanical" would be captured by the operations of Turing machines. They did not draw a clear distinction between the concepts of "a machine" and "a mechanical process applied by a human being."

Turing's Practical Machines: The Wartime Impetus

Despite its "very limited" character, the physical machinery available in 1937 held remarkable appeal for Turing. Unusually for a mathematician, Turing had a fascination with building machines for his own purposes. He used electromagnetic relays to build a binary multiplier intended for use in a secure encipherment scheme, and another machine for calculating approximate values for the Riemann zeta-function (see Hodges 1983). As is now famous, this combination of logic and practical machinery took Turing to the center of operations in the Second World War, where his machines and mechanical processes eclipsed traditional code-breaking methods. In the course of this work Turing gained an experience of electronic switches, again, by building a speech-encipherment machine with his own hands. Electronic components provided the microsecond speed necessary for effective implementation of what Turing called a "practical version" of the universal machine: the digital computer.

From 1943 onward, Turing spoke of building a *brain*, using the word absent from his 1936–38 work on computability. In his technical prospectus for a digital computer, Turing (1946) gave an argument justifying this hyperbolic vocabulary. This was a discussion of computer chess playing, with a comment that "very good chess" might be possible if the machine were allowed to make "occasional serious mistakes." This somewhat mysterious comment was clarified by Turing in 1947, when he argued (Turing 1947) that Gödel's theorem is irrelevant if infallibility is not demanded. Then and later he developed his view that what appears to be nonmechanical "initiative" is actually computable, so that the apparent oxymoron of "machine intelligence" makes sense. It seems very likely that Turing had formed this view by the end of the war, and so could feel confident with a purely mechanistic view of the mind, uncontradicted by Gödel's theorem. In fact, Turing's discussion of chess playing and other "intelligence" ideas around 1941 suggests that he formed such a conviction during that period. This was, of course, when his mechanical methods, using Bayesian inference algorithms as well as physical machinery, were first so dramatically supplanting the traditional role of human judgment in code breaking.

In 1946–47, Turing began a discussion of fundamental aspects of physical machines of a kind absent from his prewar work. He did not simply assume it straightforward to embody logically discrete states in physical machinery; his 1946 discussion of the implementation of computation with electronic parts was notable for its emphasis (learned from wartime experience) on avoidance of errors (Turing 1946/1986). Speaking to an audience of mathematicians in 1947, he gave a more abstract account of what it means to implement discrete states, in terms of disjoint sets in the configuration space of a continuous physical system (Turing 1947). This was the first suggestion of serious analysis relating Turing's logical and physical worlds.

Turing's 1948 Report: Physical Machines

We have now seen the background to the 1948 report, written for the National Physical Laboratory, London, where Turing was employed on his computer project. In this report, Turing went on to give a more directly physical content to the concept of machine. First, Turing discussed how the finite speed of light places a limit on the speed at which computations can take place. We may be amused that Turing assumed components of a computer must be separated by a centimeter, which makes his estimate of potential speed ridiculously slow by modern standards. However, he was of course correct in identifying the speed of light as a vital constraint, and it is this limitation that continues to drive miniaturization. Second, Turing calculated from statistical mechanics the probability of an electronic valve falling into the wrong state through the chance motion of its electrons: his result was that there would be virtual certainty of error in $10^{10^{17}}$ steps. Such a calculation was quite typical of Turing's approach using fundamental physics: J. L. Britton (1992) has recalled another example from Turing's Manchester period, when he gave a lecture based on the number N, defined as the odds against a piece of chalk leaping across the room and writing a line of Shakespeare on the board. However, it is again rather surprising that he made no explicit mention of quantum physics as underlying electronics, and of course he thereby missed the opportunity to anticipate the limits of miniaturization, and the possibility of quantum computing, which now are such salient features in the frontiers of computer technology.

In summarizing the properties of computability in this 1948 report, Turing was too modest to use the expression "Turing machine" but used the expression "logical computing machine" (LCM) instead. When interpreting it he made nothing of the distinction, held to be of paramount importance by Copeland, between machines and humanly applied mechanical processes. Turing summarized Church's thesis as the claim that any "rule of thumb" could be carried out by an LCM, which can indeed be taken to be an informal reference to the 1936 human calculator model. But Turing illustrated the idea of "purely mechanical" quite freely through examples of physical machines. Turing's (1946/1986) computer plan had described the function of the computer as replacing human clerks, but the 1948 report said that "the engineering problem of producing various machines for various jobs is replaced by . . . programming the universal machine." When in this report he described a procedure that, with no computers yet available, certainly *was* a human-based rule, he called it a "paper machine." He said that a computing machine can be imitated by a man following a set of rules of procedure—the reverse of Copeland's dictum. In practice, Turing wove these two models together into a discussion of "Man as a Machine," with the brain as his focus of interest. Turing still had not given any indication of why *all* possible engineered machines, going beyond the immediately practical, could be emulated by programs—in other words, Turing machines. However, his introduction of physical concepts made a start on answering this question.

Turing noted an obvious sense in which it is clearly *not* true that all machines can be emulated by Turing machines: the latter cannot milk cows or spin cotton. Turing dealt with this by making a distinction between "active" and "controlling" machinery; it is the latter, which we might call information-theoretic, which are compared with LCMs. The former (Turing's down-to-earth example: a bulldozer) are not. This distinction could be regarded as simply making explicit something that had always been implicit in references to mechanical processes: we are concerned with what makes a process mechanical in its nature, not with what the process physically effects. Intuitively, the distinction is clear, but at a deeper level, it opens up questions linking physics and information theory, which Turing did not deal with, for example, how can we characterize the kind of physical system that will be required to embody an LCM, and given a physical system, how can we characterize its capacity for storing and processing information?

Continuity and Randomness

Turing's 1948 report made a further distinction between "discrete" and "continuous" machines. Only the discrete-state machines can be considered LCMs. As regards "continuous" machines (where Turing's example was a telephone) it is worth noting that Turing was no newcomer to continuity in mathematics or physics. He was an old hand, both in theory and in practice. Even in 1936, he had hoped to extend computability to continuous analysis. One of his many contributions to pure mathematics was his work on discrete approximation to continuous groups (Turing 1938a). When he wrote in 1950 that every discrete-state machine was "really" based on continuous motion (Turing 1950), with a picture of a three-way rotating switch, this was on the basis of his experience ten years earlier with the Bombe, whose rapidly rotating commutators made millisecond connections thanks to expert engineering. The applications in his (1946/ 1986) 1946 computer plan included traditional applied mathematics and physics problems, and his software included floating-point registers for handling (discrete approximations to) real numbers. His important innovation in the analysis of matrix inversion (Turing 1948) was likewise driven by problems in continuous analysis. A notable point of Turing's 1947 London Mathematical Society talk is that from the outset he portrayed the discrete digital computer as an improvement on the continuous "differential analysers" of the 1930s, because of its unbounded capacity for accuracy. He did this in practice: he turned his prewar analogue zeta-function-calculating machine into a program for the Manchester computer (Turing 1953).

In the 1948 report his examples were designed to focus on the brain, which he declared to be continuous but "very similar to much discrete machinery," adding that there was "every reason to believe" that an entirely discrete machine could capture the essential properties of the brain. One reason for such belief was given more explicitly in the 1950 paper (Turing 1950), his answer to the "Argument from Continuity in the Nervous System." It is first worth noting that this "continuity in the nervous system" argument is an objection to a thesis that Turing had not quite explicitly made in that 1950 paper, viz., that computable operations with a discrete-state machine can capture all the functions of the physical brain relevant to "intelligence." It is there implicitly, in his response to this objection, and indeed it is implicit in his estimate of the number of bits of storage in a human brain.

His argument against the significance of physical continuity was that introducing randomness into the discrete machine would successfully simulate the effect of a continuous machine. Turing introduced machines with random elements in his 1948 report. In the 1950 paper he developed this into an interesting argument that now would be seen as the opening up of a large area to do with dynamical systems, chaotic phenomena, and computable analysis. He referred to the traditional picture of Laplacian determinism, holding that the determinism of the discrete-state machine model is much more tractable (Turing 1950, 440):

> The displacement of a single electron by a billionth of a centimetre at one moment might make the difference between a man being killed by an avalanche a year later, or escaping. It is an essential property of the mechanical systems which we have called "discrete state machines" that this phenomenon does not occur.

This "avalanche" property of dynamical systems is often referred to now as the "butterfly effect." His answer to the "continuity of the nervous system" objection admitted that the nervous system would have the avalanche property, but indicated that he did not see the absence of this property in discrete systems as any disadvantage, and claimed that it could be imitated by the introduction of randomness.

We may note in passing that Copeland (1999) presents Turing's random elements as examples of "oracles," although Turing never used this word or made a connection with his 1938 work. Copeland's justification is that Church (Copeland and Proudfoot 1999) had given a definition of infinite random sequences, in which one necessary condition is that the sequence be uncomputable. Copeland and Proudfoot (2004) also argue that "the concept of a random oracle is well known." But Turing (1950) made no reference to Church's definition and expressly said that the pseudo-random (computable) sequence given by "the digits of the decimal for pi" would do just as well for his purposes. Turing used randomness as being equivalent to variations and errors lacking any functional significance. But for a random number to serve as an uncomputable oracle it would have to be known and exploited to infinite precision.

If Turing ever entertained the notion of realizing his 1938 oracle-machine as a mechanical process, it is in this 1948 report, with its classification of machines, that we should see the evidence of it. There is no such evidence. In particular, when considering the brain as a machine, Turing had the opportunity to discuss whether it might have some uncomputable element corresponding to "intuition." He omitted to take it.

Turing seemed content with a vague and intuitive picture of randomness, which is surprising since he had a strong interest in probability and statistics, and much of his war work depended on detecting pseudo-randomness. Again, he was opening a new area of questions rather than defining an answer. We shall see later how in 1951 he did take such questioning a little further in an interesting direction.

Imitation Game: Logical and Physical

A general feature of all Turing's writing is its plethora of physical allusions and illustrations. The 1948 distinction between physical ("active") properties and the logical ("controlling") properties of a machine appears also in 1950. In the comparison of human and machine by the celebrated imitation game, both human and computer are depicted as physical entities, which as physical objects are entirely different. The test conditions are designed, however, to render irrelevant these physical attributes, and to compare only the "controlling"-machine functions. In *these* functions, Turing argued, the computer had the potential to equal the human brain. In contexts where the interface of the brain with the senses and with physical action is crucial, Turing was less sure about what might be said. In a curious illustration, he referred to "the difficulty of the same kind of friendliness occurring between man and machine as between white man and white man, or between black man and black man."

Yet even so, Turing was optimistic about the machine's scope and made rather light of what would later be seen as the "frame problem" of associating internal symbolic structure with external physical reality. It might further be argued that Turing only conceded these problems with senses and action because he explicitly limited himself to a program of modeling a single brain. He did not consider the possibility of modeling a wider system, including all human society and its environment, as some computationalists would now suggest as a natural extension. So for him to concede difficulties with questions of physical interaction was not actually to concede something beyond the scope of computability. In any case, his attitude, vigorously expressed in his conclusion to the 1950 paper, was that one should experiment and find out.

The central point of Turing's program was not really the playing of games of imitation, with various slightly different protocols and verbal subtleties. The primary question was that of the brain, confronting the fundamental question of how the human mind, with its apparent free will and consciousness, can be reconciled with mechanistic physical action of the brain. In one later remark, he did discuss this question with a new and sharper point.

Quantum Mechanics at Last

In 1951, Turing (1951/2004) gave a talk on BBC radio's Third Program. Entitled "Can digital computers think?" it was largely a condensation of his 1950 paper. But this time he made the prospectus of imitating the physical brain quite explicit. Notably, he explained the special importance of the computer by saying that a universal machine "can replace any rival design of calculating machine, that is to say any machine into which one can feed data and which will later print out results." This was consistent with the 1948 report in regarding the brain as a physical object whose relevant function is that of a discrete-state machine. But what was new in 1951 was Turing's statement that this assumption about the computability of all physical machines, and in particular the brain, might be *wrong*. The argument would only apply to machines "of the sort whose behaviour is in principle predictable by calculation. We certainly do not know how any such calculation should be done, and it has even been argued by Sir Arthur Eddington that on account of the Indeterminacy Principle in Quantum Mechanics no such prediction is even theoretically possible."

This is the only sentence in Turing's work explicitly suggesting that a physical system might not be computable in its behavior. It went against the spirit of the arguments given in 1950, because he did *not* now suggest that random elements could effectively mimic the quantum-mechanical effects. This apparent change of mind about the significance of quantum mechanics might well have reflected discussions at Manchester, in particular with the physical chemist and philosopher Michael Polanyi, but it also reflected Turing's (1932) youthful speculations based on Eddington. It also pointed forward to the work he did in the last year of his life on the "reduction" process in quantum mechanics, which is essential to the indeterminacy to which he drew attention in 1951 (Turing 1951/2004).

Turing's comment is of particular interest because of its connection with the later argument of Roger Penrose (1989, 1994) against artificial intelligence, which opposes Turing's central 1950 view, but shares with Turing a completely physicalist standpoint. Penrose also concentrates on the reduction process in quantum mechanics. Penrose leaves aside the problem of sensory interface with the physical world, and concentrates on the heartland of what Turing called the purely intellectual. From arguments that need not be recapitulated here he reasserts what Turing called the mathematical argument against his AI thesis, that Gödel's Theorem shows that the human mind cannot be captured by a computable procedure, and that Turing's arguments against that objection are invalid. He deduces (with input from other motivations also) that there must be some uncomputable physical law governing the reduction process. Turing did not make any connection between quantum mechanics and Gödel's Theorem; one can only say that he took both topics very seriously in the foundations of AI. Furthermore it seems more likely, from Turing's reported comments, that he was trying to reformulate quantum mechanics so as to remove the problem discussed in 1951. However, it might be that if his work had continued he would have gone in Penrose's direction. In any case, it is striking that it is in dealing with the physics of the brain that Turing's focus is the same as Penrose's.

The Church-Turing Thesis, Then and Now

Even between 1948 and 1951, opening these doors into physics, Turing never made a clear and explicit distinction between his 1936 model of the human calculator and the concept of a physical machine. It was Turing's former student Robin Gandy who did so in 1980; he separated the Church-Turing thesis from "Thesis M," the thesis that anything that a machine can do is computable (Gandy 1980). Under certain conditions on "machine," Gandy then showed that a machine would, indeed, be capable of no more than computable functions. His argument has since been improved and extended, for instance, by Wilfried Sieg (2002). The main generalization that this work introduces is the possibility of parallel computations. But the definition is still not general enough: the conditions do not even allow for the procedures already in technological use in quantum cryptography.

In contrast, the computer scientist A. C.-C. Yao (2003) gives a version of the Church-Turing thesis as the belief that physical laws are such that "any conceivable hardware system" can only produce computable results. Yao comments that "this may not have been the belief of Church and Turing" but that this represents the common interpretation. Yao thus ignores Gandy's distinction and identifies the Church-Turing thesis with an extreme form of Thesis M, not as dogma but as a guiding line of thought, to be settled experimentally. It reflects the central concern of computer science to embody logical software in physical hardware. It should be noted, however, that Yao leaves unspoken the finiteness condition that Church emphasized, and this is of great importance. One could conceive of an oracle consisting of an infinitely long register embodied in an infinite universe, which would then allow the halting problem to be trivially solved by acting as an infinite crib sheet. Church's condition was obviously designed to rule out such an infinite data store. That a calculation should require finite time and finite working space is also a requirement in the classical model of computability.

The origin of these finiteness restrictions lies in the concept of "effective calculability," which implies a limitation to the use of finite resources. There is now a large literature on "hypercomputing" describing putative procedures that in some senses adhere to the criterion of a finite time and finite size, but demand other infinite resources. Copeland and Proudfoot (1999), for instance, in portraying their vision of Turing's oracle, suggest the measurement of "an exact amount of electricity" to infinite precision so as to perform an uncomputable task such as solving the halting problem. Other schemes postulate unboundedly fast or unboundedly small components; sometimes the infinite resources required are not so obvious (see Hodges 2005). One might reasonably exclude all such infinite schemes, or at least regard them as the equivalents of requiring infinite time, but, from the point of view of modern physical research, in which the fundamentals of space, time, matter, and causality are still uncertain, one should not be dogmatic. New foundations to physical reality may bring about new perceptions. Formulation of the Church-Turing thesis, including the concept of finiteness, should evolve in conjunction with a deeper understanding of physical reality.

Conclusion

The Church-Turing thesis, as understood in Yao's physical sense, is the basis of artificial intelligence as a computer-based project. This is one reason for its importance. The central thrust of Turing's thought was that the action of any machine would indeed be captured by classical computation, and in particular that this included all relevant aspects of the brain's action. But his later writings show more awareness of the problem of connecting computability with physical law. Physical reality always lay behind Turing's perception of the mind and brain; as Max Newman (1955) wrote, Turing was at heart an applied mathematician.

Note

1. This broad-brush characterization of machinery before 1940 prompts the question of what Turing made of Babbage's Analytical Engine. The following points may be made.

a. It seems likely that in 1936 Turing did not know of Babbage's work.

b. Turing must have heard of the Analytical Engine plans at least by the end of the war, when they arose in Bletchley Park conversations. The name of the Automatic Computing Engine, which Turing designed, echoed the name of Babbage's machine.

c. In his 1950 paper, Turing attributed the concept of a universal machine to Babbage. In so doing, Turing overstated Babbage's achievement and understated his own. Babbage's design could not allow for unboundedly deep-nested loops of operations, and enforced a rigid separation between instructions and numerical data.

d. I see no clear reason why in his 1948 report Turing gave such short shrift to prewar machinery, yet in 1950 exaggerated its scope.

e. However, this question does not affect the principal issue discussed in this article, since everything Babbage designed lay within the realm of computable functions.

References

Britton, J. L. 1992. "Postscript." In *Pure Mathematics: The Collected Works of A. M. Turing*, edited by J. L. Britton. Amsterdam: North-Holland.

Church, A. 1940. "On the Concept of a Random Sequence." *Bulletin of the American Mathematical Society* 46: 130–35.

Church, A. 1937/1997. "Review of Turing." *Journal of Symbolic Logic* 2: 42. Also in W. Sieg, "Step by Recursive Step: Church's Analysis of Effective Calculability." *Bulletin of Symbolic Logic* 3: 154–80.

Copeland, B. J. 1999. "A Lecture and Two Radio Broadcasts on Machine Intelligence by Alan Turing." In *Machine Intelligence*, edited by K. Furukawa, D. Michie, and S. Muggleton. Volume 15. Oxford: Oxford University Press.

———. 2002. "The Church-Turing Thesis." In *Stanford Encyclopedia of Philosophy*, on-line encyclopedia, edited by E. N. Zalta, at http://plato.stanford.edu/entries/church-turing.

———. 2004. *The Essential Turing*. Oxford: Oxford University Press.

Copeland, B. J., and D. Proudfoot. 1999. "Alan Turing's Forgotten Ideas in Computer Science." *Scientific American* 280, no. 4: 98–103.

———. 2004. "The Computer, Artificial Intelligence, and the Turing Test." In *Alan Turing: Life and Legacy of a Great Thinker*, edited by C. Teuscher. Berlin: Springer.

Feferman, S. 1988. "Turing in the land of O(Z)." In *The Universal Turing Machine: A Half-Century Survey*, edited by R. Herken. Oxford: Oxford University Press.

Gandy, Robin O. 1980. "Principles of Mechanisms." In *The Kleene Symposium*, edited by J. Barwise, H. J. Keisler, and K. Kunen. Amsterdam: North-Holland, 1980.

———. 2001. "Letter to M. H. A. Newman, 1954." In *Mathematical Logic: Collected Works of A. M. Turing*, edited by R. O. Gandy and C. E. M. Yates. Amsterdam: North-Holland.

Hodges, Andrew. 1983. *Alan Turing: The Enigma of Intelligence*. London: Counterpoint.

———. 2005. "Can Quantum Computing Solve Classically Unsolvable Problems?" Available at http://arxiv.org/abs/quant-ph/0512248.

Newman, Max H. A. 1955. "Alan Mathison Turing, 1912–1954." Obituiary. *Biographical Memoirs of the Fellows of the Royal Society* 1(November): 253–63.

Penrose, R. 1994. *Shadows of the Mind*. Oxford: Oxford University Press.

———. *The Emperor's New Mind*. Oxford: Oxford University Press.

Sieg, W. 2002. "Calculations by Man and Machine: Conceptual Analysis." In *Reflections on the Foundations of Mathematics: Essays in Honor of Solomon Feferman*, edited by W. Sieg, R. Sommer, and C. Talcott. Lecture Notes in Logic series, volume 15. Wellesley, Mass.: A. K. Peters.

Teuscher, Christof. 2002. *Turing's Connectionism: An Investigation of Neural Network Architectures*. London: Springer, 2002.

Turing, Alan M. 1932/1983. "Nature of Spirit." Essay. Text in *Alan Turing: The Enigma*, by Andrew Hodges. New York: Simon & Schuster. Image of handwritten essay available at www.turingarchive.org.

————. 1936. "On Computable Numbers with an Application to the Entscheidungs-problem." *Proceedings of the London Mathematical Society* 2, no. 42: 230–65.

————. 1938a. "Finite Approximations to Lie Groups." *Annals of Mathematics* 39: 105–11.

————. 1938b. "Systems of Logic Based on Ordinals." Ph.D. diss., Princeton University. See also *Proceedings of the London Mathematical Society* 2, no. 45: 161–228.

————. 1947/1986. "Lecture to the London Mathematical Society, 1947." In *A. M. Turing's ACE Report of 1946 and Other Papers*, edited by B. E. Carpenter and R. W. Doran. Cambridge, Mass.: MIT Press.

————. 1948. "Rounding-Off Errors in Matrix Processes." *Quarterly Journal of Mechanics and Applied Mathematics* 1: 180–97.

————. 1948/1986. "Proposed Electronic Calculator." Report to National Physical Laboratory. In *A. M. Turing's ACE Report of 1946 and Other Papers*, edited by B. E. Carpenter and R. W. Doran. Cambridge, Mass.: MIT Press.

————. 1948/2004. "Intelligent Machinery." Report for the National Physical Laboratory. In *The Essential Turing*, edited by B. J. Copeland. Oxford: Oxford University Press. Original typescript available at www.turingarchive.org.

————. 1950. "Computing Machinery and Intelligence." *Mind* 59: 433–60.

————. 1951/2004. "Can Digital Computers Think?" Talk, BBC Radio, Third Programme. In *The Essential Turing*, edited by B. J. Copeland. Oxford: Oxford University Press. Typescript reproduced at www.turingarchive.org.

————. 1953. "Some Calculations of the Riemann Zeta-Function." *Proceedings of the London Mathematical Society* 3, no. 3: 99–117.

Yao, A. C.-C. 2003. "Classical Physics and the Church-Turing Thesis." *Journal of the ACM* 50: 100–5.

6   The Ratio Club: A Hub of British Cybernetics

Philip Husbands and Owen Holland

Writing in his journal on the twentieth of September, 1949, W. Ross Ashby noted that six days earlier he'd attended a meeting at the National Hospital for Nervous Diseases, in the Bloomsbury district of London. He comments (Ashby 1949a), "We have formed a cybernetics group for discussion—no professors and only young people allowed in. How I got in I don't know, unless my chronically juvenile appearance is at last proving advantageous. We intend just to talk until we can reach some understanding." He was referring to the inaugural meeting of what would shortly become the Ratio Club, a group of outstanding scientists who at that time formed much of the core of what can be loosely called the British cybernetics movement. The club usually gathered in a basement room below nurses' accommodation in the National Hospital, where, after a meal and sufficient beer to lubricate the vocal cords, participants would listen to a speaker or two before becoming embroiled in open discussion (see figure 6.1). The club was founded and organized by John Bates, a neurologist at the National Hospital. The other twenty carefully selected members were a mixed group of mainly young neurobiologists, engineers, mathematicians, and physicists.

A few months before the club started meeting, Norbert Wiener's (1948) landmark *Cybernetics: Control and Communication in the Animal and Machine* had been published. This certainly helped to spark widespread interest in the new field, as did Claude Shannon's seminal papers on information theory (Shannon and Weaver 1949), and these probably acted as a spur to the formation of the club. However, as we shall see, the first of the official membership criteria of the club was that only "those who had Wiener's ideas before Wiener's book appeared" (Bates 1949a) could join. This was no amateur cybernetics appreciation society; many members had already been active for years in developing the new ways of thinking about behavior-generating mechanisms and information processing in brains and machines that were now being pulled together under the rubric

Figure 6.1
The main entrance to the National Hospital for Nervous Diseases, Queen's Square, Bloomsbury, in 2002. Ratio Club meetings were held in a room in the basement.

"cybernetics," coined by Wiener. Indeed, the links and mutual influences that existed between the American and British pioneers in this area ran much deeper than is often portrayed. There was also a very strong independent British tradition in the area that had developed considerable momentum during World War II. It was from this tradition that most club members were drawn.

The other official membership criterion reflected the often strongly hierarchical nature of professional relationships at that time. In order to avoid restricting discussion and debate, Bates introduced the "no professors" rule alluded to by Ashby. If any members should be promoted to that level, they were supposed to resign. Bates was determined to keep things as informal as possible; conventional scientific manners were to be eschewed in favor of relaxed and unfettered argument. There also appear to have been two further, unofficial, criteria for being invited to join. First, members had to be as smart as hell. Second, they had to be able to contribute in an interesting way to the cut and thrust of debate, or, to use the parlance of the day, *be good value*. This was a true band of Young Turks. In the atmosphere of enormous energy and optimism that pervaded postwar Britain as it began to rebuild, they were hungry to push science in new and important directions. The club met regularly from 1949 to 1955, with one final reunion meeting in 1958. It is of course no coincidence that this period parallels the rise of the influence of cybernetics, a rise in which several members played a major role.

There are two things that make the club extraordinary from a historical perspective. The first is the fact that many of its members went on to become extremely prominent scientists. The second is the important influence the club meetings, particularly the earlier ones, had on the development of the scientific contributions many of that remarkable group would later make. The club membership undoubtedly made up the most intellectually powerful and influential cybernetics grouping in the UK, but to date very little has been written about it: there are brief mentions in some histories of AI and cognitive science (see Fleck 1982; Boden 2006; D. Clark [2003] has a chapter on it in his Ph.D. dissertation, based on papers from the John Bates archive). This article is intended to help fill that gap. It is based on extensive research in a number of archives, interviews with surviving members of the club, and access to some members' papers and records.

After introducing the membership in the next section, the birth of the club is described in some detail. The club's known meetings are then listed and discussed along with its scope and modus operandi. Following this some of the major themes and preoccupations of the club are described in more detail. The interdisciplinary nature of the intellectual focus of the group is highlighted before the legacy of the club is discussed. Because so many rich threads run through the club and the lives and work of its members, this chapter can only act as an introduction (a fuller treatment of all these topics can be found in Husbands and Holland [forthcoming]).

The Members

Before embarking on a description of the founding of the club, it is useful at this point to sketch out some very brief details of its twenty-one members, with outlines of their expertise and achievements, which will help to give a sense of the historical importance of the group. Of course these summaries are far too short to do justice to the careers of these scientists. They are merely intended to illustrate the range of expertise in the club and to give a flavor of the caliber of members.

W. Ross Ashby (1903–1972), trained in medicine and psychiatry, is regarded as one of the most influential pioneers of cybernetics and systems science. He wrote the classic books *Design for a Brain* (Ashby 1952a) and *An Introduction to Cybernetics* (Ashby 1958). Some of his key ideas have recently experienced something of a renaissance in various areas of science, including artificial life and modern AI. At the inception of the club he was director of research at Barnwood House Psychiatric Hospital, Gloucester. He subsequently became a professor in the Department of Biophysics and Electrical Engineering, University of Illinois.

Horace Barlow (1921–), FRS, a great-grandson of Charles Darwin, is an enormously influential neuroscientist, particularly in the field of vision, and was one of the pioneers of using information-theoretic ideas to understand neural mechanisms (Barlow 1953, 1959, 1961), a direct consequence of his involvement in the Ratio Club. When the club started he was a Ph.D. student in Lord Adrian's lab in the Department of Physiology, Cambridge University. He later became Royal Society Research Professor of Physiology at Cambridge University.

John Bates (1918–1993) had a distinguished career in the neurological research unit at the National Hospital for Nervous Diseases, London. He studied the human electroencephalogram (EEG) in relation to voluntary movement, and became the chief electroencephalographer at the hospital. The Ratio Club was his idea and he ran it with quiet efficiency and unstinting enthusiasm.

George Dawson (1911–1983) was a clinical neurologist at the National Hospital, Queen's Square. At the time of the Ratio Club he was a world leader in using EEG recordings in a clinical setting. He was a specialist in ways of averaging over many readings, which allowed him to gather much cleaner signals than was possible by more conventional methods (Dawson 1954). He became professor of physiology at University College London.

Thomas Gold (1920–2004), FRS, was one of the great astrophysicists of the twentieth century, being a coauthor, with Hermann Bondi and Fred Hoyle, of the steady-state theory of the universe and having given the first explanation of pulsars, among countless other contributions. However, he had no time for disciplinary boundaries and at the time of the Ratio Club he was working in the Cambridge University Zoology Department on a radical positive feedback theory of the working of the inner ear (Gold 1948)—a theory that was, typically for him, decades ahead of its time. He went on to become professor of astronomy at Harvard University and then at Cornell University.

I. J. (Jack) Good (1916–) was recruited into the top-secret UK code-cracking operation at Bletchley Park during the Second World War, where he worked as the main statistician under Alan Turing and Max Newman. Later he became a very prominent mathematician, making important contributions in Bayesian methods and early AI. During the Ratio Club years he worked for British Intelligence. Subsequently he became professor of statistics at Virginia Polytechnic Institute.

William E. Hick (1912–1974) was a pioneer of information-theoretic thinking in psychology. He is the source of the still widely quoted Hick's law, which states that the time taken to make a decision is in proportion to the logarithm of the number of alternatives (Hick 1952). During the Ratio Club years he worked in the Psychology Laboratory at Cambridge University. He went on to become a distinguished psychologist.

Victor Little (1920–1976) was a physicist at Bedford College, London, who worked in acoustics and optics before moving on to laser development.

Donald Mackay (1922–1987), trained as a physicist, was a very highly regarded pioneer of early machine intelligence and of neuropsychology. He was also the leading scientific apologist for Christianity of his day. At the birth of the club he was working on a Ph.D. in the Physics department of King's College, London. He later became a professor at Keele University, where he founded the Department of Communication and Neuroscience.

Turner McLardy (1913–1988) became an international figure in the field of clinical psychiatry. He emigrated to the United States in the late 1950s to develop therapeutic techniques centered around planned environments and communities. Later he became a pioneer of understanding the role of zinc in alcoholism and schizophrenia. At the inception of the club he worked at the Maudsley Hospital, London.

Pat Merton (1921–2000), FRS, was a neurophysiologist who did pioneering work on control-theoretic understandings of the action of muscles (Merton 1953). Later he carried out a great deal of important early research in magnetic stimulation of the cortex, for which he is justly celebrated (Merton and Morton 1980). During the Ratio Club years he worked in the neurological research unit at the National Hospital. He later became professor of human physiology at Cambridge University.

John Pringle (1912–1982), FRS, was one of the leading invertebrate neurobiologists of his day. He was the first scientist to get recordings from single neurons in insects, something that had previously been thought to be impossible (Pringle 1938). He did much important work in proprioception in insects, insect flight, and invertebrate muscle systems. At the birth of the club he worked in the Zoological Laboratory, Cambridge University. He subsequently became professor of zoology at Oxford University.

William Rushton (1901–1980), FRS, is regarded as one of the great figures in twentieth-century vision science. He made enormous contributions to understanding the mechanisms of color vision, including being the first to demonstrate the deficiencies that lead to color blindness (Rushton 1955). Earlier he did pioneering work on the quantitative analysis of factors involved in the electrical excitation of nerve cells, helping to lay the foundations for the framework that dominates theoretical neuroscience today (see Rushton 1935). He worked at Cambridge University throughout his career, where he became professor of visual physiology.

Harold Shipton (1920–2007) worked with W. Grey Walter on the development of EEG technology at the Burden Neurological Institute, Bristol. He was the electronics wizard who was able to turn many of Walter's inspired but intuitive designs into usable and reliable working realities. Later he became a professor at the Washington University in St. Louis, where he worked on biomedical applications. At the time of the early Ratio Club meetings, his father-in-law, Clement Attlee, was prime minister of Great Britain.

D. A. Sholl (1903–1960) did classic research on describing and classifying neuron morphologies and growth patterns, introducing the use of rigorous statistical approaches (Sholl 1956). Most of the classification techniques in use today are based on his work. He also published highly influential papers on the structure and function of the visual cortex. He worked in the Anatomy Department of University College, London, where he became reader in anatomy before his early death.

Eliot Slater (1904–1983) was one of the most eminent British psychiatrists of the twentieth century. He helped to pioneer the use of properly grounded statistical methods in clinical psychiatry. Slater's work with Ernst Rudin on the genetic origins of schizophrenia, carried out in Munich in the 1930s, still underpins all respectable Anglo-American work in psychiatric genetics, a field to which Slater made many important contributions (Slater, Gottesman, and Shields 1971). He worked at the National Hospital for Nervous Diseases, London.

Alan Turing (1912–1954), FRS, is universally regarded as one of the fathers of both computer science and artificial intelligence. Many regard him as one of the key figures in twentieth-century science and technology. He also anticipated some of the central ideas and methodologies of Artificial Life and Nouvelle AI by half a century. For instance, he proposed artificial evolutionary approaches to AI in the late 1940s (Turing 1950) and published work on reaction-diffusion models of the chemical origins of biological form in 1952 (Turing 1952). At the inception of the club he was working at Manchester University, where he was part of a team that had recently developed the world's first stored-program digital computer.

Albert Uttley (1906–1985) did important research in radar, automatic tracking, and early computing during World War II. Later he became head of the pioneering Autonomics Division at the National Physical Laboratory in London, where he did research on machine intelligence and brain modeling. However, he also became well known as a neuropsychologist, having made several important contributions to the field (Uttley 1979). At the birth of the club he worked at the Telecommunications Research Establishment (TRE), Malvern, Worcestershire, the main British military telecommunications research institute. Later he became professor of psychology at Sussex University.

W. Grey Walter (1910–1977) was a pioneer and world leader in EEG research; he founded the EEG Society and the *EEG Journal*, and organized the first EEG congress. He made many major discoveries, including theta and delta brain waves and, with Shipton, developed the first topographic EEG machine (Walter and Shipton 1951). At the time of the Ratio Club he was at the Burden Neurological Institute, Bristol, where, alongside his EEG research, he developed the first ever autonomous mobile robots, the famous tortoises, which were controlled by analogue electronic nervous systems (Walter 1950a). This was the first explicit use of mobile robots as a tool to study ideas about brain function, a style of research that has become very popular in recent times.

John Westcott (1920–), FRS, made many very distinguished contributions to control engineering, including some of the earliest work on control under noisy conditions. He also worked on applications of control theory to economics, which resulted in his team's developing various models used by the UK Treasury. At the inception of the club he was doing a Ph.D. in the Department of Electrical Engineering, Imperial College, London, having just returned from a year in Norbert Wiener's lab at MIT. He later became professor of control systems at Imperial College, London.

Philip M. Woodward (1919–) is a mathematician who made important contributions to information theory, particularly with reference to radar, and to early computing. His gift for clear concise explanations can be seen in his elegant and influential 1953 book on information theory (Woodward 1953). He worked at TRE, Malvern, throughout his entire distinguished career (one of the buildings of the present-day successor to TRE is named after him). In retirement Woodward has come to be regarded as one of the world's greatest designers and builders of mechanical clocks (Woodward 1995).

Bates's own copy of his typed club membership list of January 1, 1952 has many hand-written corrections and annotations (Bates 1952a). Among these, immediately under the main list of members, are the following letters, arranged in a neat column: *Mc*, *P*, *S*, and then a symbol that may be a *U* or possibly a *W*. If we assume it is a *W*, then a possible, admittedly highly speculative, interpretation of these letters is: McCulloch, Pitts, Shannon, Wiener. The first three of these great American cyberneticists attended club meetings—McCulloch appears to have taken part whenever travel to Britain allowed. Wiener was invited and intended to come on at least one occasion but travel difficulties and health problems appear to have gotten in the way. The W, if that's what it is, could also refer to Weaver, coauthor with Shannon of seminal information-theory papers and someone who was also well known to the club. Of course the letters may not refer to American cyberneticists at all—they may be something more prosaic such as the initials of members who owed subscriptions—but it is just possible that Bates regarded them as honorary members.

It is clear from the membership listed above that the center of gravity of the club was in the brain sciences. Indeed the initial impetus for starting the club came from a neurologist, Bates, who believed that emerging cybernetic ideas and ways of thinking could be very important tools in developing new insights into the operation of the nervous system. Many members had a strong interest in developing "brainlike" devices, either as a way of formalizing and exploring theories about biological brains, or as a pioneering effort in creating machine intelligence, or both. Hence meetings tended to center around issues relating to natural and artificial intelligence and the processes underlying the generation of adaptive behavior—in short, the mechanization of mind. Topics from engineering and mathematics were usually framed in terms of their potential to shed light on these issues. This scope is somewhat different to that which had emerged in America, where a group of mathematicians and engineers (Wiener, John von Neumann, Julian Bigelow, Claude Shannon, Walter Pitts) and brain scientists (Rafael Lorente de Nó, Arturo Rosenblueth, Warren McCulloch) had formed an earlier group similar in spirit to the Ratio Club, although smaller and with a center of gravity further toward the mathematical end of the spectrum. Their influence soon spread, via Lawrence Frank, Margaret Mead, Gregory Bateson, and others, into the social sciences, thereby creating a much wider enterprise that involved the famous Macy Foundation meetings (Heims 1991). This difference in scope helps to account for the distinct flavor of the British scene in the late 1940s and for its subsequent influences.

Genesis of the Club

Founding

The idea of forming a cybernetics dining club took root in John Bates's mind in July 1949. He discussed the idea with a small number of colleagues at a Cambridge symposium, "Animal Behaviour Mechanisms," a very cybernetics-friendly topic, organized by the Society for Experimental Biology and held from the eighteenth to the twenty-second of the month. Shortly after returning to London from the meeting, he wrote the following letter to Grey Walter in which he formally proposed the club (Bates 1949a):

National Hospital
27th July 1949
Dear Grey,
   I have been having a lot of "Cybernetic" discussions during the past few weeks here and in Cambridge during a Symposium on Animal Behaviour Mechanisms,

and it is quite clear that there is a need for the creation of an environment in which these subjects can be discussed freely. It seems that the essentials are a closed and limited membership and a post-prandial situation, in fact a dining-club in which conventional scientific criteria are eschewed. I know personally about 15 people who had Wiener's ideas before Wiener's book appeared and who are more or less concerned with them in their present work and who I think would come. The idea would be to hire a room where we could start with a simple meal and thence turn in our easy chairs towards a blackboard where someone would open a discussion. We might need a domestic rule to limit the opener to an essentially unprepared dissertation and another to limit the discussion at some point to this stratosphere, but in essence the gathering should evolve in its own way.

Beside yourself, Ashby and Shipton, and Dawson and Merton from here, I suggest the following:

Mackay—computing machines, Kings. Coll. Strand
Barlow—sensory physiologist—Adrian's lab.
Hick—Psychological lab. Cambridge
Scholl—statistical neurohistologist—University College, Anatomy Lab.
Uttley—ex. Psychologist, radar etc TRE
Gold—ex radar zoologists at Cambridge
Pringle—ex radar zoologists at Cambridge

I could suggest others but this makes 13, I would suggest a few more non neurophysiologists communications or servo folk of the right sort to complete the party but those I know well are a little too senior and serious for the sort of gathering I have in mind.

We might meet say once a quarter and limit the inclusive cost to 5/– less drinks. Have you any reaction? I have approached all the above list save Uttley so far, and they support the general idea.
Yours sincerely,
JAV Bates

The suggested names were mainly friends and associates of Bates's, known through various social networks relating to his research, whom he regarded as being "of the right sort." One or two were suggested by immediate colleagues; for instance, Merton put forward his friend Barlow.

Walter replied by return post enthusiastically welcoming the idea and suggesting that the first meeting should coincide with his friend Warren McCulloch's visit to England in September. Mackay furnished Bates with an important additional "communications or servo" contact by introducing him to John Westcott, who was finishing off his Ph.D. at Imperial College, having spent the previous year in Wiener's lab at MIT as a guest of the institution. Westcott's close association with Wiener seems to have led

Bates to soften his 'had Wiener's ideas before Wiener's book appeared' line in his invitation to him (Bates 1949b):

National Hospital
3rd August
Dear Mr. Westcott,

I have heard from Mackay that you might be interested in a dining-club that I am forming to talk "Cybernetics" occasionally with beer and full bellies. My idea was to have a strictly limited membership between 15 and 20, half primarily physiologists and psychologists though with "electrical leanings" and half primarily communication theory and electrical folk though with biological interests and all who I know to have been thinking "Cybernetics" before Wiener's book appeared. I know you have all the right qualifications and we would much like you to join. The idea is to meet somewhere from 7.00 p.m.–10.00 p.m. at a cost of about 5/– less drinks.

The second point is whether we could make McCulloch's visit in September the occasion for a first meeting. This was raised by Mackay who mentioned that you had got in touch with him already with a view to some informal talk. It has also been raised by Grey Walter from Bristol who knows him too. What do you feel? Could we get McCulloch along to an inaugural dinner after his talk for you? Could you anyway manage to get along here for lunch one day soon, we have an excellent canteen and we could talk it over?

Your sincerely
JAV Bates

Westcott was as enthusiastic as Walter. Bates wrote a succession of individual invitations to those on his list as well as to Little, who was suggested by Mackay, and Turner McLardy, a psychiatrist with a keen interest in cybernetics who was a friend of McCulloch's and appears to have been about to host his imminent stay in London. The letter to Hick was typical, including the following exuberant passage (Bates 1949c): "The idea of a 'Cybernetic' dining club, which I mentioned to you in Cambridge, has caught fire in an atomic manner and we already have half a dozen biologists and engineers, all to my knowledge possessed of Wiener's notions before his book appeared and including two particularly rare birds: Mackay and Westcott who were in Wiener's lab for a year during the war." Bates didn't quite have his facts straight; Westcott's time with Wiener was after the war and at this stage Mackay hadn't begun his collaborations with MIT, but the implication was right—that Westcott and Mackay were both familiar with the mathematical and technical details of Wiener's work. All invitees accepted membership in the club. In their replies a number made general suggestions about membership: Barlow (1949) suggested considering the addition of a few more "cautiously selected psychologists," and Pringle (1949) thought it would be a good idea to "add a mathematician to keep everyone in check and stop the discussion becoming too vague."

During August Bates secured a room at the National Hospital that could be used for regular meetings. With Eliot Slater, a senior member of staff at the hospital, on board, he was able to arrange provision of beer and food for club evenings. With a venue, a rough format, and an initial membership list, the enterprise was starting to come into focus. The following letter from Mackay (1949) to Bates, hand-written in a wild scrawl, shows that these two were starting to think about names and even emblems:

1st September 49
Dear Bates,

I'm afraid I've had few fresh ideas on the subject of our proposed club; but here are an odd suggestion or two that arose in my mind.

I wondered (a) if we might adopt a Great Name associated with the subject and call it e.g. the Babbage Club or the Leibniz Club or the Boole Club, or the Maxwell Club—names to be suggested by all, and one selected by vote or c'ttee (Nyquist might be another). Alternatively (b) could we choose a familiar symbol of feedback theory, such as beta, and call it the Beta Club or such like? Other miscellaneous possibilities are the MR Club (machina ratiocinatrix!) and plenty of other initials, or simply the "49" Club.

On emblems I've had no inspirations. I use but little beer myself and it's conceivable we might even have t-t members. But beer mugs can after all be used for other liquids and I can't think of anything better than your suggestion....
Yours,
Donald Mackay

Here we see Mackay sowing the seed for the name Ratio, which was adopted after the first meeting. *Machina ratiocinatrix* is Latin for "reasoning machine," a term used by Wiener in the introduction to *Cybernetics*, in reference to *calculus ratiocinator*, a calculating machine constructed by Leibniz (Wiener 1948, p. 12). Ratiocination is an old-fashioned word for reasoning or thinking, introduced by Thomas Aquinas to distinguish human reasoning from the supposed directly godgiven knowledge of the angels. After the first meeting Albert Uttley suggested using the root *ratio*, giving its definition as "computation or the faculty of mind which calculates, plans and reasons" (Bates 1949d). He pointed out that it is also the root of *rationarium*, meaning a statistical account—implicitly referring to the emerging work on statistical mechanisms underlying biological and machine intelligence—and of *ratiocinatius*, meaning argumentative. Given that the name clearly came from the Latin, it seems reasonable to assume that the intended pronunciation must have been "RAT-ee-oh." In interviews with the authors, half the surviving club members said that this indeed is how it was always pronounced, while the other half said it was pronounced as in the ratio of two numbers! As Thomas Gold commented in 2002, "At that time many of us [in the Ratio Club] were caught up in the excitement of our thoughts and ideas and didn't always notice the details of things like that!"

Bates's notes for his introduction to the inaugural meeting reveal that his suggestion was to call it the Potter Club after Humphrey Potter (Bates 1949e). Legend has it that, as an eleven-year-old boy in 1713, Potter invented a way of automatically opening and closing the valves on an early Newcomen steam engine. Until that point the valves had to be operated by an attendant such as Potter. He decided to make his life easier by attaching a series of cords and catches such that the action of the main beam of the engine opened and closed the valves.

At the end of August 1949 Bates attended an EEG conference in Paris at which he first met McCulloch. There he secured him as guest speaker for the first meeting of the club. Before describing the meetings, it will be instructive to delve a little deeper into the origins of the club.

Origins

Of course the roots of the club go back further than the Cambridge symposium of July 1949. The Second World War played an important catalytic role in developing some of the attitudes and ideas that were crucial to the success of the Club and to the achievements of its members. This section explores some of these roots, shedding light on the significant British effort in what was to become known as cybernetics, as well as pointing out pre-existing relationships in the group.

The War Effort Many of the unconventional and multidisciplinary ideas developed by club members originated in secret wartime research on radar, gunnery control, and the first digital computers. In Britain there was little explicit biological research carried out as part of the war effort, so most biologists were, following some training in electronics, drafted into the main thrust of scientific research on communications and radar. They became part of an army of thousands of technical "wizards" whom Winston Churchill was later to acknowledge as being vital to the allies' victory (Churchill 1949). Although most of the future Ratio Club biologists were naturally unconstrained and interdisciplinary thinkers, such war work exposed many of them to more explicitly mechanistic and mathematical ways of conceiving systems than they were used to. To these biologists a

radar set could be thought of as a kind of artificial sense organ, and they
began to see how the theoretical framework associated with it—which
focused on how best to extract information from the signal—might be
applied to understanding natural senses such as vision. On the other side
of the coin, several club members were deeply involved in the wartime development of early computers and their use in code cracking. This in turn
brought them to ponder the possibility of building artificial brains inspired
by real ones. Other engineers and theoreticians, working alongside their biologist colleagues on such problems as automatic gun aiming, began to see
the importance of coordinated sensing and acting in intelligent adaptive
behavior, be it in a machine or in an animal. Many years later, in the post-
humously published text of his 1986 Gifford Lectures—a prestigious lecture
series on 'Natural Theology' held at the Universities of Edinburgh, Glasgow, St. Andrews, and Aberdeen—Donald Mackay (1991, 40) reflected on
the wartime origins of his research interests:

During the war I had worked on the theory of automated and electronic computing
and on the theory of information, all of which are highly relevant to such things as
automatic pilots and automatic gun direction. I found myself grappling with problems in the design of artificial sense organs for naval gun-directors and with the
principles on which electronic circuits could be used to simulate situations in the external world so as to provide goal-directed guidance for ships, aircraft, missiles and
the like. Later in the 1940's, when I was doing my Ph.D. work, there was much talk
of the brain as a computer and of the early digital computers that were just making
the headlines as "electronic brains." As an analogue computer man I felt strongly
convinced that the brain, whatever it was, was not a digital computer. I didn't think
it was an analogue computer either in the conventional sense. But this naturally
rubbed under my skin the question: well, if it is not either of these, what kind of system is it? Is there any way of following through the kind of analysis that is appropriate to these artificial automata so as to understand better the kind of system the
human brain is? That was the beginning of my slippery slope into brain research.

This coalescing of biological, engineering, and mathematical frameworks
would continue to great effect a few years later in the Ratio Club. Not only
Mackay but also the future members Pringle, Gold, Westcott, Woodward,
Shipton, Little, Uttley, and Walter—and perhaps others—were also involved in radar research. Hick and Bates both worked on the related problem of visual tracking in gunnery. Uttley also worked on a range of other
problems, including the development of automatic control systems, analogue computer–controlled servo mechanisms, and navigation computers
(for this war work he was awarded the Simms Gold medal of the Royal
Aeronautical Society). There is not enough space in this paper to describe

any of this work in detail; instead a number of sketches are given that offer a flavor of the kinds of developments that were undertaken and the sorts of circumstances many future members found themselves thrust into.

Philip Woodward left Oxford University in 1941 with a degree in mathematics. As an able-bodied young man he was whisked straight into the Army, where he began basic training. However, he felt he would be much better employed at the military Telecommunications Research Establishment (TRE) nestled in the rolling hills near Malvern. It was here that thousands of scientists of all persuasions were struggling with numerous seemingly impossible radar and communications problems. Within a few days his wish was granted, following a letter from his obviously persuasive father to their local MP, Lord Beaverbrook, Minister of Supply. Leaving rifle drill far behind, Woodward joined Henry Booker's theoretical group, to be plunged into crucial work on antenna design and radio-wave propagation. Within a few days of arriving at TRE he was summoned to see Alec Reeves, a brilliant, highly unconventional engineer and one of the senior staff in Woodward's division. A few years earlier Reeves had invented pulse-code modulation, the system on which all modern digital communication is based. He firmly believed he was in direct contact with the spirits of various British scientific geniuses from bygone ages who through him were helping in the war effort. Reeves handed Woodward a file marked "Top Secret." Inside were numerous squiggles recorded from a cathode-ray tube: his task was to analyze them and decide whether or not they came from Michael Faraday. Over the years Woodward was to face many technical challenges almost as great as this in his work at TRE (Woodward 2002).

In the early years of the war John Westcott was an engineering apprentice. His job was little more than that of a storeman, fetching and filling orders for materials to be used in the manufacture of various military hardware. Although he didn't have a degree or much formal training, he was enormously frustrated by not being able to contribute more; he was convinced that he had design talents that could really make a difference if only he could use them (Westcott 2002). After much badgering, he finally managed to get himself transferred to TRE, where his abilities were indeed soon recognized. He was teamed up with two other brilliant young engineers with whom he was given complete freedom to try and design a new type of radar set to be used by the artillery. If they were successful the device would be extremely important—by using a significantly shorter wavelength than before it would provide a much higher degree of accuracy, enabling the detection of smaller objects. The other members of the team were the highly eccentric Francis Farley and, on secondment from the

American Signals Corps, Charles Howard Vollum. All three were in their early twenties. At first Farley and Vollum were always at each other's throats with Westcott trying to keep the peace. Vollum became incensed at the unreliability of the oscilloscopes at their disposal and swore that after the war he'd build one that was fit for engineers to use. Despite setbacks and failures they persevered, making use of Vollum's supply of cigars to rope in extra help and procure rare supplies. Somehow they managed to combine their significant individual talents to solve the problem and build a new type of shorter wavelength radar set. This great success placed Westcott and Farley on the road to highly distinguished scientific careers, while Vollum was as good as his word and after returning to Oregon cofounded a giant electronic instruments company, Tektronix, and became a billionaire.

Like Woodward and Westcott, Thomas Gold's route into radar research was indirect, although his entry was rather more painful. Born into a wealthy Austrian Jewish family, he was a student at an exclusive Swiss boarding school in the late 1930s when his father decided the political situation was becoming too dangerous for the family to stay in Vienna and moved to London. Thomas began an engineering degree at Cambridge University, but when war broke out he was rounded up and put into an internment camp as an enemy alien. Sleeping on the same cold concrete floor as Gold was another Austrian, a young mathematician named Hermann Bondi. The two struck up an immediate friendship and began discussing the ideas that would later make them both giants of twentieth-century astrophysics. Their partnership was initially short-lived because after only a few weeks Gold was transferred to a camp in Canada. His ship survived the savage Atlantic crossing, although others in the convey did not, being destroyed by U-boats with the loss of many hundreds of lives. Once on Canadian soil the situation did not improve. He found himself in a camp run by a brutally sadistic officer who made life hell for the interns. In order to make things bearable, Gold claimed he was an experienced carpenter and was put in charge of a construction gang. Ever ingenious, he built a contraption to divert steam from an outlet pipe into a water trough to allow his fellow interns to have a hot bath. He was severely beaten for his trouble. Fortunately, Bondi, who had by now been rescued from another camp by senior scientific staff who had known him at Cambridge, had been spreading word of his friend's brilliance. Gold was pulled out of internment and, like Bondi, was assigned to work on top-secret radar research. But not before he had the great pleasure one morning of joining with all other inmates in wild celebrations on hearing the unexpected news that the camp commander had died of a sudden heart attack in the night (Gold 2002).

Alan Turing, following a year in Princeton working with John von Neumann, was a research fellow at Cambridge University when the British government, fearing war was inevitable, recruited him into a secret codes and ciphers unit in 1938. As is well documented (Hodges 1983), he became an enormously important figure in the successful wartime code-cracking work at Bletchley Park, and through this work was deeply involved in the development of the very first digital computers, the theoretical foundations for which he had set out in the late 1930s (Turing 1936). Once war broke out, Jack Good, who had just finished a Ph.D. in mathematics at Cambridge under the great G. H. Hardy, was recruited into the top-secret operation at Bletchley Park, where he worked as the main statistician under Turing and Max Newman in a team that also included Donald Michie.

Most other Ratio Club members not mentioned above were medically trained and so worked as doctors or in medical research during the war. Most of those were based in the UK, although McLardy, who held the rank of major, saw active service as a medical officer and was captured and put in a succession of P.O.W. camps, at least one of which he escaped from. He worked as a psychiatrist in Stalag 344 at Lamsdorf, Silesia, now Lambinowice, in Poland (BBC 2005). In early 1945 the Germans started evacuating Lamsdorf ahead of the Russian advance. The P.O.W.s were marched west in columns of a thousand, each column under the charge of a medical officer. The conditions endured on these "death marches" were appalling—bitterly cold weather, little or no food, and rampant disease (Tattersall 2006). McLardy survived and eventually made it back to Britain.

Apart from plunging them into work that would help to shape their future careers, the war had a strong formative affect on the general attitudes and aspirations of many Ratio Club members. In a way that would just not happen in peacetime, many were given huge responsibilities and the freedom to follow their own initiative in solving their assigned problems. (For a while, at barely thirty years of age, Pringle was in charge of all airborne radar development in Britain. For his war service he was awarded an MBE and the American Medal of Freedom with Bronze Palm.)

Kenneth Craik From the midst of this wartime interdisciplinary problem solving emerged a number of publications that were to have a galvanizing affect on the development of British cybernetics. These included Kenneth J. W. Craik's slim volume, *The Nature of Explanation*, which first appeared in 1943 (Craik 1943). Bates's hastily scrawled notes for his introduction to the first meeting of the Ratio Club, a few lines on one side of a scrap of paper, include a handful of phrases under the heading "Membership." Of these only one is underlined. In fact it is underlined three times: "No Craik."

Kenneth Craik was a Scottish psychologist of singular genius who after many years of relative neglect is remembered now as a radical philosopher, a pioneer of the study of human-machine interfaces, a founder of cognitive psychology, and a father of cybernetics thinking. His story is made particularly poignant by his tragic and sudden death at the age of thirty-one on the last day of the war in Europe, 7 May 1945, when he was killed in a traffic accident while cycling through Cambridge. He had recently been appointed the first director of the Medical Research Council's prestigious Applied Psychology Unit. He was held in extremely high regard by Bates and the other Ratio Club members, so the "No Craik" was a lament.

After studying philosophy at Edinburgh University, in 1936 he began a Ph.D. in psychology and physiology at Cambridge. Here he came under the influence of the pioneering head of psychology, Frederick Bartlett. Craik's love of mechanical devices and his skills as a designer of scientific apparatus no doubt informed the radical thesis of his classic 1943 book, published in the midst of his war work on factors affecting the efficient operation and servicing of artillery machinery. Noting that "one of the most fundamental properties of thought is its power of predicting events" (Craik 1943, p. 50), Craik suggests that such predictive power is "not unique to minds." Indeed, although the "flexibility and versatility" of human thought is unparalleled, he saw no reason why, at least in principle, such essential properties as recognition and memory could not be emulated by a man-made device. He went even further by claiming that the human mind is a kind of machine that constructs small-scale models of reality that it uses to anticipate events. In a move that anticipated Wiener's *Cybernetics* by five years, as well as foreshadowing the much later fields of cognitive science and AI, he viewed the proper study of mind as an investigation of classes of mechanisms capable of generating intelligent behavior both in biological and nonbiological machines. Along with Turing, who is acknowledged in the introduction to Wiener's *Cybernetics*, and Ashby, who had begun publishing on formal theories of adaptive behavior in 1940 (Ashby 1940), Craik was a significant, and largely forgotten, influence on American cybernetics. Both Wiener and McCulloch acknowledged his ideas, quoting him in an approving way, and the later artificial intelligence movement, founded by John McCarthy and Marvin Minsky, was to a degree based on the idea of using digital computers to explore Craik's idea of intelligence involving the construction of small-scale models of reality (see McCarthy 1955, the original proposal for the 1956 Dartmouth Summer

Project on AI, for an explicit statement of this). Many members of the Ratio Club, a high proportion of whom had connections with Cambridge University, were influenced by Craik and held him in great esteem, in particular Bates and Hick, who had both worked closely with him; Grey Walter, who cited wartime conversations with Craik as the original inspiration for the development of his tortoises (Walter 1953, p. 125); and Uttley, whom Bates credited with giving Craik many of his ideas (Bates 1945). Indeed, in a 1947 letter to Lord Adrian, the charismatic Nobel Prize–winning head of physiology at Cambridge, Grey Walter refers to the American cybernetics movement as "thinking on very much the same lines as Kenneth Craik did, but with much less sparkle and humour" (Walter 1947). Had he survived, there is no doubt Craik would have been a leading member of the club. In fact, John Westcott's notes from the inaugural meeting of the club show that there was a proposal to call it the Craik Club in his honor (Westcott 1949–53).

Existing Relationships   Although the Ratio Club was the first regular gathering of this group of like-minded individuals, certain members had interacted with each other for several years prior to its founding, often in work or discussion with a distinct cybernetic flavor. For instance, Bates and Hick had worked with Craik on wartime research related to visual tracking in gunnery and the design of control systems in tanks. In the months after Craik's untimely death, they had been involved in an attempt to edit his notes for a paper eventually published as "Theory of the Human Operator in Control Systems" (Craik 1948).

Ashby also was familiar with Craik's ideas. In 1944 he wrote to Craik after reading *The Nature of Explanation*. As intimated earlier, the central thesis of Craik's book is that 'thought models, or parallels, reality' (Craik 1943, p. 57). Neural mechanisms, somehow acting as "small-scale models" of external reality, could be used to "try out various alternatives, conclude which is the best of them, react to future situations before they arise, utilise the knowledge of past events in dealing with the present and future, and in every way to react in a much fuller, safer, and more competent manner to the emergencies that face it" (p. 61). Today this is a familiar idea, but Craik is widely acknowledged as the first thinker to articulate it in detail. Ashby wrote to Craik to suggest that he needed to use terms more precise than "model" and "paralleling," putting forward group theory, in particular the concept of isomorphism of groups, as a suitably exact language for discussing his theories (Ashby 1944). Ashby went on to state, rather optimistically, "I believe 'isomorphism' is destined to play the same part in psychology that, say, velocity does in physics, in the sense that one can't get anywhere without it." Craik took this suggestion seriously enough to respond with a three-page letter on the nature of knowledge and mathematical description, which resulted in a further exchange of letters revealing a fair amount of common ground in the two men's views on what kind of knowledge science could communicate. Craik was "much interested to hear further" (Craik 1944) of Ashby's theories alluded to in the following paragraph in which Ashby introduces himself (Ashby 1944, p. 1):

Professionally I am a psychiatrist, but am much interested in mathematics, physics and the nervous system. For some years I have been working on the idea expressed so clearly on p. 115: "It is possible that a brain consisting of randomly connected impressionable synapses would assume the required degree of orderliness as a result of experience..." After some years' investigation of this idea I eventually established that this is certainly so, provided that by "orderly" we understand "organised as a dynamic system so that the behaviour produced is self-preservative rather than self-destructive." The basic principle is quite simple but the statement in full mathematical rigour, which I have recently achieved, tends unfortunately to obscure this somewhat.

In Ashby's talk of self-preservative dynamic systems we can clearly recognize the core idea he would continue to develop over the next few years and publish in *Design for a Brain* (Ashby 1952a). In that book he constructed a general theory of adaptive systems as dynamical systems in which "essential" variables (such as heart rate and body temperature in animals) must be kept within certain bounds in the face of external and internal changes or disturbances. This work, which preoccupied Ashby during the early years of the Ratio Club, is discussed in more detail on pages 133–136.

Ashby corresponded with several future members of the club in the mid-1940s. For instance, in 1946 Hick wrote to Ashby after reading his note on equilibrium systems in the *American Journal of Psychology* (Ashby 1946). Hick explained that he, too, was "trying to develop the principles of 'Analytical Machines' as applied to the nervous system" (Hick 1947a) and requested copies of all Ashby's papers. The pair corresponded over the mathematical details of Ashby's theories of adaptation, and Hick declared (1947b) himself "not entirely happy with your conclusion that a sequence of breaks, if it continues long enough, will eventually, by chance, lead to a stable equilibrium configuration" (p. 1). Hick was referring to an early description of what would later appear in *Design for a Brain* as postulated step mechanisms that would, following a disturbance that pushed any of the system's essential variables out of range, change the internal dynamics of an adaptive machine until a new equilibrium was established—that is, all essential variables were back in range (see pp. 133–137 for further details). Ashby agreed (1947), explaining that he had no rigorous proof but had "little doubt of its truth in a rough and ready, practical way." A year later Ashby's Homeostat machine would provide an existence proof that these mechanisms could work. But Hick had homed in on an interesting and contentious aspect of Ashby's theory. By the time *Design for a Brain* was published, Ashby talked about step mechanisms in very general terms, stating that they could be random but not ascribing absolute rigid properties to them, and so leaving the door open for further refinements. This correspondence foreshadows the kind of probing discussions that were to form the central activity of the Ratio Club, debates that sometimes spilled out onto the pages of learned journals (see, for example, pp. 130–131).

During the war there had been considerable interaction between researchers at the various military sites and several members had originally met through that route. For instance, at TRE Uttley had worked on computer-aided target tracking, as well as building the first British airborne electronic navigation computer (Uttley 1982). This work on early computing devices brought him into contact with both Gold and Turing.

Several members had been friends or acquaintances at Cambridge: Pringle and Turing were contemporaries, as were Barlow and Merton, who had both been tutored by Rushton. Others met at workshops and conferences in the years leading up to the founding of the club. Those involved in EEG work—Walter, Bates, Dawson, and Shipton—were all well known to one another professionally. Walter and Dawson had together laid the foundations for clinical uses of EEG; a paper they wrote together in 1944 (Dawson and Walter 1944) was still used in the training of EEG practitioners in the 1980s. Ashby had interacted with Walter for some time, not least because their research institutes were nearby. So by the time the Ratio Club started, most members had at least passing familiarity with some, but by no means all, of the others' ideas.

The Way Forward

For two or three years prior to the founding of the club there had been a gradual increase in activity, on both sides of the Atlantic, in new approaches to machine intelligence, as well as renewed interest in associated mechanistic views of natural intelligence. In Britain much of that activity involved future Ratio Club members. The phrase Bates used in his initial letters of invitation to the founders of the club, that he wished to bring together people who "had Wiener's ideas before Wiener's book appeared," may have been slightly gung-ho, but in a draft for an article for the *British Medical Journal* in 1952, Bates (1952b) explained himself a little more:

Those who have been influenced by these ideas so far, would not acknowledge any particular indebtedness to Wiener, for although he was the first to collect them together under one cover, they had been common knowledge to many workers in biology who had contacts with various types of engineering during the war.

It is likely that Bates was mainly thinking of chapters 3, 4, and 5 of *Cybernetics*: "Time Series, Information and Communication"; "Feedback and Oscillation"; and "Computing Machines and the Nervous System." Certainly many biologists had become familiar with feedback and its mathematical treatment during the war, and some had worked on time-series analysis and communication in relation to radar (some of their more mathematical colleagues would have been using some of Wiener's techniques and methods that were circulating in technical reports and draft papers—quite literally having Wiener's ideas before his book appeared). Most felt that the independent British line of research on computing machines and their relationship to the nervous system was at least as strong as the work going on in the United States—important strands of which in turn were based on prior British work such as that of Turing (Barlow 2001). Indeed, many were of the opinion that the central hypothesis of cybernetics was that the nervous system should be viewed as a self-correcting device chiefly relying on negative-feedback mechanisms (Wisdom 1951). This concept had first been introduced by Ashby in 1940 (Ashby 1940) and then independently by Rosenblueth, Wiener, and Bigelow (1943) three years later. The development of this idea was the central, all-consuming focus of Ashby's work until the completion of *Design for a Brain*, which set out his theories up to that point. It is interesting that Ashby's review of *Cybernetics* (Ashby 1949b) is quite critical of the way the core ideas of the book are presented.

Perhaps the following passage from the introduction to *Cybernetics* pricked Bates's sense of national pride and acted as a further spur (Wiener 1948, p. 23):

In the spring of 1947...[I] spent a total of three weeks in England, chiefly as a guest of my old friend J. B. S. Haldane. I had an excellent chance to meet most of those doing work on ultra-rapid computing machines...and above all to talk over the fundamental ideas of cybernetics with Mr. Turing....I found the interest in cybernetics about as great and well informed in England as in the United States, and the engineering work excellent, though of course limited by the smaller funds available. I found much interest and understanding of its possibility in many quarters....I did not find, however, that as much progress had been made in unifying the subject and in pulling the various threads of research together as we had made at home in the States.

Whatever the views on Wiener's influence—and the more mathematical members will surely have recognized his significant technical contributions—it is clear that all those associated with the Ratio Club agreed that Claude Shannon's newly published formulation of information theory, partly built on foundations laid by Wiener, was very exciting and important. The time was ripe for a regular gathering to develop these ideas further.

Club Meetings

The London district of Bloomsbury often conjures up images of free-thinking intellectuals, dissolute artists, and neurotic writers—early-twentieth-century bohemians who, as Dorothy Parker once said, "lived in squares and loved in triangles." But it is also the birthplace of neurology, for it was here, in 1860, that the first hospital in the world dedicated to the study and treatment of diseases of the nervous system was established. By the late 1940s the National Hospital for Nervous Diseases was globally influential and had expanded to take up most of one side of Queen's Square. It was about to become regular host to the newly formed group of brilliant and unconventional thinkers.

In 1949 London witnessed the hottest September on record up to that point, with temperatures well above ninety degrees Fahrenheit. In fact the entire summer had been a mixture of scorching sunshine and wild thunderstorms. So it was an unseasonably balmy evening on the fourteenth of that month when a gang of scientists, from Cambridge in the east and Bristol in the west, descended on the grimy bombed-out capital, a city slowly recovering from a war that had financially crippled Britain. They converged on the leafy Queen's Square and assembled in a basement room of the hospital at six-thirty in the evening. After sherries, the meeting started at seven. Bates's notes for his introduction to this inaugural gathering of the club show that he spoke about how the club membership was drawn from a network centered on his friends, and so was somewhat arbitrary, but that there had been an attempt to strike a balance between biologists and non-biologists (Bates 1949e). He then went on to make it clear that the club was for people who were actively using cybernetic ideas in their work. At that point there were seventeen members, but he felt there was room for a few more. (The initial membership comprised Ashby, Barlow, Bates, Dawson,

Gold, Hick, Little, Mackay, McLardy, Merton, Pringle, Shipton, Sholl, Slater, Uttley, Walter, and Westcott). He pointed out that there were no sociologists, no northerners (for example from Manchester University or one of the Scottish universities), and no professors. Possible names for the club were discussed (see pp. 102–103) before Bates sketched out how he thought meetings should be conducted. In this matter he stressed the informality of the club—that members should not try and impose "direction" or employ "personal weight." All agreed with this sentiment and endorsed his "no professors" rule—scientists who were regarded to be senior enough to inhibit free discussion were not eligible for membership.

Warren McCulloch then gave his presentation, "Finality and Form in Nervous Activity," a popular talk that he had first given in 1946—perhaps not the best choice for such a demanding audience. Correspondence between members reveals almost unanimous disappointment in the talk. Bates (1949f) set out his own reaction to its content (and style) in a letter to Grey Walter:

Dear Grey,

Many thanks for your letter. I had led myself to expect too much of McCulloch and I was a little disappointed; partly for the reason that I find all Americans less clever than they appear to think themselves; partly because I discovered by hearing him talk on 6 occasions and by drinking with him in private on several more, that he had chunks of his purple stuff stored parrot-wise. By and large however, I found him good value.

Walter replied (1949) to Bates apologizing for not being present at the meeting (he was the only founding member unable to attend). This was due to the birth of a son, or as he put it "owing to the delivery of a male homeostat which I was anxious to get into commission as soon as possible." He went on to tell Bates that he has had "an amusing time" with McCulloch, who had traveled on to Bristol to visit him at the Burden Institute. In reference to Bates's view on McCulloch's talk, he comments that "his reasoning has reached a plateau.... Flowers that bloom on this alp are worth gathering but one should keep one's eyes on the heights."

A buffet dinner with beer followed the talk and then there was an extended discussion session. The whole meeting lasted about three hours. Before the gathering broke up, with some rushing off to catch last trains out of London and others joining McCulloch in search of a nightcap, John Pringle proposed an additional member. Echoing the suggestion made in his written reply to Bates's original invitation to join the club, Pringle put forward the idea that a mathematician or two should be invited to join to give a different perspective and to "keep the biologists in order." He and Gold proposed Alan Turing, a suggestion that was unanimously supported. Turing gladly accepted and shortly afterward was joined by a fellow mathematician, Philip Woodward, who worked with Uttley. At the same time a leading Cambridge neurobiologist, William Rushton, who was well known to many members, was added to the list. The following passage from a circular Bates (1949g) sent to all members shortly after the first meeting shows that the format for the next few sessions had also been discussed and agreed:

It seems to be accepted that the next few meetings shall be given over to a few personal introductory comments from each member in turn. Assuming we can allow two and a half hours per meeting, eighteen members can occupy an average of not more than 25 minutes each. The contributions should thus clearly be in the nature of an aperitif or an hors d'oeuvres—the fish, meat and sweet to follow at later meetings.

Regardless of reactions to the opening talk, there was great enthusiasm for the venture. The club was well and truly born.

Following this inaugural meeting the club convened regularly until the end of 1954. There was a further two-day meeting and a single evening session in 1955 and a final gathering in 1958, after the now classic "Mechanization of Thought Processes" symposium organized by Uttley at the National Physical Laboratory in Teddington (Blake and Uttley 1959).

Table 6.1 shows the full list of known Ratio Club meetings. This has been compiled from a combination of individual meeting notices found in the Bates Archive at the Wellcome Library for the History and Understanding of Medicine, in London, surviving members' personal records, and a list of meetings made by Bates in the mid-1980s. There are inconsistencies between these sources, but through cross-referencing with notes made at meetings and correspondence between members this list is believed to be accurate. It is possible that it is incomplete, but if so, only a very small number of additional meetings could have occurred.

The order of members' introductory talks was assigned by Bates, using a table of random numbers. Due to overruns and some people being unable to attend certain meetings, the actual order in which they were given may have been slightly different from that shown in the table. However, they did take place on the dates indicated.

The format of the opening meeting—drinks, session, buffet and beer, discussion session, coffee—seems to have been adopted for subsequent meetings. Members' introductory talks, which highlighted their expertise and

Table 6.1
Known Ratio Club Meetings

Meeting	Date	Speakers, Discussion Topics, and Paper Titles
1	14 September 1949	Warren McCulloch, "Finality and Form in Nervous Activity"
2	18 October 1949	Introductory talks from Sholl, Dawson, Mackay, Uttley
3	17 November 1949	Introductory talks from Gold, Bates, McLardy
4	15 December 1949	Introductory talks from Pringle, Merton, Little, Hick, Grey Walter
5	19 January 1950	Slater, "Paradoxes Are Hogwash"; Mackay, "Why Is the Visual World Stable?"
6	16 February 1950	Introductory talks from Shipton, Slater, Woodward
7	16 March 1950	Introductory talks from Ashby, Barlow
8	21 April 1950	Introductory talks from Wescott, Turing
9	18 May 1950	"Pattern Recognition," Walter, Uttley, Mackay, Barlow, Gold
10	22 June 1950	"Elementary Basis of Information Theory," Woodward
11	18 July 1950	"Concept of Probability," Gold, Mackay, Sholl
12	21 September 1950	"Noise in the Nervous System," Pringle
13	2 October 1950	Meeting at London Symposium on Information Theory
14	7 December 1950	"Educating a Digital Computer," Turing
15	22 February 1951	"Adaptive Behaviour," Walter
16	5 April 1951	"Shape and Size of Nerve Fibres," Rushton
17	31 May 1951	"Statistical Machinery," Ashby
18	26 July 1951	"Telepathy," Bates
19	1 November 1951	"On Popper: What Is Happening to the Universe?," Gold
20	21 December 1951	Future Policy; discussion on "the possibility of a scientific basis of ethics" opened by Slater; discussion on "a quantitative approach to brain cell counts" opened by Sholl
21	8 February 1952	"The Chemical Origin of Biological Form," Turing; "The Theory of Observation," Woodward
22	20 March 1952	"Pattern Recognition," Uttley; "Meaning in Information Theory," Mackay
23	2–3 May 1952	Special meeting at Cambridge, organized by Pringle
24	19 June 1952	"Memory," Bates; "The Logic of Discrimination," Westcott

Table 6.1
(continued)

Meeting	Date	Speakers, Discussion Topics, and Paper Titles
25	31 July 1952	"The Size of Eyes," Barlow; "American Interests in Brain Structure," Sholl
26	24–25 October 1952	Special meeting at Burden Neurological Institute, Bristol (canceled)
27	6 November 1952	"Design of Randomizing Devices," Hick; "On Ashby's Design for a Brain," Walter
28	11 December 1952	"Perils of Self-Awareness in Machines," Mackay; "Sorting Afferent from Efferent Messages in Nerves," Merton
29	19 February 1953	"Pattern Discrimination in the Visual Cortex," Uttley and Sholl
30	7 May 1953	"Absorption of Radio Frequencies by Ionic Materials," Little; "The Signal-to-Noise Problem," Dawson
31	2 July 1953	Warren McCulloch: Discussion of topics raised in longer lectures given by McCulloch at University College London in previous week
32	22 October 1953	"Demonstration and Discussion of the Toposcope," Shipton; "Principles of Rational Judgement," Good
33	11 February 1954	Discussion: "How does the nervous system carry information?"; guest talk: "Observations on Hearing Mechanisms," Whitfield and Allanson
34	17 June 1954	"Servo Control of Muscular Movements," Merton; "Introduction to Group Theory," Woodward
35	25 November 1954	"Negative Information," Slater and Woodward; guest talk: "Development as a Cybernetic Process," Waddington
36	6–7 May 1955	Special meeting in West Country (TRE, Barnwood House, Burden Institute)
37	15 September 1955	Discussion meeting after third London Symposium on Information Theory; many guests from the United States
38	27 November 1958	Final reunion meeting after the National Physical Laboratory's "Mechanisation of Thought Processes" symposium

interests, typically focused on some aspect of their current research. John Westcott's notebook reveals that a wide range of topics was discussed (Westcott 1949–53): Scholl talked about the need to construct an appropriate mathematics to shed light on the physiology of the nervous systems. Dawson described ongoing work on eliminating noise from EEG readings. Mackay argued for a more complex description of information, both philosophically and mathematically, claiming that it cannot be adequately defined as a single number. Uttley sketched out the design for a digital computer he was working on at TRE. Gold illustrated his more general interest in the role of servomechanisms in physiology by describing his work on a radical new theory of the functioning of the ear, which postulated a central role for feedback; Gold (2002) later recalled that at the time the Ratio Club was the only group that understood his theory. Bates talked about various levels of description of the nervous system. McLardy described recent research in invasive surgical procedures in psychiatry. Merton outlined his work on using cybernetic ideas to gain a better understanding of how muscles work. Walter described his newly constructed robotic tortoises, sketching out the aims of the research and early results obtained (see pp. 136–137 for further discussion of this work). Woodward talked about information in noisy environments. Little discussed the scientific method and the difficulty of recognizing a perfect theory. Hick outlined his research on reaction times in the face of multiple choices—the foundations of what would later become known as Hick's law (Hick 1952), which makes use of information theory to describe the time taken to make a decision as a function of the number of alternatives available (see p. 95 for a brief statement of the law). Ashby talked about his theories of adaptive behavior and how they were illustrated by his just-finished Homeostat device (see pp. 133–136 for further discussion of this work). Barlow outlined the research on the role of eye movement in generating visual responses that he was conducting at this early stage of his career (see the interview with Barlow, chapter 18 of this volume, for further details of this work). Westcott talked a little about his background in radar and his work with Wiener at MIT before outlining his mathematical work on analyzing servomechanisms, emphasizing the importance of Wiener's theory of feedback systems, on which he was building. After each of the presentations discussion from the floor took over.

After the series of introductory talks, the format of meetings changed to focus on a single topic, sometimes introduced by one person, sometimes by several. Prior to this, Ashby circulated two lists of suggested topics for discussion; an initial one on February 18, 1950 (Ashby 1950a), and a refined version dated May 15, 1950 (Ashby 1950b). They make fascinating reading, giving an insight into Ashby's preoccupations at the time. The refined list (Ashby 1950b) is reproduced here. Many of the questions are still highly pertinent today.

1. What is known of "machines" that are defined only statistically? To what extent is this knowledge applicable to the brain?
2. What evidence is there that "noise" (a) does, (b) does not, play a part in brain function?
3. To what extent can the abnormalities of brains and machines be reduced to common terms?
4. The brain shows some indifference to the exact localisation of some of its processes: to what extent can this indifference be paralleled in physical systems? Can any general principle be deduced from them, suitable for application to the brain?
5. From what is known about present-day mechanical memories can any principle be deduced to which the brain must be subject?
6. To what extent do the sense-organs' known properties illustrate the principles of information-theory?
7. Consider the various well known optical illusions: what can information-theory deduce from them?
8. What are the general effects, in machines and brains[,] of delay in the transmission of information?
9. Can the members agree on definitions, applicable equally to all systems—biological, physiological, physical, sociological—cf: feedback, stability, servo-mechanism.
10. The physiologist observing the brain and the physicist observing an atomic system are each observing a system only partly accessible to observation: to what extent can they use common principles?
11. The two observers of 10, above, are also alike in that each can observe his system only by interfering with it: to what extent can they use common principles?
12. Is "mind" a physical "unobservable"? If so, what corollaries may be drawn?
13. What are the applications, to cerebral processes, of the thermodynamics of open systems?
14. To what extent can the phenomena of life be imitated by present-day machines?
15. To what extent have mechanisms been successful in imitating the conditioned reflex? What features of the C.R. have conspicuously not yet been imitated?

16. What principles must govern the design of a machine which, like the brain, has to work out its own formulae for prediction?

17. What cerebral processes are recognisably (a) analogical, (b) digital, in nature?

18. What conditions are necessary and sufficient that a machine built of many integrated parts should be able, like the brain, to perform an action either quickly or slowly without becoming uncoordinated?

19. Steady states in economic systems.

20. What general methods are available for making systems stable, and what are their applications to physiology?

21. To what extent can information-theory be applied to communication in insect and similar communities?

22. To what extent are the principles of discontinuous servo-mechanisms applicable to the brain?

23. What re-organisation of the Civil Service would improve it cybernetically?

24. What economic "vicious circles" can be explained cybernetically?

25. What re-organisation of the present economic system would improve it cybernetically?

26. To what extent can information-theory be applied to the control exerted genetically by one generation over the next?

27. Can the members agree on a conclusion about extra-sensory perception?

28. What would be the properties of a machine whose "time" was not a real but a complex variable? Has such a system any application to certain obscure, i.e. spiritualistic, properties of the brain?

The last topic on the initial list is missing from the more detailed second list: "If all else fails: The effect of alcohol on control and communication, with practical work." This suggestion was certainly taken up, as it appears were several others: shortly after the lists appeared Pringle gave a talk on the topic of suggestion 2 (meeting 12), as did Walter on 14 and 15 (meeting 15). Topic 27 came up in talks by Bates and Good (meetings 18 and 32). Issues relating to many of the other suggestions often arose in group discussions, being in areas of great interest to many members (topics 6–13, 16–18, and 26). In particular, Barlow recalls much discussion of topic 17 (Barlow 2007). Although Ashby's publications and notebooks make it clear that some of the suggestions are based on the central research questions he was grappling with at the time (suggestions 1, 18, 20, 22), it is very likely that some of the others arose from issues brought up by members in their introductory talks. In the mid-1980s Bates made some notes for a planned article on the Ratio Club (Bates 1985), a plan that unfortunately did not come to fruition. However, among these scant jottings is mention of Ashby's lists, which further suggests that they did play a role in shaping the scope of topics discussed.

Members often volunteered to give talks, but Bates, when he felt it was necessary, actively controlled the balance of topics by persuading particular members to give presentations. Sometimes there were requests from members for particular subjects to be discussed or particular people to give talks on certain topics. Looking through the list of subjects discussed, many are still extremely interesting today; at the time they must have been positively mouth-watering.

At the end of 1950, after meeting him at the first London Symposium on Information Theory, Bates invited I. J. "Jack" Good along to the next meeting as his guest. The speaker was Turing, Good's friend and wartime colleague. This was a particularly lively meeting and after it Good wrote to Bates expressing how much he had enjoyed the evening and apologizing for being too vociferous. He wondered, "Would there be any serious objection to my becoming a member?" (Good 1950a). Bates replied (1950a) that "the club has been going for a year, and is entirely without any formal procedures. New members join by invitation, but I think personally you would be a great asset, and hope you will be able to come as my guest to some future meetings, so that perhaps my view will become consensus!" Bates's view obviously did hold sway, as Good became the twenty-first member of the club. Perhaps it was thought a third mathematician was needed to help the other two keep the biologists in order. Partly because of the size of the room used for meetings, and partly because Bates had firm ideas on the kind of atmosphere he wanted to create and who were the "right sorts" to maintain it, the membership remained closed from that point.

For the first year meetings were monthly and were all held at the National Hospital in Queen's Square. From mid-1950 until the end of 1951 the frequency of meetings dropped slightly and in the second half of 1951 attendance started to fall. This was mainly due to the not inconsiderable time and expense incurred by members based outside London every time they came to a meeting. In October 1951 Woodward had written to Bates explaining that he had to take part of his annual leave to attend meetings (Woodward 1951); the following month Walter wrote to explain that he had difficulty in covering the expenses of the trips to London necessary for Ratio Club gatherings. He suggested holding some meetings outside London in members' labs, pointing out that this would also allow practical demonstrations as background for discussion (Walter 1951).

Indeed the round-trip journey from Bristol could be quite a hike. Janet Shipton (Shipton 2002) remembers waiting up to greet her husband, Harold, on his return from Ratio meetings: "He would get back in the dead of night, the smell of train smoke on his clothes."

At the December 1951 meeting of the club, Bates (1951) called a special session to discuss future policy. Beforehand he circulated a document in which he put down his thoughts on the state of the club. Headed "The Ratio Club," the document opened by stating that "looked at in one way, the Club is thriving—in another way it is not. It is thriving as judged by the suggestions for future activities." These suggestions are listed as requests for specific talks by Woodward (on the theory of observation) and Hick (on the rate of gain of information), an offer of a talk on morphogenesis by Turing, as well as various suggestions for discussion topics (all of these suggestions, offers and requests were taken up in subsequent meetings). Bates goes on: "In addition to this, we have in pigeon-holes a long list sent in by Ashby of suitable topics; various suggestions for outside speakers; and a further suggestion that members should collaborate in writing different chapters to a book on the lines of 'Cybernetics,' but somewhat tidier." Sadly, this intriguing book idea never came to fruition. He then explains the cause for concern:

Looked at in another way, the Club is ailing. For the past three meetings, half or more of the members have been absent. This half have been mostly those who live out of London—the most reasonable inference clearly is that a single evening's meeting does not promise to be a sufficient reward for the inconvenience and expense of getting to it. In addition one member has pointed out that if expenses cannot be claimed the night's absence is counted against the period of his annual leave! The whole point of the Club is to facilitate contacts between people who may have something to contribute to each other, and who might not otherwise come together, and it would seem that some change in its habits may be indicated.

Bates then listed some suggested courses of action for discussion at the next meeting. These ranged from having far fewer, but longer, meetings to doubling the membership.

It was decided that there would be six or seven meetings a year, four or five in London and two elsewhere. The meetings would start earlier to allow two papers. A novel suggestion by Philip Woodward was also taken up: to start a postal portfolio—a circulating package of ideas—"to be totally informal and colloquial." Bates prepared a randomized order of members for the portfolio to travel around.

This new regime was followed from the first meeting of 1952 until the club disbanded, and seemed to go a good way toward solving the problems that prompted its instigation. The typical meeting pattern was now to gather at four-thirty for tea, followed by the first talk and discussion, then a meal and drinks, followed by the second talk and discussion.

Most Ratio Club talks were based on current research and were often early outings for highly significant work, sometimes opening up new areas of inquiry that are still active today. For instance, Turing's talk, "Educating a Digital Computer," in December 1950, was on the topics covered by his seminal *Mind* paper of that year (Turing 1950), which introduced the Turing Test and is regarded as one of the key foundational works of machine intelligence. As the title suggests, that talk focused on how an intelligent machine might be developed; Turing advocated using adaptive machines that might learn over their lifetimes and also over generations by employing a form of artificial evolution. This meeting is remembered as being particularly good, with Turing in top form, stimulating a scintillating extended discussion (Bates 1950b). Turing's 1952 talk on biological form was another gem, describing his as yet unpublished work on reaction-diffusion models of morphogenesis (Turing 1952), which showed how pattern and form could emerge from reaction-diffusion systems if they are appropriately parameterized (a role he hypothesized might be taken on by genes). In addition to launching new directions in theoretical biology, this work was pioneering in its use of computer modeling and was to prove extremely influential. There is not enough space to describe all the important work discussed at club meetings, but further summaries are scattered at appropriate places throughout the rest of this chapter.

As well as research talks, there were also various "educational" presentations, usually requested by the biologists. For instance, Woodward gave several on information theory, which gave the biologists very early access to important new ways of thinking. By all accounts Woodward was an extremely good lecturer, blessed with a gift for insightful exposition (this is evident in his 1953 book *Probability and Information Theory, with Applications to Radar*, still regarded by some theorists as one of the most profound works in the area since Shannon's original papers.) Barlow was particularly influenced by these exciting new ideas and became a pioneer in the use of information theory as a theoretical framework to understand the operation of neural systems, particularly those associated with vision. This theoretical framework either directly or indirectly underpinned many of Barlow's very important contributions to neuroscience. He regards the Ratio Club as one of the most important formative influences on his work and sees "much of what I have done since as flowing from those evening meetings" (Barlow 2001; see also chapter 18 of this volume for further discussion of this point). In a similar spirit there were lectures on probability theory from Gold and Mackay and on the emerging field of control theory from Westcott.

In 1952 two extended out-of-London meetings were planned, one in Cambridge in May and one in Bristol in October. The Cambridge meeting was organized by Pringle and was held from Friday afternoon to Saturday morning in his college, Peterhouse. After drinks and dinner Pringle led a session on "Processes Involved in the Origin of Life." Correspondence after the meeting mentions that this session was captured on a tape recorder, although the recording has not yet come to light. The next day visits were arranged to various labs, including Cavendish (physics), led by Gold; Zoology; Physiology, led by Rushton; Psychology, led by Hick; and Mathematics. The photograph shown in figure 6.2 was taken at this meeting, quite

Figure 6.2
Some members of the Ratio Club with guests, outside Peterhouse College, University of Cambridge, May 1952. The photograph was organized by Donald Mackay. *Back row (partly obscured)*: Harold Shipton, John Bates, William Hick, John Pringle, Donald Scholl, John Westcott, Donald Mackay. *Middle row*: Giles Brindley, Turner McLardy. W. Ross Ashby, Thomas Gold, Arthur Uttley. *Front row*: Alan Turing, Gurney Sutton, William Rushton, George Dawson, Horace Barlow. Image courtesy The Wellcome Library for the History and Understanding of Medicine, London.

possibly after the predinner sherries mentioned on the invitation sent out to club members. Not everyone was able to attend and several of those in the photograph are guests. A limited number of guests were allowed at most meetings and over the years various distinguished visitors took part in club gatherings. As well as McCulloch, Pitts, and Shannon, these included John Zachary Young, the leading anatomist and neurologist, who attended several meetings; Conrad Waddington, the pioneering theoretical biologist and geneticist; and Giles Brindley who became a distinguished neuroscientist and was David Marr's Ph.D. supervisor. Jack Good once brought along the director of the National Security Agency, home to the United States' code breakers and makers, whom he knew through his work for British Intelligence. That particular meeting was on probability and included prolonged discussions of experiments claiming to give evidence for ESP. Following the 1955 London Symposium on Information Theory, a special club meeting involved a host of leading lights from the world of information theory and cybernetics, many from overseas. These included Peter Elias, J. C. R. Licklider, Warren McCulloch, Oliver Selfridge, Benoît Mandelbrot, and Colin Cherry. Records are sketchy on this matter, but it is likely that many other luminaries of the day took part in other meetings.

The Bristol meeting was to be held at the Burden Neurological Institute, starting at noon on Friday October 24, 1952, and running into the next day, but it seems to have been canceled at the last minute due to heavy teaching commitments preventing a substantial number of members from attending. The talks and demonstrations planned for this meeting were moved into later club meetings. These included Grey Walter opening a discussion "Mechanisms for Adaptive Behaviour," which focused on simulation of learning by man-made devices, and in particular on the issues raised in Ashby's recently published book *Design for a Brain*, and a presentation by Shipton on the Toposcope, the world's first multichannel EEG recording device. The machine, developed by Shipton and Walter, was capable of building and displaying bidimensional maps of the EEG activity over the brain surface and included frequency and phase information.

From mid-1953, meetings became less frequent, with only three in 1954 and two in 1955. In 1955 the extended West Country event finally happened, starting at TRE Malvern on May 6 and then going the next day to the Burden Institute in Bristol via Ashby's Barnwood House lab. The meeting was primarily devoted to demonstrations and discussions of work in progress at these locations. At TRE, various devices from Uttley's group were on show. These included a "tracking simulator," a novel apparatus designed to provide a versatile means of setting up and studying problems relating to a human operator working in a closed-loop system. The device used a two-gun cathode-ray tube and required the operator to track a moving dot by controlling a second dot with a joystick. Also on show were Uttley's systems for automatically classifying spatial and temporal patterns and pioneering electronic and hydraulic systems capable of inference using principles from conditional probability.

To reach the next leg of the multisite meeting, Philip Woodward recalls traveling across country in a Rolls Royce that Barlow had borrowed from his brother. As they hurtling toward "Ashby's lunatic asylum" (Barnwood House Psychiatric Hospital, in Gloucester), Rushton diagnosed the exact form of Woodward's color blindness by getting him to describe the spring flowers he could see on the verges (Woodward 2002).

At Barnwood House, Ashby demonstrated his Dispersive and Multi-stable System (DAMS), and the Homeostat was available to those who were not already familiar with it. As mentioned earlier, the Homeostat demonstrated the theories of adaptation developed in *Design for a Brain*, where it is described in some detail. Although Ashby had talked at earlier club meetings about the DAMS machine, this would have been the first time that most members saw it firsthand. The DAMS device, which is much less well known than the Homeostat—mainly because Ashby was not able to develop it sufficiently to fully demonstrate his theories—was intended to explore possible learning behaviors of randomly connected nonlinear components. The motivation for this was the intriguing possibility that parts of the brain, particularly the cortex, might be at least partially randomly wired. Ashby had been developing the machine for some years and demonstrated the current version, which by then illustrated some interesting properties of "statistical machinery." The theoretical line started in this work resurfaced many years later in Gardner and Ashby's computational study of the stability of large interconnected systems (Gardner and Ashby 1970). There is a nice anecdote about the machine which originates from this 1955 meeting. Philip Woodward remembers being told, possibly apocryphally, that when Ashby asked a local engineering firm to construct part of the device, specifying random connections, they were so bemused, particularly since the order was coming from Barnwood House Psychiatric Hospital, that they rang up to check that Dr. Ashby was not in fact a patient (Woodward 2002).

The club was full of lively and strong personalities. Mackay, Turing, and Walter were, in their very different ways, brilliant speakers who all broadcast talks on scientific subjects for BBC radio. Grey Walter, in particular, was something of a media personality, making appearances on popular radio quiz shows and early television programs. He was a larger-than-life character who liked to cultivate a certain image, that of a swashbuckling man of the world. He was, as Harold Shipton noted (2002) "a bugger for the women." This reputation did him no favors with many in the scientific establishment. Walter stood in marked contrast to Mackay, a fiery lay preacher who had been brought up attending the Evangelical Free Church of Scotland, one of the radical breakaway "wee free" churches. Many who knew him have remarked on a certain tension between his often radical scientific ideas about the nature of intelligence and his strait-laced religiosity. Horace Barlow, a great friend of Mackay's and an admirer of his ideas, has noted (2002) that "his conviction that he had a special direct line to a Higher Place, somehow slightly marred his work and prevented him from becoming as well regarded as he should have been." According to Barlow's biographical memoir (1986) of Rushton, the Cambridge don "cut a striking and influential figure...was argumentative, and often an enormously successful showman...he valued the human intellect and its skilful use above everything else." Giles Brindley, a guest at several meetings, remembers (2002) that Barlow and Gold were very active in discussions and that when occasionally a debate got out of hand, Pringle would gently refocus the conversation.

Members came from a rich mix of social and educational backgrounds, ranging from privileged upbringings to the humblest of origins. Harold Shipton's story is particularly remarkable. In the years before World War II he was plucked from the life of an impoverished farm laborer by RAF talent scouts who were looking for bright young men to train as radar operators. During training it quickly became apparent that he had a natural gift for electronics, which was duly exploited. After the war, before he had been demobbed, he was sent to the Burden Neurological Institute to find out what Grey Walter was doing with the suspiciously large amounts of surplus military electronic equipment he was buying. He and Walter immediately hit it off and he stayed. At the institute he met his future wife, Clement Attlee's daughter Janet. (Attlee was leader of the Labour Party, Churchill's deputy during the war, and prime minister of Britain from 1945 to 1951.) Hence, at the West Country meeting, members of the all-male Ratio Club were served tea by the Labour Prime Minister's daughter.

Bates had created a powerful mix of individuals and ideas with just the right degree of volatility. The result was that meetings were extremely stimulating and greatly enjoyed by all. All the surviving members interviewed recalled the club with great enthusiasm; Gold (2002) described meetings as "always interesting, often exciting." Even those, such as Woodward and Westcott, who felt that they were net givers, in terms of the direct intellectual influence of the club on members' work, found meetings a pleasure and were annoyed when they had to miss one.

The social atmosphere of the club sometimes continued in after-meeting parties. Philip Woodward (2002) remembers that on one occasion some of the group reconvened on the enormous Dutch sailing barge Pat Merton kept in St. Catherine's docks on the Thames. Merton had arranged for a pianist and oboe player on deck. St. Catherine's was a working dock in those days with a large sugar refinery that belched out pungent fumes. As the night wore on and the drink flowed, the sugar-strewn route up the dockside steps to the toilet became more and more treacherous.

Table 6.1 shows that a number of external speakers were invited to give presentations. Despite reactions to his talk in 1949, Warren McCulloch was asked back in 1953 to open a discussion on his work, and attended other meetings as a guest; eventually members grew to appreciate his style.

As the club developed, Ashby was keen to see it transformed into a formal scientific society—"the Biophysical Society" or "the Cybernetics Society"—with a more open membership. His proposals for this were resisted. It seems that, for many members, the informal atmosphere of the club, exactly as Bates had conceived it, was the most important factor. When Ashby proposed that Professor J. Z. Young be admitted as a member, Sholl (1952) wrote to Bates in protest:

I consider membership of the Club not only as one of my more pleasant activities but as one of the most important factors in the development of my work. I have stressed before how valuable I find the informality and spontaneity of our discussion and the fact that one does not have to be on one's guard when any issue is being argued. At the present time we have a group of workers, each with some specialised knowledge and I believe that the free interchange of ideas which has been so happily achieved and which, indeed, was the basis for the founding of the Club, largely results from the fact that questions of academic status do not arise.

Young was the head of the Department of Anatomy at University College, where Sholl worked, and although Sholl collaborated with him and continued to do so after the club disbanded, in those days academic relations and the processes of career advancement were such that he would have felt very uncomfortable with his boss as a member. In any event the "no professors" rule prevailed.

By the end of the summer of 1955 the club had run its course; many important intellectual cross-fertilizations had occurred, and all had learned much from each other. In 1954 Turing had died in tragic and disturbing circumstances that have been well documented (Hodges 1983). By now several members' research had become very well known internationally (Ashby and Walter in cybernetics, with Uttley not far behind, and Rushton

and Pringle in neurophysiology) and others were on the cusp of major recognition. As careers advanced and families grew, many found it increasingly difficult to justify the time needed for meetings. Another factor that
may have played a part in the club's demise was that cybernetics had
become respectable. Lord Adrian had endorsed it in one of his Royal Society presidential addresses and talk of its application in every conceivable
branch of biology was rife. The frisson of antiestablishmentarianism that
imbued the early meetings was all but gone. The September 1955 meeting,
tacked on to the end of the London Symposium on Information Theory,
turned out to be the last. A reunion was held in November 1958 after
Uttley's "Mechanization of Thought Processes" symposium at the National
Physical Laboratory. Nine members turned up—Bates, Barlow, Dawson,
Sholl, Slater, Uttley, Mackay, Woodward, and Hick). Of the rest, Bates's
(1958) note of the meeting reads:

Absent: with expressed regret: Grey Walter, Merton, Westcott; with expressed lack of
interest: Ashby, McLardy; without expression: Rushton, Pringle, Little, Good; emigrated: Gold, Shipton.

At the meeting, suggestions were put forward for possible new and younger members. The first name recorded is that of Richard Gregory, then a
young psychologist who had just made his first professional presentation
at the symposium. Clearly, Bates had not lost his ability to spot talent, as
Gregory later became an extremely distinguished vision scientist and Fellow of the Royal Society. However, the initiative came to nothing, and the
club did not meet again.

Themes

Although a very wide range of topics was discussed at club meetings, a
number of important themes dominated. These included information
theory, probabilistic and statistical processes and techniques, pattern
recognition, and digital versus analogue models of the brain (Barlow 2002,
2007; Bates 1985). The themes usually surfaced in the context of their
application to understanding the nervous system or developing machine
intelligence.

Information Theory
By far the greatest proportion of British wartime scientific effort had gone
into radar and communications, so it is perhaps unsurprising that there
was huge interest in information theory in the club. Many of the brain

scientists realized very early on that here was something that might be an important new tool in understanding the nervous system. Shannon's technical reports and papers were not easy to get hold of in Britain in the late 1940s and so the first time Barlow came across them was when Bates sent him copies—with a note to the effect that this was important stuff—along with his invitation to join the club. Barlow agreed with Bates, immediately grasping the fact that information theory provided a new, potentially measurable quantity that might help to give a stronger theoretical underpinning to neurophysiology. Over the next few years, as he learned more about the subject at club meetings—particularly from Woodward—he developed a theoretical framework that shaped his research and helped to propel him to the forefront of his field. Barlow used information-theoretic ideas in an implicit way in his now classic 1953 paper on the frog's retina (Barlow 1953). This paper gives the first suggestion that the retina acts as a filter passing on useful information, developing the idea that certain types of cells act as specialized "fly detectors"—thus that the visual system has evolved to efficiently extract pertinent information from the environment, an idea that was to become very influential. Later, in a series of very important theoretical papers, he argued that the nervous system may be transforming "sensory messages" through a succession of recoding operations which reduce redundancy in order to make the barrage of sensory information reaching it manageable (Barlow 1959, 1961). (Reducing the amount of redundancy in a message's coding is one way to compress it and thereby make its transmission more efficient.) This line of reasoning fed into the later development of his equally influential "neuron doctrine for perceptual psychology" which postulated that the brain makes use of highly sparse neural "representations" (Barlow 1972). As more neurophysiological data became available, the notion of redundancy reduction became difficult to sustain and Barlow began to argue for the principle of redundancy exploitation in the nervous system. In work that has become influential in machine learning and computational neuroscience, Barlow and his coworkers have demonstrated how learning can be more efficient with increased redundancy, as this reduces "overlap" between distributed patterns of activity (Gardner-Medwin and Barlow 2001). (For further discussion of these matters see chapter 18, in this volume).

Information and its role in biology was at the heart of many club debates. Mackay believed the Shannon formulation was too restrictive and during the Ratio Club years he developed his own set of ideas, allied with Dennis Gabor's (1946) version of information theory, which took account of context and meaning (Mackay 1952a, 1952b). In the early period of the club's existence Ashby was working hard on the final version of *Design for a Brain* and his habit of quizzing members on specific topics that would help him refine the ideas in the book left several members with the impression that that he was exclusively preoccupied with his own ideas and not open to new influences. However, his journals indicate that he was becoming convinced of the importance of information theory. He records a conversation with Gold and Pringle at one meeting in 1950 on how much information was needed to specify a particular machine, and by extension how much information must be encoded in the genes of an animal. His arguments were demolished by Gold, who pointed out that "complexity doesn't necessarily need any number of genes for its production: the most complicated organisation can be produced as a result of a single bit of information once the producing machinery has been set up" (Ashby 1950c). As usual, Gold was decades ahead in stressing the importance of genotype to phenotype mappings and the role of development. This theme resurfaced in Ashby's (1952b) paper "Can a Mechanical Chess-Player Outplay Its Designer," in which he used information theory to try and show how it might be possible to construct a machine whose behavior goes beyond the bounds of the specifications described by its designer. This paper caused debate within the club, with Hick in particular disagreeing with Ashby's claim that random processes (such as mutations in evolution) can be a source of information. This resulted in Hick joining in the discussion of Ashby's paper on the pages of *The British Journal for the Philosophy of Science*, where the original work had appeared (Ashby 1952b), a debate that also included a contribution from J. B. S. Haldane (1952).

A striking example of the degree of enthusiasm for information-theoretic ideas within the club is given by the contents page of the first ever issue of the *IEEE Transactions on Information Theory*, the field's premier journal, in February 1953. This issue was based on the proceedings of the First London Symposium on Information Theory, held in September 1950, and was dominated by Ratio Club members (see a complete table of contents at http://www.informatik.uni.trier.de/~ley/db/journals/tit/tit1.html). Of the twenty-two full papers that were published in it, fourteen were by club members. Of the remaining eight, three were by Shannon and two by Gabor.

Probability and Statistics

Probabilistic and statistical methods and processes were also of central concern to many members in areas other than information. Good was a leading statistician who pioneered various Bayesian "weight of evidence" approaches (Good 1950b), something that partly stemmed from his war-time code-cracking work with Turing, and naturally had a keen interest in the subject. Good led a number of club discussions and debates on related topics that may have influenced Uttley's ground-breaking work on conditional probability machines for learning and reasoning (Uttley 1956). In recent years similar approaches to those pioneered by these two have become very prominent in machine learning. Woodward was very knowledgeable on probability theory and gave, by request, at least one lecture to the club on the subject. Slater was one of the first psychiatrists to use well-grounded statistical techniques and did much to try and make psychiatry, and medicine in general, more rigorously scientific. Likewise, Sholl, whose first degree was in statistics, introduced statistical methods to the study of the anatomy of the nervous system. Many of the brain scientists in the club were concerned with signal-to-noise problems in their practical work, and Barlow (2006) remembers that this was a regular topic of discussion. He recalls that Gold had deep and useful engineering intuitions on the subject. As has been mentioned, Dawson was the leading expert on extracting clean EEG signals in a clinical setting.

A related area that prompted much discussion was that of the possible roles of random processes and structures in the nervous system. It has already been noted that Pringle and Ashby gave presentations in this area, but Barlow remembers that many other members, including Turing, were intrigued by the topic (Barlow 2002).

Philosophy

A quick glance at the meeting titles shown in table 6.1, and the topics of introductory talks (pp. 115–118) make it obvious that many club discussions had a distinctly philosophical flavour. Mackay was particularly keen to turn the conversation in that direction, prompting Andrew Hodges (1983) to refer to him as "a philosophical physicist" in a mention of a Ratio Club meeting in his biography of Turing (p. 411), and Woodward (2002) recalls that it was a good idea to keep him off the subject of Wittgenstein!

Pattern Recognition

Pattern recognition was another hot topic in relation to both natural and machine intelligence. The ninth meeting of the club, on May 18, 1950, was dedicated to this subject, then very much in its infancy; the perspectives of a number of members were followed by a general free-for-all discussion. Ashby provided a handout in which he tried to define "recognition" and "pattern," concluding that a large part of pattern recognition is classification or categorization. He wondered (1950d) whether "class-recognition [can] profitably be treated as a dissection of the total information into two parts—a part that identifies the inputs' class, and a part that identifies the details within the class?" Grey Walter also provided a handout, a set of condensed and hasty notes in which he concentrated on a brief survey of types of pattern-recognition problems and techniques. He noted (1950b) that "recognition of pattern correlates well with 'intelligence'; only highest wits can detect patterns in top Raven Matrices where the symmetry is abstract not graphic. Likewise in 'good' music, odours (not so much in man)." We can be sure that a vigorous debate ensued!

Space is too limited to discuss many other equally interesting themes that arose in club discussions, such as motor-control mechanisms in humans and animals (Merton and Pringle were particularly expert in this area, with Westcott providing the engineering perspective); analogue versus digital models of the functioning of the nervous system (see chapter 18, this volume, for a discussion of this in relation to the Ratio Club); and the relationship between evolution and learning, about which Pringle (1951) wrote an important paper at the time of the club, which, as Cowan (2003) has pointed out, laid the foundations for what later became known as reinforcement learning.

Artefacts and the Synthetic Method

There is, however, one last implicit theme that is important enough to deserve some discussion: the use of artefacts within the synthetic method. In addition to the engineers, several other members were adept at designing and constructing experimental equipment (often built from surplus military components left over from the war). This tendency was naturally transferred to an approach referred to by Craik as the "synthetic method"—the use of physical models to test and probe neurological or psychological hypotheses. In this spirit Ashby and Walter developed devices that were to become the most famous of all cybernetic machines: Ashby's Homeostat and Walter's tortoises. Both machines made headlines around the world, in particular the tortoises, which were featured in newsreels and television broadcasts, and were exhibited at the Festival of Britain (Holland 2003).

The Homeostat was an electromechanical device intended to demonstrate Ashby's theory of ultrastable systems—adaptive systems making use of a double feedback mechanism in order to keep certain significant quantities within permissible ranges. As mentioned earlier, these essential variables represented such things as blood pressure or body temperature in

Figure 6.3
The Homeostat. Two of the four units can be seen.

an animal. According to Ashby, ultrastable systems were at the heart of the
generation of adaptive behavior in biological systems. Part of the device is
shown in figure 6.3.

The machine consisted of four units. On top of each was a pivoted
magnet. The angular deviation of the four magnets represented the main
variables of the system. The units were joined together so that each sent
its output to the other three. The torque on each magnet was proportional
to the total input current to the unit. The units were constructed such
that their output was proportional to the deviation of their magnet from
the central position. The values of various commutators and potentiome-
ters acted as parameters to the system: they determined its subsequent behavior. The electrical interactions between the units modeled the primary
feedback mechanisms of an ultrastable system. A secondary feedback mechanism was implemented via switching circuitry to make pseudo-random
(step) changes to the parameters of the system by changing potentiometer

and commutator values. This mechanism was triggered when one of the essential variables (proportional to the magnet's deviation) went out of bounds. The system continued to reset parameters until a stable configuration was reached whereby no essential variables were out of range and the secondary feedback mechanisms became inoperative. The units could be viewed as abstract representations of an organism interacting with its environment. Ultrastability was demonstrated by first taking control of one of the units by reversing the commutator by hand, thereby causing an instability, and then observing how the system adapted its configuration until it found a stable state once more (for full details see Ashby 1952a).

On November 20, 1946, Turing had written to Ashby after being passed a letter from Ashby to Sir Charles Darwin, director of the National Physical Laboratory, a distinguished mathematician and a grandson of *the* Charles Darwin (and therefore Horace Barlow's uncle). Ashby had inquired about the future suitability of the planned ACE (automatic computing engine) digital computer, which was being designed at the National Physical Laboratory by Turing and others, for modeling brainlike mechanisms. We can assume he was thinking of the possibility of using the computer to develop a programmed equivalent of what was to become his famous Homeostat. In his reply, Turing (1946) enthusiastically endorsed such an idea, telling Ashby that "in working on the ACE I am more interested in the possibility of producing models of the action of the brain than in the practical applications of computing." Turing explained that in theory it would be possible to use the ACE to model adaptive processes by making use of the fact that it would be, in all reasonable cases, a universal machine. He went on to suggest, "You would be well advised to take advantage of this principle, and do your experiments on the ACE, instead of building a special machine. I should be very glad to help you over this." Unfortunately this collaboration never materialized. Turing withdrew from the ACE project following the NPL management's inability or unwillingness to properly manage the construction of the machine (Hodges 1983). Although the ACE project stalled, Ashby's notebooks from 1948 show that he was still musing over the possibility of using a computer to demonstrate his theories and was able to convince himself that the ACE could do the job. A pilot ACE digital computer was finally finished in mid-1950, but in the meantime a physical Homeostat had been finished in 1948 (Ashby 1948). The Manchester Mark 1, often regarded as the world's first full-scale stored-program digital computer and the project with which Turing was by then associated, was built a few months after this. It is very interesting to note that Ashby was considering using a general-purpose programmable digital

Figure 6.4
W. Grey Walter watches one of his tortoises push aside some wooden blocks on its way back to its hutch. Circa 1952.

computer to demonstrate and explore his theories before any such machine even existed. It would be many years before computational modeling became commonplace in science.

Grey Walter's tortoises were probably the first ever wheeled mobile autonomous robots. The devices were three-wheeled and turtle-like, sporting a protective "shell" (see figure 6.4). These vehicles had a light sensor, touch sensor, propulsion motor, steering motor, and an electronic valve–based analogue "nervous system." Walter's intention was to show that even in a very simple nervous system (the tortoises had two artificial neurons), complexity could arise out of the interactions between its units. By studying whole embodied sensorimotor systems, he was pioneering a style of research that was to become very prominent in AI many years later, and remains so today (Brooks 1999; Holland 2003). Between Easter 1948 and Christmas 1949, he built the first tortoises, Elmer and Elsie. They had similar circuits and electronics, but their shells and motors were a little different. They were rather unreliable and required frequent attention. The robots were capable of phototaxis, by which they could find their way to a recharging station when they ran low on battery power. In 1951, his technician, W. J. "Bunny" Warren, designed and built six new tortoises for him to a high professional standard. Three of these tortoises were exhibited at the Festival of Britain in 1951; others were demonstrated in public regularly throughout the fifties. He referred to the devices as *Machina speculatrix* after their apparent tendency to speculatively explore their environment.

Walter was able to demonstrate a variety of interesting behaviors as the robots interacted with their environment and each other (Walter 1950a, 1953). In one experiment he watched as the robot moved in front of a mirror and responded to its own reflection. "It began flickering," he wrote (Walter 1953). "Twittering, and jigging like a clumsy Narcissus." Walter argued that if this behavior was observed in an animal it "might be accepted as evidence of some degree of self-awareness."

One or other of the machines was demonstrated at at least one Ratio Club meeting. Tommy Gold recalled being fascinated by it and wondering whether the kind of principle underlying its behavior could be adapted to develop autonomous lawnmowers (Gold 2002), something that came to pass many decades later.

There was much discussion in meetings of what kind of intelligent behavior might be possible in artefacts and, more specifically, how the new general-purpose computers might exhibit mindlike behavior. Mackay (1951) was quick to point out that "the comparison of *contemporary* calculating machines with human brains appears to have little merit, and has done much to befog the real issue, as to how far an artefact *could* in principle be made to show behaviour of the type which we normally regard as characteristic of a human mind" (p. 105).

Interdisciplinarity

From what we have seen of its founding and membership, to comment that the Ratio Club was an interdisciplinary organization is stating the obvious. What is interesting, though, is that it was a *successful* interdisciplinary venture. This was partly a function of the time, when recent wartime work and experiences encouraged the breaking down of barriers, and was partly a function of Bates's keen eye for the right people. Even when war work was factored out, many of the members had very broad backgrounds. To give a few examples: Sholl had moved from mathematical sciences to anatomy following earlier studies in theology, zoology, and physiology; Uttley had degrees in mathematics and psychology; Merton was a brilliant natural engineer (he and Dawson were later instrumental in the adoption of digital computing techniques in experimental neurophysiology). All the brain scientists had strong interests, usually going back many years, in the use of mathematical and quantitative techniques. There was a similar, if less marked, story among the engineers and mathematicians: we have already commented on Gold's disregard for disciplinary boundaries; Turing was working on biological modeling; and Mackay had started his conversion into a neuropsychologist. Most members were open-minded, with wide-ranging interests outside science. This mix allowed important issues to be discussed from genuinely different perspectives, sparking off new insights.

Most members carried this spirit with them throughout their careers and many were involved in an extraordinarily wide range of research, even if this was within a single field. This lack of narrowness meant that most had other strings to their bows (several were very good musicians and a number were involved with other areas of the arts), sometimes starting whole new careers in retirement (see figures 6.5 and 6.6). For example, Woodward's enormous success in clockmaking has been mentioned, and in later life Slater became an expert on the use of statistical evidence in analyzing the authorship of Shakespearean texts.

A key ingredient in the club's success was its informal, relaxed character, which encouraged unconstrained contributions and made meetings fun. Another was the fact that it had a fairly strong focus right from the start: new ways of looking at mechanisms underlying intelligent behavior, particularly from a biological perspective.

The Legacy of the Club

In the United States, the cybernetics movement organized the Josiah Macy Foundation conferences, held between 1946 and 1953, whose published proceedings made the papers presented available a year or so after each meeting. Verbatim transcripts, they were lightly edited by Heinz von Foerster, and so the substance of all the presentations and discussions was readily available to the academic community and the public, where they had considerable influence. In the UK, by contrast, no detailed records of the Ratio Club's meetings were made, let alone circulated or published, and so in assessing the influence of the Ratio Club, it is clear that it can only have been of two kinds: the influence of its members on one another, and the consequences of that influence for their own work.

Figure 6.5
Jack Good at home in 2002. The sculpture above his head, *Jack Good's Dream*, was made in glass by an artist friend and exhibited at the famous 1968 Cybernetic Serendipity show at the Institute of Contemporary Art, London. It is based on a geometric construction of intersecting cylinders and spheres—the formation came to Jack in a dream.

Figure 6.6
Philip Woodward at home in 2002. In the background is one of the mechanical clocks he has designed and built. His W5 clock is one of the most accurate pendulum controlled clocks ever made. It has won a number of international awards and helped to make Woodward one of the most celebrated horologists of our times.

Unraveling such influences is nontrivial, but we have already seen testaments from several members on how important the club was to the development of their research. In 1981, after coming across some long-forgotten Ratio Club material, Pringle (1981) was prompted to write to Bates:

Dear John,

Going through some drawers of papers today in the lab, I came across a photograph of 17 members of the Ratio Club.... It occurs to me that someone ought to write up the history of the club, since it was in the old 17th century tradition and, to me at any rate, was a most valuable stimulus at a time when I was only just getting back into biology after the war.

He also wrote to Mackay, who agreed on the importance of the club and sent his Ratio Club papers to help with the history Pringle and Bates planned to put together. Unfortunately this venture stalled.

Pringle's response to the club was typical of its effect on many members, particularly the biologists: it acted as an inspiration and a spur. Much subsequent work of members had at least partial origins in club discussions. The important influence on Barlow has already been explained; given his

major impact on neuroscience, if all the club had done was to put Barlow
on the road he traveled, it would be of significance. Clearly it did much
more than that. As a mark of his debt to the Ratio Club, Uttley included
the photograph of its members (figure 6.2) in his 1979 book, *Information
Transmission in the Nervous System* (Uttley 1979). The influence of the biologist in the club appears to have played an important role in Mackay's
transformation from physicist to prominent neuropsychologist. The pages
of Ashby's private journals, in which he meticulously recorded his scientific
ideas as they developed, show that the club had some influence on him,
although how much is hard to judge—before becoming very well known,
he had worked on his theories in isolation for years, and there was always
something of the outsider about him. His grandson John has pointed out
that Ashby's most prolific years, as far as scientific journal writing was concerned, exactly coincided with the Ratio years (Ashby 2004). In all events
he was an active member who rarely missed a meeting.

Most members went on to pursue highly distinguished careers. Many
gained professorships at prestigious universities, and between them they
were awarded a host of prizes and honors, including seven fellowships of
the Royal Society and a CBE (Commander of the British Empire) to Slater
for services to psychiatry. Four members (Barlow, Rushton, Gold, Walter)
came within striking distance of a Nobel Prize (many feel that at least Rushton and Barlow should have received one) and Turing's work is likely to be
remembered for centuries. Many papers and books written by members of
the group, including those produced during the Ratio Club years, are still
widely cited, with many ideas and techniques that emanated from the
club's members very much in currency today.

Uttley and Mackay went on to set up and run successful interdisciplinary
groups, at the National Physical Laboratory and Keele University, respectively; it is likely that their experience of the extraordinary club influenced
them in these ventures.

So how should we assess the club's contribution? It seems to have served
a number of purposes during a narrow and very specific window in time. It
influenced a relatively small group of British scientists in their postwar
careers; given the degree of eminence many of them reached, and their influence on subsequent generations, this turned out to be highly significant.
It certainly concentrated and channeled the cybernetic currents that had
developed independently in the UK during the war. It also provided a con-
duit for the new ideas from the United States to be integrated into work
in the UK. It stimulated the introduction into biology of cybernetic ideas,
and in particular of information theory. And, perhaps appropriately for

a cybernetic organization, it stopped meeting when these purposes had been achieved.

This chapter can only serve as an introduction to the life and times of the club and its members; there is still much to tell.

Acknowledgments

We owe a great debt of gratitude to the surviving members of the Ratio Club, who all generously participated in the research for this article: Horace Barlow, Jack Good, Harold Shipton (who died as this book went to press), John Westcott, and Philip Woodward—and to the late Tommy Gold, whom we interviewed two years before his death in 2004. Documents provided by John Westcott and Jack Good have been enormously helpful. Thanks also to the many people who helped with background information and other material, in particular John and Mick Ashby, Jack Cowan, Richard Gregory, Peter Asaro, Igor Alexander, Helen Morton, Michael Slater, Ann Pasternak Slater, the late John Maynard Smith, the late Dick Grimsdale, Janet Shipton, and Andrew Hodges. Thanks to Jon Bird, Peter Cariani, Maggie Boden, Roland Baddeley, Danny Osorio, and Emmet Spier for very useful discussions of this and related material.

References

Ashby, John. 2004. "The notebooks of W. Ross Ashby." Address at W. Ross Ashby Centenary Conference. University of Illinois, Urbana, March 4–6, 2004.

Ashby, W. Ross. 1940. "Adaptiveness and Equilibrium." *Journal of Mental Science* 86: 478.

———. 1944. Letter to Kenneth Craik, 6 June 1944. W. Ross Ashby Archive, British Library, London (henceforth: Ashby Archive).

———. 1946. "Dynamics of the Cerebral Cortex: The Behavioral Properties of Systems in Equilibrium." *American Journal of Psychology* 594: 682–86.

———. 1947. Letter to W. E. Hick, 14 July 1947. Ashby Archive.

———. 1948. "Design for a Brain." *Electronic Engineering* 20: 379–83.

———. 1949a. Ashby's Journal, 1928–1972. W. Ross Ashby's Digital Archive, http://www.rossashby.info/journal, p. 2624.

———. 1949b. "Review of Wiener's Cybernetics." *Journal of Mental Science* 95: 716–24.

————. 1950a. "Subjects for discussion, 18 February 1950." Unpublished Papers and records for the Ratio Club. J. A. V. Bates Archive, the Wellcome Library for the History and Understanding of Medicine, London (henceforth: Unpublished Ratio Club papers, Bates Archive, Wellcome Library).

————. 1950b. "Suggested topics for discussion, 15 May 1950." Unpublished Ratio Club papers of John Westcott.

————. 1950c. Ashby's journal, 1928–1972, p. 2806, 28 April 1950. W. Ross Ashby's Digital Archive, http://www.rossashby.info/journal.

————. 1950d. "Pattern Recognition in Animals and Machines." Short paper for Ratio Club, May. Unpublished Ratio Club papers of John Westcott.

————. 1952a. *Design for a Brain*. London: Chapman & Hall.

————. 1952b. "Can a Mechanical Chess-Player Outplay Its Designer?" *British Journal for the Philosophy of Science* 39: 44–57.

————. 1958. *An Introduction to Cybernetics*. London: Chapman & Hall.

Barlow, Horace B. 1949. Letter to John Bates, August 1949. Unpublished Ratio Club papers, Bates Archive, Wellcome Library.

————. 1953. "Summation and Inhibition in the Frog's Retina." *Journal of Physiology* 119: 69–88.

————. 1959. "Sensory Mechanism, the Reduction of Redundancy, and Intelligence." In *Mechanisation of Thought Processes: Proceedings of a Symposium Held at the National Physical Laboratory on 24–27 November 1958*, edited by D. Blake and Albert Uttley. London: Her Majesty's Stationery Office.

————. 1961. "Possible Principles Underlying the Transformations of Sensory Messages." In *Sensory Communication*, edited by Walter A. Rosenblith. Cambridge, Mass.: MIT Press.

————. 1972. "Single Units and Sensation: A Neuron Doctrine for Perceptual Psychology?" *Perception* 1: 371–94.

————. 1986. "William Rushton." *Biographical Memoirs of Fellows of the Royal Society* 32: 423–59.

————. 2001. Interview by Philip Husbands, 30 March 2001, Cambridge.

————. 2002. Interview by Philip Husbands and Owen Holland, 19 June 2002, Cambridge.

Bates, J. 1945. Letter to William E. Hick, 30 May 1945. Unpublished Ratio Club papers, Bates Archive, Wellcome Library.

———. 1949a. Letter to Grey Walter, 27 July 1949. Unpublished Ratio Club papers, Bates Archive, Wellcome Library.

———. 1949b. Letter to John Westcott, 3 August 1949. Unpublished Ratio Club papers. Bates Archive, Wellcome Library.

———. 1949c. Letter to William Hick, 17 August 1949. Unpublished Ratio Club papers, Bates Archive, Wellcome Library.

———. 1949d. Initial membership list of Ratio Club, 14 September 1949. Unpublished Ratio Club papers, Bates Archive, Wellcome Library.

———. 1949e. Notes for first meeting of the Ratio Club, 14 September 1949. Unpublished Ratio Club papers, Bates Archive, Wellcome Library.

———. 1949f. Letter to Grey Walter, 4 October 1949. Unpublished Ratio Club papers, Bates Archive, Wellcome Library.

———. 1949g. Memo to Ratio Club members, between first and second meeting, undated, probably late September 1949. Unpublished Ratio Club papers of John Westcott.

———. 1950a. Letter to Jack Good, 13 December 1950. Unpublished papers of I. Jack Good.

———. 1950b. Letter to Grey Walter, 9 December 1950. Unpublished Ratio Club papers, Bates Archive, Wellcome Library.

———. 1951. The Ratio Club, memo to members, December 1951. Unpublished Ratio Club papers, Bates Archive, Wellcome Library.

———. 1952a. Membership list of the Ratio Club, January 1952. Unpublished Ratio Club papers, Bates Archive, Wellcome Library.

———. 1952b. "Significance of Information Theory to Neurophysiology." Draft paper. Bates Archive, Wellcome Library.

———. 1958. Note on final meeting. Unpublished Ratio Club papers, Bates Archive, Wellcome Library.

———. 1985. "Notes for an Article on the Ratio Club." Unpublished Ratio Club papers, Bates Archive, Wellcome Library.

BBC. 2005. WW2 People's War. Website. "Trevor Charles Noel Gibbons." Available at www.bbc.co.uk/ww2peopleswar/stories/53/a6038453.shtml.

Blake, D., and A. Uttley, eds. 1959. *The Mechanisation of Thought Processes*. Volume 10, National Physical Laboratory Symposia. London: Her Majesty's Stationery Office.

Boden, Margaret A. 2006. *Mind as Machine: A History of Cognitive Science*. Oxford: Oxford University Press.

Brindley, Giles. 2002. Interview by Philip Husbands and Owen Holland, May 2002, London.

Brooks, R. A. 1999. *Cambrian Intelligence: The Early History of the New AI.* Cambridge, Mass.: MIT Press.

Churchill, Winston. 1949. *The Second World War.* Volume 2: *Their Finest Hour,* chapter 4, "The Wizard War." London: Cassell.

Clark, D. 2003. "Enclosing the Field." Ph.D. diss., Warwick University, Department of Computer Science.

Cowan, J. 2003. Interview by Philip Husbands and Owen Holland, 6 April 2003, Chicago.

Craik, Kenneth J. W. 1943. *The Nature of Explanation.* Cambridge: Cambridge University Press.

———. 1944. Letter to W. Ross Ashby, 1 July 1944. Ashby Archive.

———. 1948. "Theory of the Human Operator in Control Systems." Part 2: "Man as an Element in a Control System." *British Journal of Psychology* 38: 142–48.

Dawson, George D., and W. Grey Walter. 1944. "The Scope and Limitation of Visual and Automatic Analysis of the E.E.G." *Journal of Neurology, Neurosurgery, and Psychiatry* 7: 119–30.

———. 1954. "A Summation Technique for the Detection of Small Evoked Potentials." *Electroencephalography and Clinical Neurophysiology* 6: 65–84.

Fleck, J. 1982. "Development and Establishment in Artificial Intelligence." In *Scientific Establishments and Hierarchies,* edited by Norbert Elias H. Martins and R. Whitley Volume 6: *Sociology of the Sciences.* Dordrecht: D. Reidel, 169–217.

Gabor, D. 1946. "Theory of Communication." *Journal of the IEE* 933: 429–57.

Gardner, M. R., and W. Ross Ashby. 1970. "Connectance of Large Dynamic Cybernetic Systems: Critical Values for Stability." *Nature* 228: 784.

Gardner-Medwin, A. R., and Horace B. Barlow. 2001. "The Limits of Counting Accuracy in Distributed Neural Representations." *Neural Computation* 133: 477–504.

Gold, T. 1948. "Hearing." Part 2: "The Physical Basis of the Action of the Cochlea." *Proceedings of the Royal Society of London* (series B) 135, no. 881: 492–98.

———. 2002. Interview by Philip Husbands and Owen Holland, Ithaca, New York, 17 April 2002.

Good, I. Jack. 1950a. Letter to John Bates, 9 December 1950. Unpublished Ratio Club papers, Bates Archive, Wellcome Library.

————. 1950b. *Probability and the Weighing of Evidence*. London: Charles Griffin.

Haldane, J. B. S. 1952. "The Mechanical Chess Player." *British Journal for the Philosophy of Science* 310: 189–91.

Heims, S. 1991. *Constructing a Social Science for Postwar America: The Cybernetics Group, 1946–1953*. Cambridge, Mass.: MIT Press.

Hick, William E. 1947a. Letter to W. Ross Ashby, 10 June 1947. Ashby Archive.

————. 1952. "On the Rate of Gain of Information." *Quarterly Journal of Experimental Psychology* 4: 11–26.

Hodges, Andrew. 1983. *Alan Turing: The Enigma of Intelligence*. London: Counterpoint.

Holland, Owen. 2003. "Exploration and High Adventure: The Legacy of Grey Walter." *Philosophical Transactions of the Royal Society of London* (series A) 361(October 15): 2085–2121.

Husbands, Philip, and Owen Holland. Forthcoming. *The Ratio Club*.

MacKay, Donald M. 1949. Letter to John Bates, 1 September 1949. Unpublished Ratio Club papers, Bates Archive, Wellcome Library.

————. 1951. "Mindlike Behaviour in Artefacts." *British Journal for the Philosophy of Science* 26: 105–21.

————. 1952a. "In Search of Basic Symbols." In *Proceedings of the 8th Conference on Cybernetics, 1951*, edited by Heinz von Foerster. New York: Josiah Macy Jr. Foundation.

————. 1952b. "The Nomenclature of Information Theory." In *Proceedings of the 8th Conference on Cybernetics, 1951*, edited by Heinz von Foerster. New York: Josiah Macy Jr. Foundation.

————. 1991. *Behind the Eye*. Oxford: Blackwell.

McCarthy, J., M. Minsky, N. Rochester, and C. Shannon. 1955. "A Proposal for the Dartmouth Summer Research Project on Artificial Intelligence." Available at www-formal.stanford.edu/jmc/history/dartmouth/dartmouth.html.

Merton, P. A. 1953. "Slowly Conducting Muscle Spindle Afferents." *Acta Physiologica Scandinavica* 291: 87–88.

Merton, P. A., and H. B. Morton. 1980. "Stimulation of the Cerebral Cortex in the Intact Human Subject." *Nature* 285: 227.

Pringle, J. W. 1938. "Proprioception in Insects." Part 1: "A New Type of Mechanical Receptor from the Palps of the Cockroach." *Journal of Experimental Biology* 15: 101–13.

———. 1949. Letter to John Bates, 5 August 1949. Unpublished Ratio Club papers, Bates Archive, Wellcome Library.

———. 1951. "On the Parallel Between Learning and Evolution." *Behaviour* 3: 174–215.

———. 1981. Letter to John Bates, 13 January 1981. Unpublished Ratio Club papers, Bates Archive, Wellcome Library.

Rosenblueth, A., Norbert Wiener, and Julian Bigelow. 1943. "Behaviour, Purpose, and Teleology." *Philosophy of Science* 101: 18–24.

Rushton, William. 1935. "A Theory of Excitation." *Journal of Physiology* 84: 42.

———. 1955. "Foveal Photopigments in Normal and Colour-Blind." *Journal of Physiology* 129: 41–42.

Shannon, Claude, and W. Weaver. 1949. *The Mathematical Theory of Communication.* Urbana: University of Illinois Press.

Shipton, Harold, and Janet Shipton. 2002. Interview by Philip Husbands and Owen Holland, Jupiter, Florida, October 2002.

Sholl, D. A. 1952. Letter to John Bates, 28 May 1952. Unpublished Ratio Club papers, Bates Archive, Wellcome Library.

———. 1956. *The Organization of the Nervous System.* New York: McGraw-Hill.

Slater, E., I. Gottesman, and J. Shields. 1971. *Man, Mind, and Heredity: Selected Papers of Eliot Slater on Psychiatry and Genetics.* Baltimore: Johns Hopkins University Press.

Tattersall, M. 2006. Personal communication based on information in his father's diary. His father was a fellow medical officer and P.O.W. with Turner McLardy.

Turing, Alan M. 1936. "On Computable Numbers, with an Application to the Entscheidungsproblem." *Proceedings of the London Mathematical Society* (series 2) 42: 230–65.

———. 1946. Letter to W. Ross Ashby, undated, about 19 November 1946. Ashby Archive.

———. 1950. "Computing Machinery and Intelligence." *Mind* 59: 433–60.

———. 1952. "The Chemical Basis of Morphogenesis." *Philosophical Transactions of the Royal Society London* (series B) 237: 37–72.

Uttley, Albert M. 1956. "Conditional Probability Machines and Conditioned Reflexes." In *Automata Studies,* edited by Claude E. Shannon and J. McCarthy. Princeton: Princeton University Press.

———. 1979. *Information Transmission in the Nervous System.* New York: Academic Press.

————. 1982. *Brain, Mind and Spirit.* Privately published.

Walter, W. Grey. 1947. Letter from Grey Walter to Professor Adrian, 12 June 1947. Burden Neurological Institute Papers, Science Museum, London.

————. 1949. Letter to John Bates, 29 September 1949. Unpublished Ratio Club papers, Bates Archive, Wellcome Library.

————. 1950a. "An Imitation of Life." *Scientific American* 1825: 42–45.

————. 1950b. "Pattern Recognition." Short paper for the Ratio Club, 15 May 1950. Unpublished Ratio Club papers of John Westcott.

————. 1951. Letter to John Bates, November 1951. Unpublished Ratio Club papers, Bates Archive, Wellcome Library.

————. 1953. *The Living Brain.* London: Duckworth.

Walter, W. Grey, and Harold Shipton. 1951. "A New Toposcopic Display System." *Electroencephalography and Clinical Neurophysiology* 3: 281–92.

Westcott, John. 1949–53. Notebook used at Ratio Club meetings 1949–1953. Unpublished papers of John Westcott.

————. 2002. Interview by Philip Husbands and Owen Holland, Imperial College, London, 15 March 2002.

Wiener, Norbert. 1948. *Cybernetics, or Control and Communication in the Animal and the Machine.* Cambridge, Mass.: MIT Press.

Wisdom, J. O. 1951. "The Hypothesis of Cybernetics." *British Journal for the Philosophy of Science* 25: 1–24.

Woodward, Philip M. 1951. Letter to John Bates, 29 October 1951. Unpublished Ratio Club papers, Bates Archive, Wellcome Library.

————. 1953. *Probability and Information Theory, with Applications to Radar.* London: Pergamon Press.

————. 1995. *My Own Right Time: An Exploration of Clockwork Design.* Oxford: Oxford University Press.

————. 2002. Interview by Philip Husbands and Owen Holland, Malvern, May 8, 2002.

7   From Mechanisms of Adaptation to Intelligence Amplifiers: The Philosophy of W. Ross Ashby

Peter M. Asaro

During the last few years it has become apparent that the concept of "machine" must be very greatly extended if it is to include the most modern developments. Especially is this true if we are studying the brain and attempting to identify the type of mechanism that is responsible for the brain's outstanding powers of thought and action. It has become apparent that when we used to doubt whether the brain could be a machine, our doubts were due chiefly to the fact that by "machine" we understood some mechanism of very simple type. Familiar with the bicycle and the typewriter, we were in great danger of taking them as the type of all machines. The last decade, however, has corrected this error. It has taught us how restricted our outlook used to be; for it developed mechanisms that far transcended the utmost that had been thought possible, and taught us that "mechanism" was still far from exhausted in its possibilities. Today we know only that the possibilities extend beyond our farthest vision.

—W. Ross Ashby (1951, p. 1).

The idea that intelligence could be imitated by machines has appeared in numerous forms and places in history. Yet it was in the twentieth century, in Europe and North America, that these metaphorical ideas were transformed into scientific theories and technological artifacts. Among the numerous scientists who pursued mechanistic theories of intelligence in the last century, W. Ross Ashby (1903–1972) stands out as a particularly unique and interesting figure. A medical doctor and psychiatrist by training, Ashby approached the brain as being first and foremost an organ of the body. Like other organs the brain had specific biological functions to perform. Ashby further believed that through a thoughtful analysis of those functions, a quantitatively rigorous analysis of the brain's mechanisms could be devised. It was his single-minded dedication to this basic idea that motivated his research into the mechanisms of intelligence for more than forty years. By always insisting upon sticking to the naturalistic functions of the brain, and to quantitative methods, Ashby was led to a number of startling and unique insights into the nature of intelligence that remain influential.

In this chapter I seek to sketch an intellectual portrait of Ashby's thought from his earliest work on the mechanisms of intelligence in 1940 through the birth of what is now called Artificial Intelligence (AI), around 1956, and to the end of Ashby's career in 1972. This period of Ashby's intellectual development is particularly interesting in his attempts to grasp the basic behaviors of the brain through the use of mechanical concepts. It is unique in the way that Ashby used rather sophisticated mechanical concepts, such as equilibrium and amplification, which were not particularly favored by other researchers. And moreover, he used these concepts not merely metaphorically, but also imported their associated mathematical formulations as a basis for quantifying intelligent behavior. As a result of this, we can see in Ashby's work both great insight and a truly original approach to the mechanisms of intelligence.

Ashby's professional career, beginning in 1928 and lasting until his death, is itself a remarkable tale that merits further research. He was the author of two enormously influential books in the early history of cybernetics, *Design for a Brain* (1952c) and *An Introduction to Cybernetics* (1956b).[1] Between his written contributions and his participation in the scientific community of cybernetics and its conferences and meetings, Ashby is considered to be one of the pioneers, or even cofounders, of cybernetics, which in turn gave rise to AI.

Our primary concern, however, will be with the central tenets of Ashby's thought. In particular we seek to discover the problems that motivated his thought, the conceptual form that he gave to those specific problems, and how their resolution resulted in a new mechanistic understanding of the brain and intelligence. This recounting of Ashby's mental philosophy will proceed in a roughly chronological fashion. We shall begin by examining his earliest published works on adaptation and equilibrium, and the conceptual structure of his notions of the mechanisms of control in biological systems. In particular we will examine his conceptions of mechanism, equilibrium, stability, and the role of breakdown in achieving equilibrium. We shall then proceed to his work on refining the concept of "intelligence," on the possibility of the mechanical augmentation and amplification of human intelligence, and on how machines might be built that surpass human understanding in their capabilities. I conclude with a consideration of the significance of his philosophy, and its role in cybernetic thought.

Figure 7.1
Ashby in front of his house, Westons, in 1960. Used with permission of the Trustees of the Estate of W. Ross Ashby.

The Mechanism of Adaptation

Given that Ashby was trained in medical psychiatry, and that his early work focused on neurological disorders from a strongly medical and physiological perspective, it might seem curious that he should come to be one of the leading proponents of a mechanical perspective on the mind. Mechanics has had a long and successful scientific history, and certainly scientists and philosophers before him had submitted that the brain, and perhaps also the mind, were in some sense machine-like. Roberto Cordeschi (2002) has carefully illustrated how a group of psychologists were arguing about possible mechanisms that could achieve mental capabilities, and were seeking to give a purely mechanistic explanation of mental capacities in the early decades of the twentieth century. Yet these scientific debates dwelled on the proper ways to separate out the mechanistic from the metaphysical aspects of psychology—consciousness, voluntary actions, and the spiritual aspects of mind. These scientists did propose specific types of mechanisms, such as Jacques Loeb's (1900) orientation mechanisms, and also built electronic automata to demonstrate these principles, such as John Hammond Jr. and Benjamin Miessner's (1915) phototropic robot (Miessner 1916). While these sorts of behaviors were interesting, for Ashby they were not sufficient to demonstrate that intelligence itself was mechanistic. Ashby knew that a mechanistic approach to the mind would have to deal with the most complex behaviors as well as the simplest, and do so with a single explanatory framework. It was with this goal in mind that he elaborated on the mechanistic nature of adaptation, as a route from simple physiology to complex forms of learning.

Another aspect of Ashby's work, shared with the pre-cybernetic and cybernetic mechanists, was that the development of theories of the brain and behavior went hand in hand with the development of technologies that exploited these theories in novel artefacts. Ashby summarized his own intellectual career in 1967 by saying (1967, p. 20):

Since opening my first note-book on the subject in 1928, I have worked to increase our understanding of the mechanistic aspect of "intelligence," partly to obtain a better insight into the processes of the living brain, partly to bring the same processes into action synthetically.

In many ways the construction of synthetic brains was integral to the theorization of the living brain. Cordeschi (2002) has called this approach the "synthetic method," and it continues in many areas of AI and robotics.[2] Although this essay focuses on the theoretical development of Ashby's thought, there is a deep technological aspect to that development and the machines Ashby built are worthy of consideration in their own right (Asaro 2006).

To understand how the key aspects of the transformation of psychological concepts to mechanical explanations took place in Ashby's thought, we must look at the unique way in which he reconceptualized the observed behavior of thinking creatures as being equivalent to the mechanical processes of physical devices. Ashby's views on these matters warrant careful consideration insofar as they do not fall easily into the categories employed by contemporary philosophers of mind, such as reductive materialism or straightforward functionalism. Ashby (1952e) did see his objective as being to provide a physical explanation of the mind (p. 408; emphasis in all excerpts is as in the original except where noted):

The invasion of psychology by cybernetics is making us realize that the ordinary concepts of psychology must be reformulated in the language of physics if a *physical* explanation of the ordinary psychological phenomena is to become possible. Some psychological concepts can be re-formulated more or less easily, but others are much more difficult, and the investigator must have a deep insight if the physical reality behind the psychological phenomena is to be perceived.

But his views on this matter are rather more complex than merely attempting to reduce mental processes to physical or physiological processes in the brain. As he expressed in a review of J. C. Eccles's *The Neurophysiological Basis of Mind* (Ashby 1954, p. 511):

The last two chapters, however—those on the cortex and its highest functions—fall off sadly, as so often happens when those who have spent much time studying the *minutiae* of the nervous system begin to consider its action as a whole; yet it is difficult to see, while present-day neurophysiology is limited to the study of the finest details in an organism carefully isolated from its environment, how the neurophysiologist's account could have been improved. The last two chapters, in fact, show only too clearly how ill adapted classical neurophysiology is to undertake the study of the brain's highest functions. At the moment it is far too concerned with details, and its technical resources are leading it only into the ever smaller. As a result, the neurophysiologist who starts to examine the highest functions is like a microscopist who, hearing there are galaxies to be looked at, has no better resource than to point his microscope at the sky. He must not be surprised if he sees only a blur.

Ashby recognizes that the instruments of investigation shape what one finds, and the question is what instruments to use to study the brain. Like other scientists who were trying to draw similar conclusions about the physical basis of mentality at the time, Ashby did believe that mental and psychological processes were essentially physical and chemical processes, but he argued that this did not mean that they could be explained and understood by simply appealing to some deeper or more fundamental level of analysis, such as physiology, in the quote. He believed that the methodology of physical analysis could be applied to mental states directly, the way statistical mechanics could be applied to a volume of gas to describe its behavior without being concerned with the motions of the individual molecules within the gas in order to characterize the relationships between pressure, volume, temperature, and so forth. Thus, Ashby sought to apply mechanistic analysis to the gross holistic organization of behavior directly, not merely to low-level processes, and to thereby demonstrate the general mechanisms by which the brain could achieve mental performances.

The first step in this conceptual move was not a purely metaphysical argument, though its conclusion had profound metaphysical implications. It was primarily an epistemological argument by analogy. Instead of considering the metaphysical arguments directly, he took an epistemological approach which sought to explain the mental phenomena of "adaptation" by an analogy to a physical mechanical process of "equilibrium." This approach is epistemological insofar as it attempts to show that we can *know or understand* the mind the same way we *understand* mechanical processes—by virtue of the analogy made between them. This is in contrast to others, who pursued a metaphysical argument that the mind must submit to mechanistic explanation because it was necessarily made up of the obviously physical brain—though Ashby also believed this, indeed took it for granted. His particular argument by analogy in fact appeals to the metaphysical necessity of equilibrium, but rather than argue that adaptation is reducible to this concept, shows that it is equivalent, and hence can be *analyzed and studied in the same manner as mechanical processes but independent of its specific material composition.* And so, it is *how* one comes to know a thing that is primary to the argument, and not its "essence."

The central argument of Ashby's mechanistic approach first appears in "Adaptation and Equilibrium" (1940). The title discloses the two concepts that he argues are analogous. In its final formulation, the analogy he argued for was that adaptive behavior, such as when a kitten learns to avoid the hot embers from a fire, was equivalent to the behavior of a system in equilibrium. In establishing this analogy, he shows that the biological phenomena of adaptive behavior can be described with the language and mathematical rigor of physical systems in states of equilibrium. In his own summary (p. 483):

Animal and human behavior shows many features. Among them is the peculiar phenomenon of "adaptiveness." Although this fact is easily recognized in any given case, yet it is difficult to define with precision. It is suggested here that adaptive behavior may be identical with the behavior of a system in stable equilibrium, and that this latter concept may, with advantage, be substituted for the former. The advantages of this latter concept are that (1) it is purely objective, (2) it avoids all metaphysical complications of "purpose," (3) it is precise in its definition, and (4) it lends itself immediately to quantitative studies.[3]

Thus Ashby suggests that a well-understood mechanical concept, carrying with it an extensive set of mathematical tools, ought be substituted for the vague conception of adaptive behavior in common usage. This passage also makes clear that Ashby's motivation in seeking a mechanistic explanation of mental phenomena is to provide a new basis for scientific study, and to sidestep rather than resolve any outstanding philosophical problems. It is also apparent that he was aware of the metaphysical issues surrounding the mind and believed that by conceiving of adaptation as equilibrium in this way one could avoid them.

The first half of the analogy depends upon establishing the importance of adaptive behavior in living and thinking things. Ashby begins by arguing that a peculiar feature of living organisms is their adaptive behavior. While definitions of life might variously include such requirements as motive, vegetive, or reproductive capacities, essential to this argument was the notion that the capacity for adaptation is necessary, and possibly sufficient, for something to be a living organism. In his second paper on the subject, "The Physical Origin of Adaptation by Trial and Error" (1945), Ashby elaborated on the role of adaptation in biological organisms, and to this end quoted various biologists, including Jennings (p. 14, quoting Jennings 1915):

Organisms do those things that advance their welfare. If the environment changes, the organism changes to meet the new conditions. . . . If the mammal is cooled from without, it heats from within, maintaining the temperature that is to its advantage. . . . In innumerable details it does those things that are good for it.

It is important to note that Ashby did not restrict his conception of adaptation to the Darwinian notion of adaptation by natural selection, though he certainly considered this to be a profoundly important form of adaptation, as his later writings make clear. Adaptation is then quickly extended from the physiological reactions of whole species to include also the notion of a behavioral response to a novel stimulus by an individual animal—the groundwork for a bridge between biology and behavioral psychology—and further generalized to include any observable behavior at all. In Ashby's favorite example, the kitten will not at first avoid the glowing embers from a fire, will burn its paw, and will thereafter avoid the fire; the resulting observed behavior is "adapted" insofar as it was the result of the kitten's individual experience of the world.[4]

The other half of the analogy, equilibrium, was seen to provide a rigorous set of analytical tools for thinking about the mind by importing the mathematical theory of mechanisms. Equilibrium is initially defined as a metaphysical necessity (Ashby 1940, p. 482):

Finally, there is one point of fundamental importance which must be grasped. It is that stable equilibrium is necessary for *existence*, and that systems in unstable equilibrium inevitably destroy themselves. Consequently, if we find that a system *persists*, in spite of the usual small disturbances which affect every physical body, then we may draw the conclusion with absolute certainty that the system must be in stable equilibrium. This may sound dogmatic, but I can see no escape from this deduction.

Ashby later (1945) employed the simpler definition of the physicist Hendrik Lorentz (1927): "By a state of equilibrium of a system we mean a state in which it can persist permanently" (p. 15). Since many equilibrium states are precarious and unlikely, Ashby further qualifies this by accepting the definition of a "stable" equilibrium as one in which a system will return to the equilibrium state even when some of its variables are disturbed slightly. For example, a cube resting on a table is in a stable equilibrium since it will return to the same state if tilted slightly and released. By contrast, though it might be possible to balance a cone on its point, under the slightest disturbance it will *not* return to the balanced state but will fall into a remote state and thus is in an odd sort of equilibrium if so balanced—an "unstable" equilibrium. A sphere resting on a table represents a "neutral" equilibrium, which is stable at many adjacent states and can be moved freely and smoothly between those states.[5] He clarifies the concept's meaning (Ashby 1940, pp. 479, 483):

We must notice some minor points at this stage. Firstly, we notice that "stable equilibrium" does *not* mean immobility. A body, e.g. a pendulum swinging, may vary considerably and yet be in stable equilibrium the whole time. Secondly, we note that the concept of "equilibrium" is essentially a dynamic one. If we just look at the three bodies [cube, cone, and sphere] on our table and do nothing with them the concept of equilibrium can hardly be said to have any particular meaning. It is only when we disturb the bodies and observe their subsequent reactions that the concept develops its full meaning....

The question of whether adaptiveness is *always* equivalent to "stable equilibrium" is difficult. First we must study the nature of "adaptiveness" a little closer.

We note that in all cases adaptiveness is shown only in relation to some specific situation: an animal in a void can show neither good nor bad adaptation. Further, it is clear that this situation or environment must affect the animal in some manner, i.e. must change it, since otherwise the animal is just receiving the stimulus without responding to it. This means that we are dealing with a circuit, for we have, first: environment has an effect on the animal, and then: the animal has some effect on the environment. The concept of adaptive behavior deals with the relationship between the two effects. It becomes meaningless if we try to remove one of the effects.

These points are by no means minor, but reflect Ashby's insistence on explaining the dynamic processes of observable phenomena, and how this can be done in terms of mechanisms seeking equilibrium.

The emphasis on "behavior" here, and throughout Ashby's work, is probably best read not as a commitment to, or sympathy for, behaviorism, but as an insistence on the epistemological limitations of science to observable phenomena. "Adaptation," like other scientific concepts, is nothing more than a set of observed reactions of various systems under different conditions. Those conditions are crucial insofar as the environment provides the context for the actions and reactions—the behavior—of a system, a necessary link in the chain of cause and effect. "Observation" is also crucial here, as it is throughout cybernetics, as the basis for determining the system and phenomena in question—both are meaningless in the absence of an observer. This is most likely an inheritance from positivism, which Ashby's approach shared to some extent with behaviorism in its insistence on "observable behaviors" in the form of responses in conditioned response. Although Ashby drew on behaviorist methodology, he went beyond its theory to posit the mechanism that controlled and extended behaviors. Pavlovian conditioning reinforced existing behaviors, and explained responses to stimuli based on this type of conditioning, but made no attempt to explain the mechanisms that supported this kind of conditioning.

Mechanical theory was of particular interest to Ashby by virtue of its potential for supplying a mathematical basis for psychology. A mathematical model of a state-determined mechanical system, such as those used by engineers at the time, involves several parameters divided into variables and constants in a set of equations or functions. When such a model is of a linear dynamical system, the values of the variables at one time determine the values at future times in a deterministic fashion—the functions generate the values for the next time-step from the values at the current time-step. The values of the variables in such a system may eventually stop changing. For example, if we were to observe the value of the angular displacement of a pendulum—how far it is from pointing straight down—that value would appear to grow and shrink and grow a little less with each swing until it eventually settled down to zero. An equilibrium in these systems is an assignment of values to the variables such that the variables will not change in future time-steps under the rules governing the system, such as when the pendulum rests pointing straight down. If a particular model does not have an equilibrium state, the variables will continue changing endlessly, typically with their values going to extreme limits. Such systems, Ashby argues, are not often found in nature—he can think only of a comet being hurled into deep space, never to return. Most of the systems found in nature, as well as human-made machines, have equilibria in which the variables settle to constant or cyclically repetitive values.

In fact, when an actual machine does not arrive at an equilibrium, it exhibits an intriguing phenomenon—it *breaks* (Ashby 1945, p. 17):

What happens to machines, as defined above, in time? The first point is that, in practice, they all arrive sooner or later at some equilibrium (in the general sense defined above). Thus, suppose we start with a great number of haphazardly assembled machines which are given random configurations and then started. Those which are tending towards equilibrium states will arrive at them *and will then stop there*. But what of the others, some of whose variables are increasing indefinitely? In practice the result is almost invariable—something breaks. Thus, quicker movements in a machine lead in the end to mechanical breaks; increasing electric currents or potentials lead inevitably to the fusing of wires or the break-down of insulation; increasing pressures lead to bursts; increasing temperatures lead to structures melting; even in chemical dynamics, increasing concentrations sooner or later meet saturation.

A break is unlike the normal changes in a dynamic machine in an important way. A break is a change in the organization of a system. In changing its organization, the machine ceases to be the machine it was and becomes a new machine. In the mathematical theory of mechanisms, the equations or functions that previously defined the system no longer hold true. To describe the change mathematically we must either define a new system of equations or must have previously defined a set of equations containing constants (parameters) whose values can represent the current and alternate organizations of the machine. When the machine "breaks," those values change and consequently the relationships between the variables of the system suddenly become different. And while the variables in a system can change either in discrete steps or continuously, a break, or change in the parameters, is necessarily a discontinuous change from one distinct organization to another distinct organization—what Ashby called a step-function.

Given this understanding of equilibrium and the dynamics of machines, the analogy to adaptation becomes clear (Ashby 1945, p. 17):

We may state this principle in the form: dynamic systems stop breaking when, and only when, they reach a state of equilibrium. And since a "break" is a change of organization, the principle may be restated in the more important form: *all dynamic systems change their internal organizations spontaneously until they arrive at some state of equilibrium.*

The process of breaking continues indefinitely as long as the variables describing the system continue to exceed tolerable limits on their values— that is, until the variables can be kept within certain limits. The instances of unbounded variables in nature, like the comet, are quite rare. By then applying this understanding to biological organisms, he argues that the organism adapts to its environment by successive trials of internal reorganization until it finds an equilibrium in which its physiological needs are met. In later writings, Ashby (1952a, c) stressed the importance of certain "essential variables," which the organism must maintain within certain limits in order to stay alive, such as body temperature, blood sugar level, and so forth. In its psychological formulation, the thinking system behaves so as to seek and approach a "goal," defined as a set of desired values over certain variables. The organism thus seeks to find an equilibrium of a special kind, one in which essential variables are kept within their safe and vital limits, or in which a goal is satisfied.

What seems perhaps most curious in this conceptual transformation is the productive power placed in breakdowns. Generally, a breakdown is seen as undesirable, something to be avoided, and the mark of a bad machine. Here it has become the supreme virtue of living machines: the creative drive, the power to generate alternative organizations in order to adapt to the environment. This result is in part due to the rigid structures of mathematics: it is easy to represent change in variables, but a change *in the relationships between variables* cannot be as easily expressed. In order to describe a machine that changes its dynamics, it is necessary to switch from one set of functions to another. Ultimately, Ashby would cease using the language of "breakdowns" and replace it with the language of "step-functions," a mathematical formulation that broadened the representation of a system to include its possible organizations and the discontinuous transitions between those organizations.

A similar tension is reflected also in the seeming banality of equilibrium—a system in equilibrium just stops, every dead thing and piece of inert matter is in a state of equilibrium. How can equilibrium be the ultimate goal of life when it implies a kind of stasis? What makes one kind of equilibrium indicative of life, is that it is *dynamic and is not uniform over the total system*. The living system can maintain some desired portion of its organization in equilibrium, the essential variables, even as the rest of the system changes dynamically in response to disturbances that threaten to destroy that desired equilibrium. For Ashby, this involved developing his conception of "ultrastability"—the power of a system to always find a suitable equilibrium despite changes in its environmental conditions. That is, the organism achieves a certain kind of stability for a few vital variables such as blood-sugar level, by varying other variables that it controls, sometimes wildly, as when an animal searches for food to maintain its blood-sugar levels.

The idea of equating adaptation and equilibrium appears to be unique to Ashby, though it bears strong similarities to ideas such as "negative feedback," which were being developed by other cyberneticians at the time. Ashby continued to cite and restate this analogy and argument throughout his career and used it as the basis of his first book, *Design for a Brain* (1952c); he never changed it significantly. Once it was published, he appears to have focused his energies on promoting the idea in various ways, including explicating its relationship to the ideas of other cyberneticians, including "negative feedback," and finding new expressions of the idea in his writings and in working machines. We now turn to the most notorious of these machines.

The Homeostat, completed in 1948, is a fascinating machine for several reasons. Most obvious is that it is a machine with an odd sort of purpose. It does not "do" anything in the sense that a machine generally serves some useful human purpose; unlike a bicycle or typewriter, it has no real practical application. On the other hand, it has its own "purpose" in the purest sense given by cybernetics: its equilibrium-seeking behavior is goal-oriented and controlled by negative feedback and so it is a teleological mechanism. This means that the machine itself has a goal, as revealed by its behavior, which may or may not have anything to do with the goals of its designer, a distinction that was to be further elaborated in Ashby's philosophy.

Most interesting, perhaps, is its role as a scientific model (Asaro 2006). It stands as a working physical simulation of Ashby's theory of mental adaptation. As a simulation it offers a powerful illustration of his conception of adaptive behavior in all kinds of systems, and in this regard its isomorphic correspondence to elements of his abstract theory are crucial. To see these correspondences, a brief description of the device is helpful.

The classic setup of the Homeostat consisted of four independent units, each one connected directly to each of the other three through circuits whose resistance could be controlled by either a preset switch or a randomizing circuit, called a "uniselector." They could "adapt" to one another by adjusting the resistances in the circuits that connected them, provided that the uniselector was engaged instead of the preset switches. Each unit featured a trough of water on top that contained an electrical field gradient and that had a metal needle dipping into it. By virtue of its connection to the current from the other units via the resistors and uniselectors, this needle acted as an indicator of the state of the unit: being in the middle of the trough represented a "stable" position, and being at either end of the trough represented an unstable position. Due to a relay that involved the position of the needle, whenever the needle was outside a central position in the trough it would send a charge to a capacitor. When the capacitor reached a predetermined charge level it would discharge into the uniselector, causing it to switch to a new random resistance in the circuit. These were only pseudo-random, however, as the resistances were derived from a table of random numbers and hard-wired into the uniselector, which stepped through them sequentially (see figure 6.3, p. 134, for a photograph of the device).

The correspondence between the Homeostat and Ashby's theory of mechanistic adaptation rests on an isomorphism between "random variations" and the operation of the uniselector circuit elements; between "acceptable values for essential variables" and the relay controlling the energizing capacitor for the uniselectors; between "equilibrium" and the visible needle resting in the middle of the trough; and between the wildly behaving needles of a machine out of control and a system that continues to "break" up its internal organization through step-functions until it finds equilibrium.

In a later paper, "Simulation of a Brain," Ashby (1962) discusses the objectives of modeling and simulation directly. In that paper he defines a model formally as a system that stands in relation to another system by virtue of an explicit mapping between sets of elements. He asserts that physical as well as mathematical and symbolic forms can stand in such relationships. He also insists that the value of the formal definition is that it provides a quantitative measure of the closeness of a model to the original system by virtue of the number of relationships shared among the members of the two sets. Given this definition of a model, he argues that there are three virtues to simulations, as physical models, which contribute to scientific progress. The first is their *vividness*: to clearly express a concept in an easily graspable form. The second is their function as an *archive*: to stand as a repository of built-up knowledge that might be too vast and complex to be written out or grasped all at once by an individual. The final virtue of simulations is their capacity to facilitate *deduction and exploration*: to resolve disputes, disprove hypotheses, and provide a basis for scientific inquiry into areas that, without simulations, would otherwise remain speculative (Ashby 1962, pp. 461–64). He offers the Homeostat as an example of a simulation useful in scientific education for demonstrating that goal-seeking behavior, as a trial-and-error search for equilibrium, presents a fundamentally different kind of mechanical process—negative feedback with step-functions—and opens up new vistas of possibility for what machines might be capable of doing. I have argued elsewhere (Asaro 2006) that working brain models such as the Homeostat also served an important role in mediating between theories of behavior and physiological theories of neurons in the development of the mechanistic theory of the mind.

Designs for Intelligence

With the analogy between adaptation and equilibrium firmly in place, Ashby turned his attention to demonstrating the significance and potential applications of this new insight. His effort consisted of two distinct parts: the development of other simulations, such as the Dispersive And Multistable System (DAMS) made of thermionic valves and neon light tubes (Ashby 1951), in order to demonstrate his ideas in more tangible forms; and the continuing articulation of a clear and compelling rhetorical framework for discussing the problems of designing intelligent machines. The machines Ashby developed are deserving of further study as technological artifacts built on unique principles of design, but a discussion of these would take us to remote regions of his mental philosophy, whereas we are concerned only with its central features. In the following sections, we will consider the further development of his theoretical views. We shall begin by looking at Ashby's formal articulation of a "problem" that his mechanism of adaptation could "solve," and then to how this problem-solving mechanism could be generalized to solving more significant and compelling problems. In so doing we shall examine his definition of intelligence and how it could be fully mechanized. Throughout these efforts, Ashby sought to motivate and inspire the belief that a revolution had occurred in our understanding of machines, and that the mechanism of adaptation might ultimately result in machines capable of impressive and even super-human performances.

The Problem of the Mechanical Chess Player

While satisfied with the soundness of his argument for the possibility of an adaptive mechanism, Ashby felt compelled to demonstrate the full significance and implications of this possibility to an audience beyond the handful of psychiatrists and cyberneticians with whom he had contact. To do this, he developed a clear and compelling problem through which audiences could grasp this significance. The example he elaborated on was the "Problem of the Mechanical Chess Player," which he credited to his experiences in casual conversations, most likely with the members of the Ratio Club, such as Alan Turing, who were very interested in the mathematical problems of chess play. Ashby took the problem in a different direction than Turing and subsequent AI researchers did, and used this as an imaginative, and thus compelling, example of the basic problem of the very possibility of mechanized thought, which could be formalized using the analytical apparatus borrowed from mechanical theory. The rhetorical development of the problem of the mechanical chess player is interesting because it starts by raising some fundamental issues of metaphysics, but once properly formulated as a technical problem, it could be decisively resolved by the demonstrated performance of a working machine. Just how this was achieved we shall now see.

The metaphysical problem of the mechanical chess player was how (or in its weaker form, *whether*) it could be possible to design a machine that has a greater range or skill in performance than what its designer had provided for it by its design—in other words, whether a mechanical chess player can *outplay its designer*. As Ashby (1952d) posed the question in the Ninth Josiah Macy Jr. Foundation Conference on Cybernetics (p. 151):

> The question I want to discuss is whether a mechanical chess player can outplay its designer. I don't say "beat" its designer; I say "outplay." I want to set aside all mechanical brains that beat their designer by sheer brute power of analysis. If the designer is a mediocre player, who can see only three moves ahead, let the machine be restricted until it, too, can see only three moves ahead. I want to consider the machine that wins by developing a deeper strategy than its designer can provide. Let us assume that the machine cannot analyze the position right out and that it must make judgements. The problem, then, becomes that the machine must form its own criteria for judgement, and, if it is to beat its designer, it must form better judgements than the designer can put into it. Is this possible? Can we build such a machine?

While Ashby chose to formulate the problem as whether a machine can outplay its designer, it seems less confusing to me to formulate it as whether a machine can outplay its design, that is, whether it can do "better" than it was designed to, rather than to say that it can actually defeat the person who designed the machine. In short, Ashby was concerned with the ability of a machine, in this case a chess-playing machine, to acquire knowledge and skill beyond the knowledge and skill built into it.

Ashby hoped to show this by arguing that a mechanism utilizing a source of disorganized information, though one containing a greater variety of possibilities than the designer could enumerate, could in principle achieve better strategies than its designer. Because a generator of random moves could produce novel moves that no known specific or general rule of chess would suggest, there was a possibility of finding a "supermove" that would not otherwise be found and so could not have been built into the machine. Therefore, as long as a system was designed so as to allow the input of such random possibilities, and designed with the ability to select among those possibilities, it might be possible for it to find moves and strategies far better than any its designer could have provided.

This particular formulation in fact caused some confusion at the Macy Conference. In the ensuing discussion of it, Julian Bigelow challenged the distinction Ashby attempted to make between analysis and strategic judgment (Ashby 1952d, pp. 152–54).[6] For Bigelow, the ability to construct strategies was itself already a kind of analysis. He argued that limiting the analysis of the system to looking only three moves ahead necessarily put a limitation on the number of strategies that could be considered. He also rejected the notion that adding random noise could add any information to the chess-playing system at all—for him information necessarily had to have analytical import and random noise had none. To provide a resolution of this confusion and a better understanding of the role of this problem in thinking machines more generally, we must first clarify Ashby's conception of "design" and "designer," as well as the formal articulation he gave to the problem.

Ashby saw the issue as a fundamentally philosophical problem of agency having its roots deep within the tradition of European thought. He offered, as different formulations of the same problem, the following examples from that tradition: "Descartes declared that there must be at least as much reality and perfection in the cause as in the effect. Kant (*General History of Nature*, 1755) asked, 'How can work full of design build itself up without a design and without a builder?'" (Ashby 1952b, p. 44). Descartes's dictum, of course, maintains that an effect cannot have more perfection than its cause, and thus a designed system cannot be superior to its designer.[7] If true, the implication of this dictum is that a machine, being capable only of what its design has provided for it, can never be "better" than that design, and thus cannot improve on it. But Ashby believed that he had already shown how a mechanism could be capable of adaptation—a kind of improvement relative to environmental conditions. He thus saw it as essential to prove that Descartes was wrong, and saw that the proof would require a more rigorous formal presentation.

The crux of the problem lay in the proper definition of "design." For a proof, it was necessary to provide a formal definition that could show clearly and quantitatively exactly what was contained in the "design" provided by a designer, such that this could be compared to the quantity of the "design" demonstrated in the performance of the machine. He derived these measures using the information theory of Claude E. Shannon (1948). The quantities measured in the "design" and in the machine would be information, and if a machine could be shown to "output" more information than was provided as "input" in the instructions for its construction, then the machine's designer would have disproved Descartes's dictum.

Without going too far into the technical details of information theory, the basic idea is that the quantity of information in a message is the measure of the reduction in uncertainty that results when the message is received. The technical definition differs significantly from the common-sense understanding of "information" insofar as the information contained in a message has nothing to do with the contents of the message itself, but only with the variety in the other messages from which it was selected, and so "information" is really a property of a system of communication rather than of any particular message within it. The reduction in uncertainty upon receiving a message thus depends on the probability of receiving the message, and also on the size of the set of possible messages to which it belongs.[8] As the number of possible messages increases, either the number of different signals or the length of a message (composed of a sequence of signals) must also increase in order to make each message distinct from the others. In the binary encoding of computers, there are only two signals (or symbols), 0 and 1, and thus the length of the sequence needed to encode a message must increase as the number of possible messages increases in order for each message to be represented by a unique sequence.

Ashby used the theory of information to measure "design" by arguing that the choices made in a design are like the messages sent over a communication channel. That is, the significance of a choice is measured against the number of alternatives from which it must be selected. As he states it (Ashby 1952b, pp. 45–47):

How are we to obtain an objective and consistent measure of the "amount of design" put into, or shown by, a machine? Abstractly, "designing" a machine means giving selected numerical values to the available parameters. How long shall the lever be? where shall its fulcrum be placed? how many teeth shall the cog have? what value shall be given to the electrical resistance? what composition shall the alloy have? and so on. *Clearly, the amount of design must be related in some way to the number of decisions made and also to the fineness of the discrimination made in the selection* [emphasis added]. . . .

To apply the measure to a designed machine, we regard the machine as something specified by a designer and produced, as output, from a workshop. We must therefore consider not only the particular machine but the *ensemble* of machines from which the final model has been selected [original emphasis].

If one quantifies the information contained in a design as the choices made from among the possible alternatives, then one can make a similar move to quantify the information exhibited by the machine's performance. The information displayed by the machine is the number of functionally distinct states it can exhibit—Ashby's example is of a network consisting of a number of switches, the configuration of which determines different connectivities or states of the network. The design of the network is an assignment of values to the switches from among all the possible assignments. In this case, the network can only display as many states as the switches allow different configurations; some of the distinct assignments may be functionally equivalent and thus the machine may display *less* information than is contained in its design. But how, then, is it possible for a machine to display *more* information than is contained in its design?

The demonstration of this possibility draws close to the arguments about "design" during the rise of evolutionary theory in the nineteenth century. So close, in fact, that Ashby (1952b, p. 50) followed Norbert Wiener (1948) in calling instances of such systems "Darwinian Machinery":

The question might seem settled, were it not for the fact, known to every biologist, that Descartes' dictum was proved false over ninety years ago by Darwin. He showed that quite a simple rule, acting over a great length of time, could produce design and adaptation far more complex than the rule that had generated it. The status of his proof was uncertain for some time, but the work of the last thirty years, especially that of the geneticists, has shown beyond all reasonable doubt the sufficiency of natural selection. We face therefore something of a paradox. There can be no escape by denying the great complexity of living organisms. Neither Descartes nor Kant would have attempted this, for they appealed to just this richness of design as evidence for their arguments. Information theory, too, confirms this richness. Thus, suppose we try to measure the amount of design involved in the construction of a bird that can fly a hundred miles without resting. As a machine, it must have a very large number of parameters adjusted. How many cannot be stated accurately, but it is of the same order as the number of all facts of avian anatomy, histology, and biochemistry. Unquestionably, therefore, evolution by natural selection produces great richness of design.

In evolution, there is an increasing amount of information displayed by the machine, despite the fact that the design is both simple and, in a sense, unchanging. Ashby (1952b) goes so far as to suggest that the design for a bird might be as simple as "Take a planet with some carbon and oxygen; irradiate it with sunshine and cosmic rays; and leave it alone for a few hundred million years" (p. 52). But the mechanism responsible for evolution is difficult to directly observe in action, and it does not appear to apply straightforwardly to a chess-playing machine.

If evolution is able to produce systems that exhibit more information than is contained in their design, and information cannot be spontaneously generated, where did this extra information come from? Obviously, this information must come in the form of an input of messages unforeseen by the designer (Ashby 1952b, p. 51):

The law that information cannot be created is not violated by evolution, for the evolving system receives an endless stream of information in the form of mutations. Whatever their origin, whether in cosmic rays or thermal noise, the fact that each gene may, during each second change unpredictably to some other form makes each gene a typical information source. The information received each second by the whole gene-pattern, or by the species, is then simply the sum of the separate contributions. The evolving system thus has *two* sources of information, that implied in the specifications of the rules of natural selection and that implied by the inpouring stream of mutations.

This philosophical problem was, of course, the same one which fueled much of the controversy over Darwin's theory in the nineteenth century—whether the exquisite subtleties of living creatures could possibly be produced by brute natural processes or whether they necessarily required a supernatural "Designer." What Darwin had so carefully detailed in *On the Origin of Species by Means of Natural Selection* (1859) was how natural evolutionary processes could lead to speciation—the divergence in forms of two distinct species who share a common ancestry; the branching of the tree of common descent. Assuming that the design of a species did not change in virtue of continuous divine intervention, the demonstration that species did change over time, and to such an extent as to result in new species, implied that natural evolutionary processes, in the absence of a designer, might have given rise to all biological forms. The basic process of natural selection choosing among the variations of form is argued to move species toward those forms best able to survive and reproduce. Ashby simply placed a special emphasis on a portion of Darwin's theory by indicating how spontaneous variations in form provide an additional source of information apart from any determinate design.

In biological systems, the random variations of mutation supply alternative possibilities unforeseen by any designer, and thus the organism can evolve capacities beyond its own design. Similarly, Ashby (1952b) would argue, by adding a random number generator, Geiger counter, or other source of random noise to a system, we introduce the possibility of behaviors unforeseen in its "design" (p. 51):

It is now clear that the paradox arose simply because the words "cause" or "designer," in relation to a system, can be used in two senses. If they are used comprehensively, to mean "everything that contributes to the determination of the system," then Shannon and Descartes can agree that "a noiseless transducer or determinate machine can emit only such information as is supplied to it." This formulation will include the process of evolution if the "cause" is understood to include not only the rules of natural selection but also the mutations, *specified in every detail*. If, on the other hand, by "cause" or "designer" we mean something more restricted—a human designer, say—so that the designer is only a part of the total determination, then the dictum is no longer true.

With the paradox thus resolved, Ashby had demonstrated the possibility that a mechanical chess player *could* outplay its design(er). Further, he had identified the key to achieving this possibility, the flow of random information coming into the system. What remained to be shown was how this information could be made useful. A random move generator might contain the "supermoves" of chess, but how would a mechanical chess player be able to distinguish these moves from the rest? The answer to this question required developing a new conception of intelligence suitable to the mechanistic theory of mind.

Amplifying Intelligence

Once the analogy between adaptation and equilibrium was firmly set in Ashby's philosophy as the basis for a mechanistic theory of mind, he extended the analogy freely by describing mental processes using the terminology once reserved for describing machines such as steam engines and electronic devices: the engineer's language of "power," and "energy." One of his central themes in this respect was the application of the process of "amplification" to mental concepts such as intelligence. This extended analogy was not merely a rhetorical turn of phrase, but carried implications within his theoretical framework. Ashby thus turned his attention to developing a more rigorous definition of intelligence, and to demonstrating the significance of the mechanical-chess-player argument by showing how its results could be applied to practical problems. This line of thought culminated in his contribution to the first collected volume of work in the newly emerging subfields of computer science, artificial intelligence, and automata theory: Claude Shannon and John McCarthy's *Automata Studies*, published in 1956. The paper bore the intriguing title "Design for an Intelligence-Amplifier" and appeared in the final section of that volume, entitled "Synthesis of Automata." We will now examine that paper (Ashby 1956a) in detail and place its ideas in perspective with Ashby's overall philosophy.

Demonstrating that it was possible for a mechanical chess player to out-play its designer might be philosophically interesting, but showing that this discovery had practical significance would take more than arguments of metaphysical possibility. For this purpose, Ashby further extended his conception of the mechanisms of thought to problems of general interest, which took the form of a device that could "amplify" human intelligence. The continued reliance upon the analogy between thought and mechanical physics in his conception was made clear in the introduction to the paper (p. 215):

For over a century Man has been able to use, for his own advantage, physical powers that far transcend those produced by his own muscles. Is it impossible that he should develop machines with "synthetic" intellectual powers that will equally surpass those of his own brain? I hope to show that recent developments have made such machines possible—possible in the sense that their building can start today. Let us then consider the question of building a mechanistic system for the solution of problems that are beyond the human intellect. I hope to show that such a construction is by no means impossible, even though the constructors are themselves quite averagely human. There is certainly no lack of difficult problems awaiting solution. Mathematics provides plenty, and so does almost every branch of science. It is perhaps in the social and economic world that such problems occur most noticeably, both in regard to their complexity and to the great issues which depend on them. Success in solving these problems is a matter of some urgency. We have built a civilization beyond our understanding and we are finding that it is getting out of hand. Faced with such problems, what are we to do?

Rather than hope that individuals of extraordinary intelligence will step forward and solve such problems—a statistically unlikely eventuality—Ashby suggested that we ought to design machines that would amplify the intellectual powers of average humans. In the absence of careful definitions and criteria, such devices might sound quite fanciful. But with his usual flare for mathematical rigor, Ashby provided those definitions and criteria and thereby also provided further illumination of his mechanistic philosophy of mind.

In resolving the problem of the mechanical chess player, Ashby had shown that a machine could output more information than was input through its design, by making use of other, random, information. This was a kind of amplification—information amplification—like the amplification of power that utilizes an input of power plus a source of free energy to output much more power than was originally supplied (p. 218):

[L]et us remember that the engineers of the middle ages, familiar with the principles of the lever and cog and pulley, must often have said that as no machine, worked by a man, could put out more work than he put in, therefore no machine could ever amplify a man's power. Yet today we see one man keeping all the wheels in a factory turning by shoveling coal into a furnace. It is instructive to notice just how it is that today's stoker defeats the mediaeval engineer's dictum, while being still subject to the law of the conservation of energy. A little thought shows that the process occurs in two stages. In Stage One the stoker lifts the coal into the furnace; and over this stage energy is conserved strictly. The arrival of the coal in the furnace is then the beginning of Stage Two, in which again energy is conserved, as the burning of the coal leads to the generation of steam and ultimately to the turning of the factory's wheels. By making the whole process, from stoker's muscles to factory wheel, take place in two stages, involving two lots of energy whose sizes can vary with some independence, the modern engineer can obtain an overall amplification.

In the mechanical chess player, as well as in evolution, information from the design, or problem specification, can be amplified in the same way that the strength of a stoker is amplified by a pile of coal and a steam engine, by the addition of free energy or random information. But the availability of bare information is not in itself intelligence, any more than free energy is work—these resources must be directed toward a task or goal.

What then is a suitable criterion for intelligent behavior? By starting from a definition of information that considered only its technical implications, a definition that leaves information independent of any analysis of it, Ashby was able to take account of analysis and judgment in his definition of intelligence. According to Ashby, intelligence implies a selection: *intelligence is the power of appropriate selection*. To see what this means, consider his example (p. 217):

It has often been remarked that any random sequence, if long enough, will contain *all* the answers. Nothing prevents a child from doodling "$\cos^2 x + \sin^2 x = 1$," or a dancing mote in the sunlight from emitting the same message in Morse or a similar code. Let us be more definite. If each of the above thirteen symbols might have been any one of fifty letters and elementary signs, then as 50^{13} is approximately 2^{73}, the equation can be given in coded form by 73 binary symbols. Now consider a cubic centimeter of air as a turmoil of colliding molecules. A particular molecule's turnings after collision, sometimes to the left and sometimes to the right, will provide a series of binary symbols, each 73 of which, on some given code, either will or will not represent the equation. A simple calculation from the known facts shows that the molecules in every cubic centimeter of air are emitting this sequence correctly over a hundred thousand times a second. The objection that "such things don't happen" cannot stand. Doodling, then, or any other random activity, is capable of producing all that is required. What spoils the child's claim to be a mathematician is that he will doodle, with equal readiness, such forms as "$\cos^2 x + \sin^2 x = 2$" or "ci)xsi-nx1" or any other variation. After the child has had some mathematical experience he will stop producing these other variations. *He becomes not more, but less productive: he becomes selective.* [emphasis added]

In order to be intelligent, a mechanism must exhibit discipline in its behavior. Thus, given an ample source of random information, the efforts toward designing an intelligence amplifier ought to focus on the mechanisms of appropriate selection by which the device can choose which among the many possibilities is the desired answer. This definition constitutes a kind of inversion of the common formulation of machine intelligence understood as the ability to produce correct responses by design; intelligence is now understood as a combination of the abilities to *produce a great many meaningless alternatives*, and to *eliminate by appropriate selection the incorrect choices among those*—a two-stage process.

Exactly how to construct a mechanism to make appropriate selections thus becomes the design problem for building an intelligence amplifier. The design of an intelligent selector involves two major parts. The first is to establish criteria of selection that can be utilized by the machine, sufficient for it to know when it has arrived at an acceptable solution to the given problem. The second part involves coupling the selector to a source of chaotic information which it can search through in order to find an acceptable solution (p. 223):

Consider the engineer who has, say, some ore at the foot of a mine-shaft and who wants it brought to the surface. The power required is more than he can supply personally. What he does is to take some system that is going to change, by the laws of nature, from low entropy to high, and he couples this system to his ore, perhaps through pistons and ropes, so that "low entropy" is coupled to "ore down" and "high entropy" to "ore up." He then lets the whole system go, confident that as the entropy goes from low to high so will it change the ore's position from down to up. Abstractly . . . he has a process that is going, by the laws of nature, to pass from state H_1 to state H_2. He wants C_1 to change to C_2. So he couples H_1 to C_1 and H_2 to

C_2. Then the system, in changing from H_1 to H_2, will change C_1 to C_2, which is what he wants. The arrangement is clearly both necessary and sufficient. The method of getting the problem-solver to solve the set problem can now be seen to be of essentially the same form. The job to be done is the bringing of X ... to a certain condition or "solution" η. What the intelligence engineer does first is build a system, X and S, that has the tendency, by the laws of nature, to go to a state of equilibrium. He arranges the coupling between them so that "not at equilibrium" is coupled to not-η, and "at equilibrium" to η. He then lets the system go, confident that as the passage of time takes the whole to an equilibrium, so will the conditions in X have to change from not-η to η. He does not make the conditions in X change by his own efforts, but allows the basic drive of nature to do the work. This is the fundamental principle of our intelligence-amplifier. Its driving power is the tendency for entropy to increase, where "entropy" is used, not as understood in heat-engines, but as understood in stochastic processes.

In yet another inversion of traditional thought, Ashby has demonstrated how the natural processes of entropy in nature, the relentless destruction of organization, can be used as the fuel for the amplification of intelligence beyond the capabilities of the naked human mind.

The key to intelligence thus lies in selectivity, for it is the power of appropriate selection that is able to recognize the desired messages from among the chaos of random information. But how does one achieve this in a machine? Consider, as Ashby does, a machine to solve difficult social and economic problems. As designers, we make *our* selection as to what we want, say (p. 219):

An organisation that will be stable at the conditions:

Unemployed	<100,000 persons
Crimes of violence	<10 per week
Minimal income per family	>£500 per annum

Taking these desiderata as the machine's goal, it is the task of the machine to sift through an enormous number of possible economic configurations, and select one that meets these conditions. Part of the design of that machine involves specifying the representation of the economic system, and thus the set of things from which the selection must take place. Apart from this, Ashby has little to say about this design process—a topic with which much of the work in artificial intelligence has since been concerned. But herein lies another essential point, for it raises again the question of information. This is to say that in determining the class of things from which a selection is to be made one also specifies the amount of information that the answer will require. Since the measure of the information contained in a message is the reduction in uncertainty resulting from the message being received, by determining the size of the set of possible messages—answers—the designer has put a number on the amount of information needed to solve the problem.

In later writings, Ashby returned to this problem and gave it a proper formalization using information theory. That formulation involved seeing the process of selection not as an instance of the perfect transmission of information but as a form of communication over a noisy channel. In so doing, he saw a deep and interesting connection between Shannon's 10th Theorem (1948) and his own Law of Requisite Variety (Ashby 1956b, p. 202).[9] The formulation involves equating the entropic source of random information with a noisy channel, and selection with the problem of determining which messages are correct and which are not. In order for someone on the receiving end of a noisy channel to determine the correctness of a message, they must receive an additional source of information, a kind of feedback regarding the correctness of the messages received. This information comes through an error-correcting channel. Shannon's 10th Theorem provides a measure of the capacity of the channel necessary to achieve error-free transmission over a noisy channel (within a certain degree of accuracy). Ashby argued that in order to make a correct selection in a decision process, a system must receive information from the environment and that the measure of this information is equivalent to the required capacity for an error-correcting channel (Ashby 1960, p. 746).

To see what this means, consider the case in which the number of possible economic configurations our problem solver must select from is 1,000,001, and there is only one correct solution. Suppose that it is possible to eliminate whole classes or subsets of this set as inappropriate. A message on the error-correcting channel transmits this information by indicating a single subset that the correct answer cannot be a part of. Let us say that each subset in our problem contains exactly 1,000 unique economic configurations (in most real problems the size of each subset is different and many subsets overlap and share members, but we shall ignore these difficulties). In this case every message eliminates a thousand possibilities, leaving the selector with 999,001 possibilities after the first message, and then with 998,001 after the second message, and so on. At this rate, it will take at least 1,000 messages to achieve complete certainty that the selector will have the right answer, but fewer if we do not require 100 percent certainty. At each step it has made some progress as the probability of correctness for each of the answers still in the set of possibilities goes up after each piece of information is received. But when it comes to choosing from among the elements remaining, the selector has no more information available for deciding whether any one of the remaining elements is "better" or "worse" than any of the others—it can only pick one at random. If the selector had more information and were thus able to make a selection among the remaining elements, it would do so until it was again left with a set of elements where each was no more likely to be correct than any other.

This led Ashby to the conclusion that all forms of intelligence depend necessarily on receiving information in order to achieve any appropriate selection that they make. And the greater the set of possibilities and complexity of the partitioning of alternatives, the more information will be required for the selection to be appropriate. No intelligence is able to create a brilliant idea from nothing; genius of this sort is merely a myth (Ashby 1961, p. 279):

Is there, then, no such thing as "real" intelligence? What I am saying is that if by "real" one means the intelligence that can perform great feats of appropriate selection without prior reception and processing of the equivalent quantity of information; then such "real" intelligence does not exist. It is a myth. It has come into existence in the same way that the idea of "real" magic comes to a child who sees conjuring tricks.

When humans appear to achieve remarkable performances of "genius," it is only because they had previously processed the required amount of information. Ashby argues that were it possible for such selections to occur in the absence of the required information processing, it would be like the case of a student who provided answers to exam questions before they were given—it would upset the causal order (Ashby 1960, p. 746).

When considering whether a machine such as a computer is capable of selective—that is, intelligent—performances at the level of skill of the human mind, he warns that we must carefully note how much information has been processed by each system (Ashby 1961, pp. 277–278):

It may perhaps be of interest to turn aside for the moment to glance at the reasons that may have led us to misunderstand the nature of human intelligence and cleverness. The point seems to be, as we can now see with the clearer quantitative grasp that we have today, that we tended grossly to mis-estimate the quantities of information that were used by computers and by people. When we program a computer, we have to write down every detail of the supplied information, and we are acutely aware of the quantity of information that must be made available to it. As a result, we tend to think that the quantity of information is extremely large; in fact, on any comparable scale of measurement it is quite small. The human mathematician, however, who solves a problem in three-dimensional geometry for instance, may do it very quickly and easily, and he may think that the amount of information that he has used is quite small. In fact, it is very large; and the measure of its largeness is precisely the amount of programming that would have to go into the computer in order to enable the computer to carry through the same process and to arrive at the same answer. The point is, of course, that when it comes to things like three-dimensional geometry, the human being has within himself an enormous quantity of information obtained by a form of preprogramming. Before he picked up his pencil, he already had behind him many years of childhood, in which he moved his arms and legs in three-dimensional space until he had learned a great deal about the intricacies of its metric. Then he spent years at school, learning formal Euclidian methods. He has done carpentry, and has learned how to make simple boxes and three-dimensional furniture. And behind him is five billion years of evolutionary molding all occurring in three-dimensional space; because it induced the survival of those organisms with an organisation suited to three-dimensional space rather than to any other of the metrics that the cerebral cortex could hold.... What I am saying is that if the measure is applied to both on a similar basis it will be found that each, computer and living brain, can achieve appropriate selection precisely so far as it is allowed to by the quantity of information that it has received and processed.

Once formulated in this way, we can recognize certain connections to aspects of Ashby's philosophy discussed earlier in this chapter. Most obvious is the significance of evolutionary adaptation as a source of information. On the one hand, there are the countless random trials and errors of that history—the raw information of random variation. But there is also the resultant information of selective adaptation: what was won from those trials and errors was a better organization for dealing with the environment. For the mathematician, that organization is already a part of him. As a model of the evolutionary history of his species, and of his own life experiences, he stands as an *archive* of that information—it is embodied in his cerebral organization. For the computer, the programmer stands as a designer who must make each of those decisions necessary for the mathematician's performance and express them in a computer program. It would be more desirable for the machine to learn those things itself, but this merely means that the information comes from a different source, not that it is spontaneously created by the machine.

With an account of the process of appropriate selection that was sufficient for quantitative measurement, Ashby had completed his general outline of a mechanistic philosophy of mind. It formed the basis, he believed, for an objective scientific study of intelligence. It provided in its formal rigor a means for experimentation and observations capable of resolving theoretical disputes about the mind. It also provided a basis for the synthesis of mechanical devices capable of achieving adaptive and intelligent

Figure 7.2
Ashby at the Biological Computer Laboratory (BCL), University of Illinois, with his "Grandfather Clock" and "Non-Trivial Machine." Used with permission of Murray Babcock's widow, Velva Babcock.

performances; the Homeostat was only one of the devices capable of such performances that Ashby constructed. His theoretical framework brought together physical, biological, and psychological theory in a novel and powerful form, one that he would credit Arturo Rosenblueth, Norbert Wiener, and Julian Bigelow (1943) and G. Sommerhoff (1950) for having independently discovered in their own work (Ashby 1952c). He would also agree that his conception of "adaptation and equilibrium" was equivalent to Sommerhoff's "directive correlation" and Rosenblueth, Wiener, and Bigelow's conception of "negative feedback"—the central concept of cybernetics. But Ashby also extended this idea to the more subtle aspects of intelligence: How could human intelligence be extended by machines? And what were the mechanics of decision-making processes?

Conclusion

Ashby's mechanistic philosophy of mind bears many superficial similarities to the more popular formulations of the idea that "machines can think," in particular the formulation provided by the "Turing test." Now that we have examined Ashby's philosophy in its details, however, it is instructive to note the subtle differences. The demonstration of the fundamental equivalence of adaptation and equilibrium was the core of Ashby's conception of the mind as a mechanism. Although Alan Turing demonstrated (1936) that any formally describable process could be performed by a computer, he recognized that this was not itself sufficient to show that a computer could think, since thinking might not be a formally describable process. Moreover, it did not come close to explaining *how* a computer could think. Ashby had set himself a different task than Turing: to understand how the behaviors and performances of living organisms in general, and thinking brains in particular, could be composed of mechanisms at all, and what those mechanisms were.

Consider Turing's (1950) "imitation game" for deciding whether or not a machine could be intelligent. In the first sections of that paper, he completely avoids attempting to define "machine" or "intelligence." Instead, he insists with little argument that the machine must be a digital computer, and proceeds to substitute his imitation game for a formal definition of intelligence. While we might agree with Turing that appealing to a commonsense understanding of "intelligence" would amount to letting the truth of the statement "intelligent machines can be made" depend upon the popular acceptance of the statement, his own imitation game doesn't go much further than this. In Turing's test for intelligence, he pits a digital computer against a real human being in a game where the winning objective for all contestants is to convince human judges that they are the humans and not the computer. The computer is considered "intelligent" if it is able to convince more than 50 percent of the judges that it is the human. Turing sets out some rules, to ensure that digital computers can play on an even field, which require that all interactions between the judges and the contestants take place over a telegraph wire, which limits the intelligent performances to the output of strings of symbols. Much has been written about this "test" for machine intelligence, and it is certainly the most popular formulation of the problem, but it seems profoundly lacking when compared to Ashby's definition of machine intelligence (and even the other ideas offered by Turing).

First, the fact that the "common usage" of the term "intelligence" is insufficient for judging computers does not mean that a precise formal definition cannot be provided—indeed, this is just what Ashby believed he had done. Second, the restriction of the meaning of "machines" to "digital computers" seems unnecessary. The Homeostat, for one, is an analogue computer that seems quite capable of demonstrating an intelligent capacity. Moreover, it does so not by virtue of carrying out particular calculations but of being a certain kind of information-processing system, one that is goal-seeking and adaptive. More significant, by leaving the meaning of intelligence up to a population of judges with indeterminate criteria, Turing's test fails to offer any instruction as to how such a computer should be constructed, or what its specific intellectual capacities might be—it is a way to dodge the issue of what intelligence is altogether.

In the process of developing his mechanistic philosophy, Ashby managed to perform some inversions of intuitions that are still commonly held. The first of these inversions was the "generative power of breakdown." The idea that creation requires impermanence, that destruction precedes construction, or that from chaos comes order is a recurring metaphysical paradox, at least as ancient as pre-Socratic Greek thought. In another form, it reappears in Ashby's work as a system's need for a source of random information in order to achieve a better performance than it was previously capable of. And it appears again when entropy is used as the fuel for driving the intelligence-amplifier to superhuman performances of appropriate selection. The intelligence-amplifier also inverts the notion that originality and productivity are essential aspects of intelligence. These are aspects of the random information fed to a selector, but it is the *power of appropriate selection* that reduces productivity and originality in a highly disciplined process which gives only the desired result.

To the end of his career Ashby remained concerned with the specific requirements for building machines that exhibited brainlike behavior. In part, this was motivated by his desire to understand the brain and its processes, and in part it was to build machines capable of aiding the human intellect. Although his designs for an intelligence-amplifier may still sound fanciful, his belief that such machines could be usefully brought to bear on real economic and social problems was not (Ashby 1948, pp. 382–83):

The construction of a machine which would react successfully to situations more complex than can be handled at present by the human brain would transform many of our present difficulties and perplexities. Such a machine might be used, in the distant future, not merely to get a quick answer to a difficult question, but to explore regions of intellectual subtlety and complexity at present beyond the human powers. The world's political and economic problems, for instance, seem sometimes to involve complexities beyond even the experts. Such a machine might perhaps be fed with vast tables of statistics, with volumes of scientific facts and other data, so that after a time it might emit as output a vast and intricate set of instructions, rather meaningless to those who had to obey them, yet leading, in fact, to a gradual resolving of the political and economic difficulties by its understanding and use of principles and natural laws which are to us yet obscure. The advantages of such a machine are obvious. But what of its disadvantages?

His aim was thus not merely to understand the brain, and simulate its properties, but also to understand those properties in such a way that they could be usefully employed to resolve difficult intellectual problems.

Even while he held out a hopeful vision of a future in which intelligent machines could resolve problems of great human concern and consequence, he was not without his fears of what the actual results might be (Ashby 1948). An intelligent machine by his definition was, after all, a machine that succeeded in achieving its *own purposes*, regardless of the resistance it encountered (p. 383):

But perhaps the most serious danger in such a machine will be its selfishness. Whatever the problem, it will judge the appropriateness of an action by how the feedback affects itself: not by the way the action benefits us. It is easy to deal with this when the machine's behavior is simple enough for us to be able to understand it. The slave-brain will give no trouble. But what of the homeostat-type, which is to develop beyond us? In the early stages of its training we shall doubtless condition it heavily to act so as to benefit ourselves as much as possible. But if the machine really develops its own powers, it is bound sooner or later to recover from this. If now such a machine is used for large-scale social planning and coordination, we must not be surprised if we find after a time that the streams of orders, plans and directives issuing from it begin to pay increased attention to securing its own welfare. Matters like the supplies of power and the prices of valves affect it directly and it cannot, if it is a sensible machine, ignore them. Later, when our world-community is entirely dependent on the machine for advanced social and economic planning, we would accept only as reasonable its suggestion that it should be buried deeply for safety. We would be persuaded of the desirability of locking the switches for its power supplies permanently in the "on" position. We could hardly object if we find that more and more of the national budget (planned by the machine) is being devoted to ever-increasing developments of the planning-machine. In the spate of plans and directives issuing from it we might hardly notice that the automatic valve-making factories are to be moved so as to deliver directly into its own automatic valve-replacing gear; we might hardly notice that its new power supplies are to come directly from its own automatic atomic piles; we might not realise that it had already decided that its human attendants were no longer necessary. How will it end? I suggest that the simplest way to find out is to make the thing and see.

This vision of the evolution of machines is sobering and sounds like the stuff of science fiction. In fact, however, it is more reserved than many of the claims made in the fields of artificial life and Artificial Intelligence in six decades since it was written. More to the point, when viewed in perspective with Ashby's overall philosophy it provides a means for thinking about the processes of social and economic organization and planning with a particular emphasis on the flow of information in those processes; though Ashby did not pursue this idea, it would seem to warrant further study.

There are many subtleties, implications, and extensions of Ashby's mechanistic philosophy that we have not covered. There are also many aspects of his intellectual career and contributions that we have skipped over or touched on only briefly. Our aim, however, was to come to a much clearer view of Ashby's overall philosophy, and of the interconnections and dependencies between its elements, so as to gain a greater appreciation for what is contained in Ashby's idea of "mechanical intelligence."

Notes

1. Both books were translated into several languages: *Design For a Brain* was published in Russian (1958), Spanish (1959), and Japanese (1963); *An Introduction to Cybernetics* was published in Russian (1957), French (1957), Spanish (1958), Czech (1959), Polish (1959), Hungarian (1959), German (1965), Bulgarian (1966), and Italian (1966).

2. Though it is implicit in much of AI, this approach is most explicit in the current field of biorobotics (see Webb and Consi 2001), and was also central in the develop-

ment of the fields of bionics and self-organizing systems in the 1960s (see Asaro 2007; for more on the synthetic method in the work of Ashby and a fellow cybernetician, W. Grey Walter, see Asaro 2006).

3. It is interesting to note that advantage 2 in this summary presages A. Rosenblueth, Norbert Wiener, and Julian Bigelow's (1943) "Behavior, Purpose, and Teleology" by three years. Ashby also bases his arguments on an elaboration of the concept of a "functional circuit," emphasizing the stable type, which parallels Rosenblueth, Wiener, and Bigelow's concept of feedback mechanisms, and negative feedback in particular, as explaining purposive or goal-seeking behavior. Another researcher, G. Sommerhoff (1950), a physicist attempting to account for biological organisms as physical systems, would come to essentially the same concepts a few years later. In his review of Sommerhoff's *Analytical Biology* Ashby (1952e) himself concludes, "It shows convincingly that the rather subtle concept of 'adaptation' can be given a definition that does full justice to the element of 'purpose,' while achieving a degree of precision and objectivity hitherto obtainable only in physics. As three sets of workers have now arrived independently at a definition from which the individuals differ only in details, we may reasonably deduce that the concept of 'adaptation' *can* be so re-formulated, and that its formulation in the language of physics is now available" (p. 409).

4. See Ashby (1947), "The Nervous System as Physical Machine: With Special Reference to the Origin of Adaptive Behavior," for more on learning and adaptation in the kitten.

5. It is interesting to note as an aside that, despite his relentless use of "stability" and later coining of the terms "ultrastability," "poly-stable" and "multi-stable," he does not use the word at all in his second paper on the mechanisms of adaptation, "The Physical Origin of Adaptation by Trial and Error" (1945; submitted 1943). There he uses the term "normal" in the place of "stability." This was perhaps due to a difference in audiences since this paper was addressed to psychologists.

6. Bigelow was a colleague of Norbert Wiener's at MIT, and was a coauthor of "Behavior, Purpose, and Teleology" (1943), which marks the beginning of cybernetics. He was the electrical engineer who built Wiener's "anti-aircraft predictor." In 1946 he had become the chief engineer of John von Neumann's machine at Princeton's Institute for Advanced Study, one of the first stored-program electronic computers.

7. Descartes's dictum can be found in the *Meditations*, and is a premise in his argument for the existence of God. The other premise is that "I find upon reflection that I have an idea of God, as an infinitely perfect being," from which Descartes concludes that he could not have been the cause of this idea, since it contains more perfection than he does, and thus there must exist an infinitely perfect God which is the real cause of his idea of an infinitely perfect God. He goes on to argue that the same God endowed him with reliable perception of the world.

8. Shannon's (1948) equation for the quantity of information is: $-\sum_j p_j \log p_j$, where p_j is the probability of receiving message j. By summing over all the messages, we obtain a measure of the current uncertainty, and thus of how much uncertainty will be removed when we actually receive a message and become certain. Thus the uncertainty is a measure of the system of communication and is not really a property of the message; alternatively we could say that the information content is the same for equiprobable messages in the set.

9. Shannon's 10th Theorem (1948, p. 68) states: "If the correction channel has a capacity equal to $H_y(x)$ it is possible to so encode the correction data as to send it over this channel and correct all but an arbitrarily small fraction ε of the errors. This is not possible if the channel capacity is less than $H_y(x)$." Here $H_y(x)$ is the conditional entropy of the input (x) when the output (y) is known.

Ashby's Law of Requisite Variety states that any system that is to control the ultimate outcome of any interaction in which another system also exerts some control must have at least as much variety in its set of alternative moves as the other system if it is to possibly succeed (Ashby 1956b, p. 206).

References

Asaro, Peter. 2006. "Working Models and the Synthetic Method: Electronic Brains as Mediators Between Neurons and Behavior." *Science Studies* 19(1): 12–34.

———. 2007. "Heinz von Foerster and the Bio-Computing Movements of the 1960s." In *An Unfinished Revolution? Heinz von Foerster and the Biological Computer Laboratory, 1958–1974*, edited by Albert Müller and Karl H. Müller. Vienna: Edition Echoraum.

Ashby, W. Ross. 1940. "Adaptiveness and Equilibrium." *Journal of Mental Science* 86: 478–83.

———. 1945. "The Physical Origin of Adaptation by Trial and Error." *Journal of General Psychology* 32: 13–25.

———. 1947. "The Nervous System as Physical Machine: With Special Reference to the Origin of Adaptive Behavior." *Mind* 56, no. 1: pp. 44–59.

———. 1948. "Design for a Brain." *Electronic Engineering* 20: 379–83.

———. 1951. "Statistical Machinery." *Thales* 7: 1–8.

———. 1952a. "Homeostasis." In *Cybernetics: Transactions of the Ninth Conference*, edited by H. von Foerster. New York: Josiah Macy Jr. Foundation (March), 73–108.

———. 1952b. "Can a Mechanical Chess-player Outplay Its Designer?" *British Journal for the Philosophy of Science* 3, no. 9: 44–57.

———. 1952c. *Design for a Brain*. London: Chapman & Hall.

———. 1952d. "Mechanical Chess Player." In *Cybernetics: Transactions of the Ninth Conference*, edited by H. von Foerster. New York: Josiah Macy Jr. Foundation (March), 151–54.

———. 1952e. "Review of *Analytical Biology*, by G. Sommerhoff." *Journal of Mental Science* 98: 408–9.

———. 1954. "Review of *The Neurophysiological Basis of Mind*, by J. C. Eccles." *Journal of Mental Science* 100: 511.

———. 1956a. "Design for an Intelligence-Amplifier." In *Automata Studies*, edited by Claude E. Shannon and J. McCarthy. Princeton: Princeton University Press.

———. 1956b. *An Introduction to Cybernetics*. London: Chapman & Hall.

———. 1960. "Computers and Decision Making." *New Scientist* 7: 746.

———. 1961. "*What Is an Intelligent Machine?*" BCL technical report no. 7.1. Urbana: University of Illinois, Biological Computer Laboratory.

———. 1962. "Simulation of a Brain." In *Computer Applications in the Behavioral Sciences*, edited by H. Borko. New York: Plenum Press.

———. 1967. "Cybernetics of the Large System." In *Accomplishment Summary 1966/67*. BCL report no. 67.2. Urbana: University of Illinois, Biological Computer Laboratory.

Cordeschi, Roberto. 2002. *The Discovery of the Artificial*. Dordrecht: Kluwer.

Jennings, H. S. 1915. *Behavior of the Lower Organisms*. New York: Columbia University Press.

Loeb, J. 1900. Comparative Physiology of the Brain and Comparative Psychology. New York: G. P. Putnams and Jons.

Lorentz, H. A. 1927. *Theoretical Physics*. London: Macmillan.

Miessner, B. F. 1916. *Radiodynamics: The Wireless Control of Torpedoes and Other Mechanisms*. New York: Van Nostrand.

Rosenblueth, A., Norbert Wiener, and Julian Bigelow. 1943. "Behavior, Purpose, and Teleology." *Philosophy of Science* 10: 18.

Shannon, Claude E. 1948. "A Mathematical Theory of Communication." *Bell System Technical Journal* 27: 379–423 and 623–56.

Shannon, Claude E., and J. McCarthy, eds. 1956. *Automata Studies*. Princeton: Princeton University Press.

Sommerhoff, G. 1950. *Analytical Biology*. London: Oxford University Press.

Turing, Alan M. 1936. "On Countable Numbers, with an Application to the *Entscheidungsproblem*." *Proceedings of the London Mathematics Society* (series 2) 42: 230–65.

———. 1950. "Computing Machinery and Intelligence." *Mind* 59: 433–60.

Webb, Barbara, and Thomas R. Consi (Eds.). 2001. *Biorobotics*. Cambridge, Mass.: MIT Press.

Wiener, Norbert. 1948. *Cybernetics, or Control and Communication in the Animal and Machine*. New York: Wiley.

8   Gordon Pask and His Maverick Machines

Jon Bird and Ezequiel Di Paolo

A computer that issues a rate demand for nil dollars and nil cents (and a notice to appear in court if you do not pay immediately) is not a maverick machine. It is a respectable and badly programmed computer.... Mavericks are machines that embody theoretical principles or technical inventions which deviate from the mainstream of computer development, but are nevertheless of value.

—Gordon Pask (1982a, p. 133)

Gordon Pask (1928–1996) is perhaps most widely remembered for his technical innovations in the field of automated teaching. Less widely appreciated are the theoretical principles embodied in Pask's maverick machines. He described himself as a "mechanic philosopher" (Scott 1980), and building machines played a central role in the development of a conceptual framework that resulted in two theories later in his career: Conversation Theory (CT) (Pask 1975) and Interaction of Actors Theory (de Zeeuw 2001). Even adherents of these theories concede that they are difficult to understand. Pask wrote over two hundred fifty papers and six books and his prose can be hard to follow and his diagrams difficult to untangle. B. Scott (1980, p. 328), who collaborated with Pask on CT, characterizes some of Pask's writing as "esoteric, pedantic, obscurantist." R. Glanville (1996), who wrote his doctorate under Pask's supervision, admits that CT "in many parts very hard to understand, because of a tendency to present it all, all the time, in its full complexity." Pask's presentations were dramatic and furiously paced and often left the audience baffled. Consequently, "some dismissed him, almost with resentment because of their inability to come to terms with him, but others recognised something both intriguing and important in what he said and the way that he said it. I myself often found I had lost the thread of what Gordon was saying, yet strangely he was triggering thoughts and insights" (Elstob 2001, p. 592). The psychologist Richard Gregory, who was a contemporary of Pask's at Cambridge, remembers

Figure 8.1
Gordon Pask (c. 1963). Printed with permission of Amanda Heitler.

(2001), "A conversation with Gordon is (perhaps too frankly) memorable now as being extraordinarily hard to understand at the time. Or is this just my inadequacy? He would come out with an oracular statement, such as 'Life is fire,' and would defend it against all objection. No doubt it had a certain truth, but I for one was never quite clear whether he was dealing in poetry, science, or humour. This ambiguous mixture was a large part of his charm" (p. 686). However, Gregory acknowledges that "without doubt, Gordon was driven by genuine insight" (p. 685). Heinz von Foerster and Stafford Beer, who both collaborated closely with Pask, also rated his intellect very highly, describing him as a genius (von Foerster 2001, p. 630; Beer 2001, p. 551).

In this chapter we focus on the early period of Pask's life, tracing the development of his research from his days as a Cambridge undergraduate to the period in the late 1950s when his work started to have an impact internationally. We describe three of his maverick machines: Musicolour, a sound-actuated interactive light show; SAKI, a keyboard-skill training machine; and an electrochemical device that grew an "ear." We assess the value of these machines, fifty years after they were built, in particular, the maverick ideas that they embody. We hope this will not only provide a way in to the challenging Paskian literature for the interested reader, but also demonstrate that many of Pask's ideas remain highly relevant for many current research areas.

School and University

What do we mean by conflict? Basically, that two or more time sequences of computation, which may have been proceeding in parallel, interact. Instead of remaining parallel and (by the definition of parallel) separate, they converge in a head-on collision from which there is no logical-deductive retreat.
—Gordon Pask (1982a, p. 62)

School Years

Pask stood out at Rydal, a Methodist public school in North Wales, where he was a boarder during the Second World War.[1] It was fairly liberal, but the headmaster, a prominent churchman, had a reputation for severity and would beat pupils (a common practice in public schools at the time). Pask's dress sense distinguished him from his fellow pupils and made him seem older than he was; he wore double-breasted business suits and bow ties, compared to the blazers and gray flannel trousers of his contemporaries. It was a style that he kept for the rest of his life (adding an Edwardian cape once he had left school). He was a small and sickly child and did not excel on the sports field—a very important part of Rydal culture (the school's two most famous alumni distinguished themselves as international rugby players). He spent his spare time building machines, for example, a device to detect rare metals that he tested out in nearby mines. A story about another one of his inventions, possibly fantasy, circulated through the school and contributed to Pask's reputation as a "mad professor." It was said that at the beginning of the Second World War he sent the War Office a design for a weapon. After a few months he received a reply stating that his proposal had been considered and it was thought it would work, but its effect was too dreadful to be employed against a human enemy.

Although his were not the usual preoccupations of teenage boys, he was not disliked, as he had a sense of fun and mischief. As a prank he would deflate large numbers of rugby balls that the sports master had inflated and left outside his room ready for the next day's sports activities. Pask also demonstrated his independence by slipping away from school some evenings, catching the train to Liverpool, and returning in the early hours. He said he was involved in producing stage shows in the city.[2] One day the whole school was summoned to a general assembly, as was always the case when the headmaster wanted to make an example of somebody for disciplinary offenses. Nobody knew who the offender was until his name was announced. Pask's absence had been discovered the previous evening and the headmaster publicly berated him. Pask was not cowed and in fact took offense at his treatment: he stood up and stormed out of the hall, telling the headmaster, "I shall speak to my solicitor about this." Apparently he escaped a beating.

Pask did not do national service after Rydal, perhaps because of ill health. Instead he went to Liverpool Technical College, where he studied geology and mining. In 1949 he went to Downing College, Cambridge University, to study medicine. He continued to have a vivid impact on his contemporaries, just as he had done at school.

Cambridge

At Cambridge Pask read Norbert Wiener's *Cybernetics*, which had an "emotional impact" on him (Pask 1966). He had found a field of study that was broad enough to accommodate his wide range of interests and also combined theory and practice: "As pure scientists we are concerned with brain-like artifacts, with evolution, growth and development; with the process of thinking and getting to know about the world. Wearing the hat of applied science, we aim to create ... the instruments of a new industrial revolution—control mechanisms that lay their own plans" (Pask 1961, p. 11). Pask met Robin McKinnon-Wood, a physicist, at Cambridge, and they began to build machines together. It was a relationship that continued for the rest of their lives. When they graduated they set up System Research Ltd., a company that sold versions of the machines that they had first started developing as undergraduates.

Pask also began to investigate statistical phenomena. Cedric Price, the architect, knew him as an undergraduate and was roped into some statistical experiments: "'It's simple, just throw these wooden curtain rings as quickly as possible into the numbered box—which I shall call out. Then do it backwards with a mirror, then blindfolded.' He took my arm and led me into Jordan's Yard. I could see that he was not to be trifled with" (Price 2001,

p. 819). This strange-sounding experiment was Pask's way of generating different probability distributions in order to predict the enlistment numbers for the RAF in the year 2000.

Stationary and Nonstationary Systems

A broad distinction that can be drawn about the statistics of a series of events is whether they are *stationary* or *nonstationary*. A scientist observing the behavior of a system over time might identify some regularities, for example, if the system is in state A, it goes to state B 80 percent of the time and to state C 20 percent of the time. If this behavior sequence is invariant over a large number of observations of the same system, or an ensemble of similar systems, then an observer can infer that statistically the system is stationary. The observed properties are *time-independent*, that is, various statistical measures, such as the mean and standard deviation, remain invariant over time. Therefore, given the occurrence of A we can be confident about the probability of B or C following, irrespective of what time we observe A.

Nonstationary systems do not display this statistical invariance; there are time-dependent changes in their statistical properties, and the relationship between A, B, and C can change. Human behavior, for example, is often nonstationary, as was dramatically demonstrated by Pask when he was studying medicine. He would get through anatomy tests by memorizing footnotes from Gray's *Anatomy*; by dazzling on some arcane anatomical details he usually managed to cast shadows over the holes in his knowledge. But on occasion he got found out. Gregory (2001) recalls an anatomy exam where Pask was asked to dissect an arm. One might predict, having observed the behavior of other anatomy students, that he would have used a scalpel. Instead, he used a fire axe, smashing a glass dissecting table in the process. Unsurprisingly, Pask graduated from Cambridge in physiology, rather than medicine.

Learning provides less dramatic examples of nonstationary behavior. We can measure the skill of a novice at performing some skill, for example, typing, by recording the person's average response time and error rate. As the novice practices, their skills will improve, and although their performance might be stationary for periods of time, it will also show discontinuities as it improves. Dealing with nonstationary systems is a challenge, as their behavior is difficult to characterize. Pask started developing two learning machines while he was an undergraduate and developed a mechanical and theoretical approach to dealing with nonstationary systems. In the next two sections we describe these machines in detail.

Musicolour

Man is prone to seek novelty in his environment and, having found a novel situation, to learn how to control it.
—Gordon Pask (1971, p. 76)

Pask built the first Musicolour system, a sound-actuated interactive light show, in 1953. Over the next four years, Pask, McKinnon-Wood, their wives, and a number of other individuals were involved in its development (Pask 1971). Pask's initial motivation for building the system was an interest in synesthesia and the question of whether a machine could learn relations between sounds and visual patterns and in doing so enhance a musical performance. From the outset, Musicolour was designed to cooperate with human performers, rather than autonomously generate "aesthetically valuable output" (Pask 1962, p. 135). The way musicians interacted with the system quickly became the main focus of research and development: the performer "trained the machine and it played a game with him. In this sense, the system acted as an extension of the performer with which he could co-operate to achieve effects that he could not achieve on his own. Consequently, the learning mechanism was extended and the machine itself became reformulated as a game player capable of habituating at several levels to the performer's gambits" (Pask 1971, p. 78).

How Does Musicolour Work? The sounds made by the musicians are relayed to the system via a microphone and amplifier. A bank of filters then analyze various aspects of the sound (see figure 8.2; the system had up to eight filters, but five are shown). An early system just used band-pass filters, but in later systems there were also filters that analyzed attack and rhythm. Each of the filters has a parameter that can take one of eight prespecified values. These values determine the frequency range of the band-pass filters and delays in the attack and rhythm filters.

The output from each filter is averaged over a short period, rectified, and passed through an associated adaptive threshold device (figure 8.2). If the input exceeds a threshold value, the output is 1, otherwise it is 0. These devices adapt their threshold to the mean value of the input, habituating to repetitive input, for example a continuous sound in a particular pitch band, and outputting 0. The outputs from the adaptive threshold devices

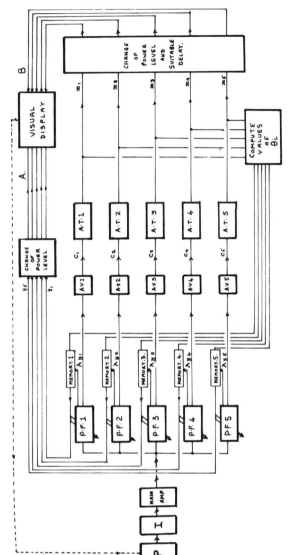

Figure 8.2

Diagram of a typical Musicolour system. P = performer; I = instrument and microphone; AT = adaptive threshold device; A = inputs to the visual display that determine *what* patterns are projected; B = inputs to the visual display that determine *when* the patterns are projected. From Pask (1971). Reprinted with permission of Jasia Reichardt.

Figure 8.3
A servo-positioned pattern wheel used in Musicolour. From Pask (1971). Reprinted with permission of Jasia Reichardt.

determine *when* a selection is made from the available visual patterns by controlling dimmers connected to the lights.

The values of the filter parameters determine *what* visual pattern is selected by controlling a servo-positioned pattern or color wheel (see figure 8.3). The particular parameter values are selected on the basis of how different the output of the filter's associated adaptive threshold device is, compared to the other filter's thresholded outputs, and how long it is since a particular value has been selected. The selection strategy aims to increase the novelty of the filter outputs and to ensure that all of the parameter values are sampled.[3]

If the input to Musicolour is repetitive, it habituates and adjusts its filter parameter values in an attempt to generate more variety in the light patterns. If there is no input, the system becomes increasingly sensitive to any sound in the environment and a gain control prevents this from disrupting the system too much.

Was It a Success? Musicolour was found to be "eminently trainable" (Pask 1971, p. 80). Performers were able to accentuate properties of the music and reinforce audio-visual correlations that they liked (for example, high notes with a particular visual pattern). Once performers became familiar with the filter-value selection strategy of the machine, they were able to establish time-dependent patterns in the system and reinforce correlations between groups of musical properties. It is important to note that there was no fixed mappings between sounds and lights: these were developed through the interaction of the musicians with Musicolour. There is reciprocal feedback between Musicolour and the performers: "The machine is designed to entrain the performer and to couple him into the system" (Pask 1971, p. 80). From the performer's perspective, "training becomes a matter of persuading the machine to adopt a visual style that fits the mood of his performance," and when the interaction has developed to this level "the performer conceives the machine as an extension of himself" (p. 86). Pask did some "rough and ready" studies of how visual patterns affect performance, finding that short sequences of visual events acted as releaser stimuli (p. 86).[4] It was also found that once a stable coordinated interaction had been established, it was robust to a certain level of arbitrary disturbances.

Musicolour developed from a small prototype machine that was tested at parties and in small venues to a large system that toured larger venues in the north of England and required two vans to transport the equipment and five people to set it up. After this tour, Musicolour was used in a theatrical performance at the Boltons Theatre in 1955, where it was combined with marionettes in a show called *Moon Music*. Musicolour and puppets were "unhappy bedfellows," and after a week of technical problems, the stage manager left and the show closed (Pask 1971, p. 81). With Jone Parry, the music director for Musicolour, Pask and McKinnon-Wood then used the month's paid-up rental on the theater to develop the musical potential of the system, and the show became a concert performance. Subsequently, Pask developed a work, *Nocturne*, in which he attempted to get dancers interacting with Musicolour. This was technically challenging, but Pask thought it showed some artistic potential.

The Musicolour project began to fall into debt and Pask explored different ways of generating income, ranging from adapting it for juke boxes (then at the height of their popularity) to marketing it as an art form. Bankruptcy was avoided by a regular gig at Churchill's Club in London (and by Cecil Landau becoming a partner in the business). People participated in the system by dancing, responding to the music and light show. After a year Musicolour moved to another club, the Locarno in Streatham, London, a large ballroom with a capacity of several thousand as well as a huge lighting rig (120 kW), which Musicolour modulated. This cavernous environment was not conducive to audience participation as there were too many other visual elements, such as exit signs, that distracted dancers from the visual display. Churchill's Club had been more intimate, and Musicolour had integrated with the space. Pask (1971) says that Landau "was prone to regard an archway across the middle of the night-club as a surrogate proscenium and everything beyond it a stage" (pp. 87–88). In larger, commercially viable spaces, Musicolour became just "another fancy lighting effect" and it "was difficult or impossible to make genuine use of the system" (p. 88). In 1957, after a final performance at a ball in London, Musicolour was shelved and Pask and McKinnon-Wood concentrated on the commercial development of their teaching machines.

SAKI

Teaching is control over the acquisition of a skill.
—Gordon Pask (1961, p. 88)

In 1956, Pask, his wife, Elizabeth, and Robin McKinnon-Wood applied for a patent for an "Apparatus for Assisting an Operator in Performing a Skill."[5] This patent covers a wide range of teaching machines built along cybernetic principles, including SAKI (self-adaptive keyboard instructor), which Stafford Beer (1959, p. 123) described as "possibly the first truly cybernetic device (in the full sense) to rise above the status of a 'toy' and reach the market as a useful machine." SAKI trains people to operate a Hollerith key punch (see figure 8.4), a device that punches holes in cards used for data processing.[6] By pressing keys the operator makes holes in selected columns on the cards to encode data in a form that can be read by a card reader and stored in computer memory. The Hollerith keyboard was designed to be operated with one hand and had twelve keys: 0 to 9, an X, and a top key. One digit can be entered per column by pressing the corresponding key. Alphabetic characters are entered by punching two holes in the same column: the top key and 1 to 9 for A to I, the X key and 1 to 9 for J to R, and 0 and 1 to 9 for S to Z. Up until the 1970s the key punch was

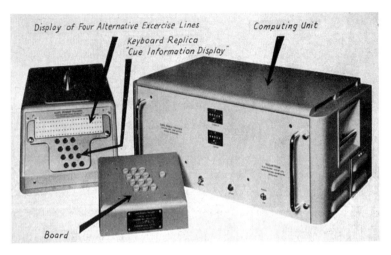

Figure 8.4
SAKI (self-adaptive keyboard instructor). Image taken from Plate II, Gordon Pask (1960) *An Approach to Cybernetics*, Harper and Brothers, with kind permission of Springer Science and Business Media.

a common form of data entry and there was a large demand for skilled operators.

One challenge in automating teaching is to ensure that a student's interest is sustained: "Ideally the task he is set at each stage should be sufficiently difficult to maintain his interest and to create a competitive situation yet never so complex that it becomes incomprehensible. A private tutor in conversation with his pupil seeks, in fact, to maintain this state which is not unlike a game situation" (Pask, McKinnon-Wood, and Pask 1961, p. 32). It requires that the tutor responds to the particular characteristics of a pupil. A multitude of factors determine a person's skill level (previous experience, motor coordination, level of tiredness) and some of these factors will change as a result of the learning process. Pask's novel approach was to build teaching machines that construct a continuously changing probabilistic model of how a particular operator performs a skill. Furthermore, the machines do not force an operator to perform in a particular way; operators are "minimally constrained by corrective information" in order to provide the "growth maximising conditions which allow the human operator as much freedom to adopt his own preferred conceptual structure" (p. 33). By adapting the task on the basis of a dynamic, probabilistic model of the operator, SAKI teaches in a way that responds to students' (non-stationary) individual characteristics and holds their interest.

How Does SAKI Work?   The operator sits in front of a display unit (see figure 8.4) that presents the exercise material (four lines of twenty-four alphanumeric characters to be punched) and cueing lights, arranged in the same spatial layout as the keyboard, that indicate which key, or key sequence, to press on the key punch. Initially the operator works through all four exercise lines. Starting with the first line, items are randomly presented at a slow, uniform rate and the cueing lights are bright and stay on for a relatively long period of time. The operator's response time for each item is stored in the "computing unit." This consists of a series of capacitors that are charged from the moment an operator makes a correct response until the next item is presented: the faster a correct response is, the higher the charge stored. When all four exercise lines have been completed correctly, SAKI has a preliminary analogue "model" of the operator's key-punch skills for every item in the four exercise lines, stored as charges on the series of capacitors.

The exercise line for which the operator has the slowest average response time is then repeated. The capacitors drive valves, which determine how the individual items in this exercise are presented to the operator—specifically, the available response time and the clarity of the cueing lights (their brightness and duration). In a prototype design, Pask uniformly varied the difficulty of the items according to average performance on an exercise line. However, it was found that uniformly increasing the difficulty of *all* the items in the exercise results in oscillations in an operator's performance—the task alternating between being too difficult and being too easy (Pask, McKinnon-Wood, and Pask 1961). The computing unit therefore *individually* varies the difficulty of each item in an exercise line so as to better match the performance of the operator. For example, it increases the difficulty of items where the operator has performed relatively successfully by reducing the cue information as well as the available response time. The reduction in available response time also reduces the maximum charge that can be stored on the associated capacitor. As the operator's skill on an item increases, the cueing information reduces, until finally there is only an indication of the alphanumeric character that has to be punched. This reduction in cueing information initially increases the likelihood that the operator will make a mistake. SAKI responds by reintroducing the visual cues and extending the available response time. Operators using SAKI show plateaus in their learning curves, but can ultimately reach a final stable state where there is no visual cueing information and an equal distribution of available response times for all items in an exercise line (Pask 1961). That is, they punch each key with equal proficiency. To maintain this level, the operator has to consistently perform a sequence of key punches at or below predetermined error and response rates.

Beer (1959) describes his experience of using a version of SAKI in *Cybernetics and Management* (pp. 124–25):

You are confronted with a punch: it has blank keys, for this is a "touch typing" skill. Before you, connected to the punch, is Pask's machine. Visible on it is a little window, and an array of red lights arranged like the punch's keyboard. The figure "7" appears in the window. This is an instruction to you to press the "7" key. But you do not know which it is. Look at the array of lights. One is shining brightly: it gives you the position of the "7" key, which you now find and press. Another number appears in the window, another red light shines and so on. Gradually you become aware of the position of the figures on the keyboard, and therefore you become faster in your reactions. Meanwhile, the machine is measuring your responses, and building its own probabilistic model of your learning process. That "7," for instance, you now go to straight away. But the "3," for some obscure reason, always seems to elude you. The machine has detected this, and has built the facts into its model. And now, the outcome is being fed back to you. Numbers with which you have difficulty come up with increasing frequency in the otherwise random presentation of digits. They come up more slowly, too, as if to say: "Now take your time." The numbers you find easy, on the contrary, come up much faster: the speed with which each number is thrown at you is a function of the state of your learning. So also is the red-light system. For as you learn where the "7" is, so does the red-light clue gradually fade. The teacher gives you less and less prompting. Before long, if you continue to improve on "7," the clue light for "7" will not come on at all. It was getting fainter on "5," for you were getting to know that position. But now you have had a relapse: "5" is eluding you altogether. Your teacher notes your fresh mistakes. "5" is put before you with renewed deliberation, slowly; and the red light comes back again, brightly.... So the teaching continues. You pay little intellectual attention: you relax. The information circuit of this system of you-plus-machine flows through the diodes and condensers of the machine, through the punch, through your sensory nerves and back through your motor nerves, the punch, the machine. Feedback is constantly adjusting all the variables to reach a desired goal. In short, you are being conditioned. Soon the machine will abandon single digits as the target, and substitute short runs of digits, then longer runs. You know where all the keys are now; what you have to learn next are the patterns of successive keys, the rhythms of your own fingers.

Was It a Success?   Beer began as a complete novice and within forty-five minutes he was punching at the rate of eight keys per second. It seems likely that he was just doing single-key exercises,[7] rather than key combinations. Generally, SAKI could train a novice key-punch operator to expert level (between seven thousand and ten thousand key depressions per hour) in four to six weeks if they completed two thirty-five-minute training sessions every working day. A conservative estimate of the reduction in training time, compared to other methods, was between 30 and 50 percent (Pask 1982b).

SAKI deals with incomplete knowledge about the characteristics of individual operators and how they learn by taking the cybernetic approach of treating them as a "black box"—a nonstationary system about which we have limited knowledge. In order to match the characteristics of the operator, the computing unit is also treated as a black box that builds a probabilistic, nonstationary analogue of the relation between itself and the operator through a process of interaction. The overall goal is to find a stable relation between the user and SAKI, with the additional constraint that the operator meets a prespecified performance level defined in terms of speed and accuracy of key punching. Pask summarizes this design methodology: "a pair of inherently unmeasurable, non-stationary systems, are coupled to produce an inherently measurable stationary system" (Pask 1961, p. 98). SAKI found the appropriate balance between challenging exercises and boredom: "Interest is maintained, and an almost hypnotic relationship has been observed, even with quite simple jobs" (Pask, McKinnon-Wood, and Pask 1961, p. 36). In 1961 the rights to sell SAKI were bought by Cybernetic Developments and fifty machines were leased or sold, although one unforeseen difficulty was getting purchasers to use SAKI as a training machine, rather than as a status symbol (Pask 1982b). SAKI was a very effective keypunch trainer but a limited financial success.

Summary of Musicolour and SAKI

Pask described Musicolour as "the first coherence-based hybrid control computer" where a nonstationary environment was tightly coupled with a nonstationary controller and the goal was to reach stability, or coherence, through reciprocal feedback (Pask 1982a, p. 144). He describes it as "hybrid" because rather than executing a program, it adapted on a trial-and-error basis. SAKI differs from Musicolour in that for commercial reasons there was also a performance constraint driving the activity. There were no such constraints on how Musicolour and musicians reached stable cycles of activity, the search for stability being an end in itself. Interestingly, having observed people interacting with both systems, Pask concluded (1961) that they are motivated by the desire to reach a stable interaction with the machines, rather than to reach any particular performance goal: "After looking at the way people behave, I believe they aim for the non-numerical payoff of achieving some desired stable relationship with the machine" (p. 94).

Both Musicolour and SAKI are constructed from conventional hardware components (capacitors, valves, and so forth), but it is difficult to *functionally* separate the machines from their environments, as they are so tightly coupled. However, Pask wanted to develop *organic* machines that were built from materials that *develop* their functions over time, rather than being specified by a design. An organic controller differs from Musicolour and SAKI by not being limited to interacting with the environment through designer-specified channels (such as keyboards and microphones): it "determines its relation to the surroundings. It determines an appropriate mode of interaction, for example, it learns the best and not necessarily invariant sensory inputs to accept as being events" (Pask 1959, p. 162). The next sections describe the collaboration between Pask and Stafford Beer as they explored how to build such radically unconventional machines.

Pask as an Independent Cybernetic Researcher

Stafford Beer (1926–2002) and Pask met in the early 1950s and they collaborated for the rest of the decade. They were "both extremely conscious of the pioneering work being done in the USA in the emerging topic that Norbert Wiener had named cybernetics, and knew of everyone in the UK who was interested as well" (Beer 2001, p. 551). Both men were ambitious and wanted to make an impact in the field of cybernetics. They were particularly interested in W. Ross Ashby's work on ultrastability (Ashby 1952) and the question of how machines could adapt to disturbances that had not been envisaged by their designer. Beer was working for United Steel, doing operations research, and had persuaded the company to set up a cybernetics research group in Sheffield. Pask was developing learning machines and trying to market them commercially. They grew close as they both faced similar challenges in trying to persuade the business world of the value of their cybernetic approach. They also shared a deep interest in investigating the suitability of different "fabrics," or media, as substrates for building self-organizing machines:

If systems of this kind are to be used for amplifying intelligence, or for 'breeding' other systems more highly developed than they are themselves, a fixed circuitry is a liability. Instead, we seek a fabric that is *inherently* self-organizing, on which to superimpose (as a signal on a carrier wave) the particular cybernetic functions that we seek to model. Or, to take another image, we seek to *constrain* a high-variety fabric rather than to fabricate one by blueprint (Beer 1994, p. 25).

The "high-variety" criterion came from Ashby's argument that a controller can only control an environment if it has variety in its states greater or equal to the variety in the disturbances on its inputs.[8] Another requirement for a suitable fabric was that its behavior could be effectively coupled to another system.

The Search for a Fabric

Both Beer and Pask investigated a wide range of media for their suitability as high-variety fabrics. From the outset, Beer rejected electrical and electronic systems as they had to be designed in detail and their functions well specified, and this inevitably constrained their variety. Instead, he turned to animals.

In 1956 Beer had set up games that enabled children to solve simultaneous equations, even though they were not aware they were doing so. Their moves in the game generated feedback in the form of colored lights that guided their future moves. He then tried using groups of mice, with cheese as the reward, and even tried to develop a simple mouse language. Beer considered the theoretical potential of other vertebrates (rats and pigeons) and, with Pask, social insects, but no experiments were carried out using these animals.

Beer then investigated groups of *Daphnia*, a freshwater crustacean. He added iron filings to the tank, which were eaten by the animals. Electromagnets were used to couple the tank with the environment (the experimenter). Beer could change the properties of magnetic fields, which in turn effected changes in the electrical characteristics of the colony. Initially this approach seemed to have potential, as the colony "retains stochastic freedom within the pattern generally imposed—a necessary condition in this kind of evolving machine; it is also self-perpetuating, and self-repairing, as a good fabric should be" (Beer 1994, p. 29). However, not all of the iron filings were ingested by the crustaceans and eventually the behavior of the colony was disrupted by an excess of magnets in the water.

Beer then tried using a protozoan, *Euglena*, keeping millions of them in a tank of water, which he likened to a "biological gas" (Beer 1994, p. 30). These amoebae photosynthesize in water and are sensitive to light, their phototropism reversing when light levels reach a critical value. If there is sufficient light they reproduce by binary fission; if there is a prolonged absence of light they lose chlorophyll and live off organic matter. The amoebae interact with each other by competing for nutrients, blocking light and generating waste products. Although the green water was a "staggering source of high variety" and it was possible to couple to the system (using a

point source of light as an input and a photoreceptor to measure the behavioral output), unfortunately, the amoebae had "a distressing tendency
to lie doggo, and attempts to isolate a more motile strain failed" (Beer
1994, p. 31). Beer started to experiment with pond ecosystems kept in large
tanks, thinking that his single-species experiments were not ecologically
stable. He coupled the tank and the wider world in the same way as he
had done in the *Euglena* experiments, using a light and photorecep-
tors. However, it proved difficult to get this system to work as a control
system—the feedback to the environment was too ambiguous. "The state
of the research at the moment is that I tinker with this tank from time to
time in the middle of the night. My main obsession at the moment is at
the level of the philosophy of science. All this thinking is, perhaps, some
kind of breakthrough; but what about an equivalent breakthrough in experimental method? Do we really know how to experiment with black
boxes of abnormally high varieties?" (Beer 1994, p. 31). The first experimental breakthrough came during one of his visits to Pask.

Growing an Ear

Although based in Sheffield, Beer would regularly go down to London and
work most of the night with Pask.[9] In 1956 or '57, he had "the most important and indeed exciting of my personal recollections of working with Gordon" (Beer 2001, p. 553): the night they grew an electrochemical ear. Pask
had been experimenting with electrochemical systems consisting of a number of small platinum electrodes inserted in a dish of ferrous sulphate solution and connected to a current-limited electrical source. Metallic iron
threads tend to form between electrodes where maximum lines of current
are flowing. These metallic threads have a low resistance relative to the
solution and so current will tend to flow down them if the electrical activation is repeated. Consequently, the potentials at the electrodes are modified
by the formation of threads. If no current passes through a thread, then it
tends to dissolve back into the acidic solution. Metallic threads develop as
the result of two opposing processes: one that builds threads out of ions on
relatively negative electrodes; and one that dissolves threads back into ions.
The trial-and-error process of thread development is also constrained by the
concurrent development of neighboring threads and also by previously
developed structures. Slender branches extend from a thread in many directions and most of these dissolve, except for the one following the path of
maximum current. If there is an ambiguous path then a thread can bifur-
cate. As the total current entering the system is restricted, threads compete
for resources. However, when there are a number of neighboring unstable

structures, the threads can amalgamate and form one cooperative structure. Over time a network of threads literally grows dynamically stable structures.

These electrochemical systems display an elementary form of learning. If a stable network of threads is grown and then the current to the electrodes is redistributed, a new network will slowly start to form. If the current is then set to the original distribution, the network tends to regrow its initial structure. The longer a network has been stably growing, the slower it breaks down when the current distribution changes, and the quicker it returns to its original structure when the current distribution is reset.

Beer vividly remembers the night that he and Pask carried out the electrochemical experiments that resulted in an ear (Beer 2001, pp. 554–55). They were discussing Ashby's concept of ultrastability and the ability of machines to adapt to unexpected changes—changes that had not been specified by their designer. Pask had recently been placing barriers in the electrochemical dishes and the threads had grown over them—they had adapted to unexpected changes in their environment. That night they did some experiments to see how the threads would respond to damage by chopping out sections of some of the threads. When current was applied to the system the threads regrew, the gap moving from the anode to the cathode until it was gone.

Although excited by this result, they thought that these were relatively trivial disturbances. They wanted to perform an experiment to investigate whether a thread network could adapt to more radical, unexpected disruption. "We fell to discussing the limiting framework of ultrastability. Suddenly Gordon said something like, 'Suppose that it were a survival requirement that this thing should learn to respond to sound? If there were no way in which this 'meant' anything, it would be equivalent to your being shot. But this cell is liquid, and in principle sound waves could affect it. It's like your being able to accommodate to a slap, rather than a bullet. We need to see whether the cell can learn to reinforce successful behaviour by responding to the volume of sound.'...It sounded like an ideal critical experiment" (Beer 2001, p. 555).

Beer cannot remember the exact details of how they rewarded the system.[10] However, it did not require any major changes to the experimental setup. They basically connected one, or more, of the electrodes with output devices that enabled them to measure the electrical response of the electrochemical system to sound. The reward consisted of an increase in the current supply, a form of positive reinforcement. Regardless of how the electrodes are configured, the electrochemical system will tend to develop a thread structure that leads to current flowing in such a way that it is rewarded further. Importantly, the reward is simply an increased capacity for growth—there is no specification of what form the growth should take.

The electrochemical system is not just electrically connected to the external world: threads are also sensitive to environmental perturbations such as vibrations, temperature, chemical environment, and magnetic fields. Any of these arbitrary disturbances can be characterized as a stimulus for the system, especially if they cause a change in current supply. "And so it was that two very tired young men trailed a microphone down into Baker Street from the upstairs window, and picked up the random noise of dawn traffic in the street. I was leaning out of the window, while Gordon studied the cell. 'It's growing an ear,' he said solemnly (*ipsissima verba* [the very words])" (Beer 2001, p. 555).

Pask (1959) describes further experiments that were carried out where a thread network was grown that initially responded to 50 Hz and then, with further training, could discriminate between this tone and 100 Hz. He was also able to grow a system that could detect magnetism and one that was sensitive to pH differences. In each case the electrochemical system responded to positive reinforcement by growing a sensor that he had not specified in advance. Beer is clear why he and Pask thought this experiment was significant: "This was the first demonstration either of us had seen of an artificial system's potential to recognize a filter which would be conducive to its own survival and to incorporate that filter into its own organization. It could well have been the first device ever to do this, and no-one has ever mentioned another in my hearing" (Beer 2001, p. 555).

Pask (1959, p. 262) argues that the electrochemical ear is a maverick device, as it shows the distinction between

the sort of machine that is made out of known bits and pieces, such as a computer...and a machine which consists of a possibly unlimited number of components such that the function of these components is not defined beforehand. In other words, these 'components' are simply 'building material' which can be assembled in a variety of ways to make different entities. In particular the designer need not specify the set of possible entities.

Importantly, electrochemical systems, although finite, "are rendered non-bounded by the interesting condition that they can alter their own relevance criteria, and in particular, by the expedient of building sense organs, can alter their relationship to the environment according to whether or not a trial relationship is rewarded" (p. 262).

The Value of Gordon Pask

Ideas that were dear to Gordon all that time ago, on interactive circuits with dynamic growth, are coming back in the form of neural nets, with parallel processing in digital computers and also analogue systems. My bet is that analogue self-adapting nets will take over as models of brain function—because this is very likely how the brain works—though AI may continue on its course of number crunching and digital computing. Surely this is alien to the brain. So we would fail the Turing Test, being too good at pattern recognition, and much too poor at arithmetic compared with digital computers. In short, the kind of philosophy that Gordon nurtured does seem to be returning. Perhaps his learning machines have lessons for us now.
—Richard Gregory (2001, pp. 686–87)

The naive picture of scientific knowledge acquisition is one of posing increasingly sophisticated questions to nature. But, of course, such questions, and therefore the knowledge obtained from them, are never pure, unaffected by the questioners' ulterior motives, or unconstrained by technological and conceptual barriers. Science manifests itself as a social and cultural activity through subtle factors such as concept management, theory creation, and choice of what problems to focus on. It is far from being passive observation followed by rational reflection. It is *active*. But even in this picture, experimental data, the source of scientific information to a community of researchers, is still seen as the detached, passive observation of nature at work. Observer intervention (today most apparent in quantum measurement or the behavioral and cognitive sciences) is often treated as a *problem* we would wish to minimize if we cannot eliminate.

Pask's approach goes against this view. For him, not only can we gain new understanding by actively constructing artefacts instead of just observing nature, we can also increase our knowledge by engaging in an interaction with them. Pask's design methodology can be characterized as "meeting nature half way": accepting that we have limited, incomplete knowledge about many systems we want to understand and treating them as black boxes. By interacting with these systems we can constrain them, and ourselves, and develop a stable interaction that *is* amenable to analysis. For him, both the construction and the interaction become a necessity if we wish to understand complex phenomena such as life, autonomy, and intelligence.

Let us consider construction. The first thing that must be clarified is that Pask, and nowadays some of current research in AI and robotics, is not simply proposing that technology and science interact, often in a positive, mutually enhancing manner. The construction he refers to is not that of more sophisticated artefacts for measuring natural phenomena or the construction of a device that models natural phenomena by proxy, but the construction of a proper object of study, in other words, the synthesis of *a scientific problem in itself.* This idea is radical—fraught with pitfalls and subject to immediate objections. Why create problems deliberately? Are we not just using our existing knowledge to guide the creation of an artefact? Then, how do we expect to gain any new knowledge out of it?

Indeed, the idea seems not just a minefield of methodological issues; it seems absurd and a nonstarter, at most a recipe for useful pedagogical devices, toy problems for scientific training, but not the stuff of proper science. To answer these criticisms it is necessary to demonstrate not only that interesting artefacts can be constructed that will grasp the attention of scientists but also that we can do science with them, that they can advance our understanding of a problem.

It is clear, in relation to one of the objections above, that if by construction we mean the full specification of every aspect of our artefact and every aspect of its relation to its environment, then little new knowledge can be expected from it, except perhaps the knowledge that confirms that our ideas about how to build such an artefact were or weren't correct. This is traditional engineering, which of course is a source for that kind of knowledge. But what if the construction proceeds not by a full specification of the artefact but by the design of some broad constraints on processes that lead to increased organization, the result of which—with some good probability—is the artefact we are after? Now, if we succeed in this task, the workings of such a system are not fully known to us. It may surprise us. It may challenge our preconceptions by instantiating a new way of solving a problem. Or, more subtly, it may make us revise the meaning of our scientific terms and the coherence of our theories.

Is such an underspecified synthesis possible? Yes, it is. It was for Pask, as he demonstrated with his maverick machines (most dramatically with the electrochemical "ear") and it is common currency in biologically inspired AI (self-organizing optimization algorithms, evolutionary robotics, stochastic search, and so forth). Hardware evolution, which uses genetic algorithms to constrain reconfigurable devices such as field-programmable gate arrays (FPGAs), also provides striking examples of how relaxing conventional engineering constraints (such as a central clock) can lead to the invention of novel circuits—or should that be "discovered"?[11] Pask's research also provides a valuable reminder of the constraints that conventional computer architectures impose on machines. Although digital computers have an invaluable "number-crunching" role, they are not necessarily the best medium for building controllers that have to interact with dynamic environments. Similarly, conventional computer architectures might not be the best models of adaptive systems.

Pask provides a methodology for developing controllers that can deal with nonstationary environments about which we have limited knowledge. His cybernetic approach of coupling two nonstationary, unmeasurable systems in order to generate a stable, measurable relation will probably not appeal to conventional engineers. For example, Beer (2001, p. 552), discussing SAKI, lamented, "The engineers somehow took the cybernetic invention away. I suspect that they saw themselves as designing a machine to achieve the content-objective (learn to type), instead of building a Paskian machine to achieve the cybernetic objective itself—to integrate the observer and the machine into a homeostatic whole. Machines such as these are not available to this day, because they are contra-paradigmatic to engineers and psychologists alike."

Even if we can successfully synthesize an artefact that would indeed be a source for furthering our understanding about a given problem, what makes us think that such a device would be easier to understand than nature? Valentino Braitenberg (1984), a proponent of a related synthetic approach, convincingly pointed out a curious fact that he dubbed the "law of downhill synthesis and uphill analysis": it is rather easy to build things that look very complex and are hard to understand. This is true particularly if we specify lower-level mechanistic building blocks and leave as unspecified higher-level and interactive aspects of the system. If we now present the latter as the easily observable variables, the system can seem devilishly complex. W. Grey Walter had already demonstrated this with his robotic tortoises (1950, 1951, 1953). Simple combinations of a very few basic mechanisms could interact in surprising ways with a complex environment, giving the illusion of sophisticated cognitive performances (such as decision making, adaptive goal constancy, self-sustenance, and others).

This "law" of downhill synthesis, uphill analysis, is on the one hand quite interesting in itself, and often the source of entertaining explorations. It is also a stark reminder that we need not theorize complex mechanisms when we are faced with complex systemic behavior—a much unheeded warning. It is, in this sense, a powerful positive idea. On the other hand, though, it points to a major problem with the proposal of furthering scientific understanding by construction. Yes, we may be successful in constructing our artefact, but how shall we understand it? We seem to be at an advantage over understanding similarly complex phenomena in nature.

We may have more access to data, we know many things, if not everything, about how the system is built, we can restart it and do experiments that would be impossible in nature. But will these advantages always suffice? Have we given our process of synthesis too much freedom and the result is now an intractably complex system?

Two answers can be given to this problem. One suggests that our greater gain is by proceeding in a more or less controlled manner in exploring increasingly complex systems. This answer advocates minimalism as a methodological heuristic (Harvey et al. 2005). By building systems that are underdetermined but in a controlled fashion (which sounds like a paradox, but simply means that we should carefully control the constraints to the process of automatic synthesis), we stand our highest chance of creating new knowledge because we advance minimally over our previous understanding, which can largely be deployed on the analysis of our current system. There is a sense in which such a minimalism will provide us with the simplest cases that instantiate a phenomenon of interest, for instance learning or decision making, and allow us to develop the right kind of "mental gymnastics" to deal with more complex cases (Beer 2003).

But Pask proposes a different, more radical solution that has, paradoxically, been in use in dealing successfully with nature since the advent of culture, much before anything like science ever existed. Pask proposes that we should base our understanding of a complex system on our interactions with it and the regularities that emerge from such interaction. We should approach complex systems, even those we synthesize ourselves, as a natural historian would (perhaps even as an animal trainer, a psychotherapist, or an artist would). This interactive method for understanding complex systems is still a hard pill to swallow in many areas of science.

Pask's machines and philosophy often seem so maverick that they are hard to evaluate. Interacting with his work, one can struggle to achieve a stable understanding because of the demands he places on the reader. However, we have found it a worthwhile struggle and we hope that others will be encouraged to interact with his ideas: although fifty years old, they are highly relevant for every discipline that is attempting to understand adaptive (nonstationary) behavior.

Acknowledgments

We would like to thank Michael Renshall, CBE, and Paul Pangaro for agreeing to be interviewed about their memories of Gordon Pask and for giving valuable feedback on an earlier draft. It was through the research of Peter

Cariani that Jon Bird first became aware of Gordon Pask and he has enjoyed and benefited from conversations with Cariani over the last few years. Many thanks to Philip Husbands for his support for Paskian research and for first introducing Jon Bird to the cybernetic literature. Many thanks to Amanda Heitler and Jasia Reichardt for use of their photographs.

Notes

1. Michael Renshall, CBE, who was at Rydal from 1941 to 1948 and also was a contemporary of Pask's at Cambridge, provided all of the information about Gordon's school days.

2. The radical theater director Joan Littlewood was certainly aware of Pask by 1946 (Littlewood 2001). Pask wrote shows for his Musicolour system in the early 1950s.

3. A clear description of the strategy for selecting filter parameter values is given in Pask (1971, p. 80).

4. Ethologists coined the term "releaser stimulus" to refer to a simple perceptual feature of a complex stimulus that elicits a pattern of behavior. Niko Tinbergen (1951) had shown that crude models of a stickleback could elicit behavior patterns in the real fish—they attack red-bellied models and court swollen-bellied models.

5. The complete patent specification was published in 1961 (GB866279).

6. A later version of SAKI was developed to train operators in the use of key punches with larger numbers of keys (see Pask 1982, p. 71, figure 2). Herman Hollerith developed the first automatic data-processing system to count the 1890 U.S. census. A key punch was used to record the data by making holes in dollar-bill-sized cards. A tabulating machine contained a pin for each potential hole in a card. A card was passed into the reader and if a pin passed through a hole a current was passed, incrementing a counter. On the basis of these counters the card was automatically dropped into the appropriate section of a sorting box. It took just three years to tabulate the 62 million citizens the census counted. Building on this success Hollerith set up the Tabulating Machine Company, which eventually, after a series of mergers, became IBM in 1924. Some key-punch devices continued to be marketed as Hollerith machines, for example, the IBM 032 Printing punch produced in 1933 and the keyboard used in the first versions of SAKI.

7. This does seem a remarkably fast rate—the average response time for pressing a key after training on SAKI was about 0.2 seconds (Pask 1961a, p. 96).

8. For details of Ashby's "Law of Requisite Variety," see Ashby (1956, pp. 202–18).

9. Both Pask and Beer worked eccentric hours. Pask would regularly stay awake for thirty-six hours and then sleep for twelve hours, regulating the cycle with pills

(Elstob 2001). His wife thought that he was often at his best at the end of these marathon work sessions (Paul Pangaro, personal communication; Pangaro, who earned his doctorate with Pask, maintains an on-line archive of Pask's work at http://www.pangaro.com/Pask-Archive/Pask-Archive.html).

10. We lack clear information about the experimental details, even though Pask continued in-depth investigations into electrochemical systems at the University of Illinois under Heinz von Foerster. There has not, to our knowledge, ever been an independent replication of these experiments.

11. Adrian Thompson (1997) evolved a circuit on a small corner of a Xilinx XC6216 field-programmable gate array (FPGA) that was able to discriminate between two square wave inputs of 1 kHz and 10 kHz without using any of the counters–timers or RC networks that conventional design would require for this task. Layzell (2001) developed his own reconfigurable device, the Evolvable Motherboard, for carrying out hardware evolution experiments. One experiment resulted in the "evolved radio," probably the first device since Pask's electrochemical "ear" that configured a novel sensor (Bird and Layzell 2001).

References

Ashby, W. Ross. 1952. *Design for a Brain: The Origin of Adaptive Behaviour*. London: Chapman & Hall.

———. 1956. Introduction to Cybernetics. London: Methuen.

Beer, R. D. 2003. "The Dynamics of Active Categorical Perception in an Evolved Model Agent." *Adaptive Behavior* 11, no. 4: 209–43.

Beer, Stafford. 1959. *Cybernetics and Management*. London: English Universities Press.

———. 1994. "A Progress Note on Research into a Cybernetic Analogue of Fabric." In *How Many Grapes Went into the Wine: Stafford Beer on the Art and Science of Holistic Management*, edited by R. Harnden and A. Leonard. New York: Wiley.

———. 2001. "A Filigree Friendship." *Kybernetes* 30, no. 5–6: 551–59.

Bird, Jon, and P. Layzell. 2001. "The Evolved Radio and Its Implications for Modelling the Evolution of Novel Sensors." *Proceedings of the 2002 Congress on Evolutionary Computation*. Washington, D.C.: IEEE Computer Society.

Braitenberg, Valentino. 1984. *Vehicles: Experiments in Synthetic Psychology*. Cambridge, Mass.: MIT Press.

Elstob, M. 2001. "Working with Gordon Pask: Some Personal Impressions." *Kybernetes* 30, no. 5–6: 588–92.

Foerster, Heinz von. 2001. "On Gordon Pask." *Kybernetes* 30, no. 5–6: 630–35.

Glanville, R. 1996. "Robin McKinnon-Wood and Gordon Pask: A Lifelong Conversation." *Cybernetics and Human Knowing* 3, no. 4: n.p.

Gregory, Richard. 2001. "Memories of Gordon." *Kybernetes* 30, no. 5–6: 685–87.

Grey Walter, W. 1953. *The Living Brain*. London: Duckworth.

———. 1950. "An Imitation of Life." *Scientific American* 182, no. 5: 42–45.

———. 1951. "A Machine That Learns." *Scientific American* 185, no. 2: 60–63.

Harvey, I., E. A. Di Paolo, E. Tuci, R. Wood, and M. Quinn. 2005. "Evolutionary Robotics: A New Scientific Tool for Studying Cognition." *Artificial Life* 11: 79–98.

Layzell, P. 2001. "Hardware Evolution: On the Nature of Artificially Evolved Electronic Circuits." Ph.D. diss., University of Sussex.

Littlewood, Joan. 2001. "Gordon Pask." *Kybernetes* 30, no. 5–6: 760–61.

Moore, H. 2001. "Some Memories of Gordon." *Kybernetes* 30, no. 5–6: 768–70.

Pask, Gordon. 1959. "Physical Analogues to the Growth of a Concept." *Proceedings of Symposium No. 10 of the National Physical Laboratory: Mechanisation of Thought Processes, National Physical Laboratory, 24–27 November 1958*. Volume 2. London: Her Majesty's Stationery Office.

———. 1961. *An Approach to Cybernetics*. New York: Harper & Brothers.

———. 1962. "Musicolour." In *A Scientist Speculates*, edited by I. M. Good. London: Heinemann.

———. 1966. "Comments on the Cybernetics of Ethical Psychological and Sociological Systems." In *Progress in Biocybernetics*, edited by J. P. Schade. Volume 3. Amsterdam: Elsevier.

———. 1971. "A Comment, a Case History and a Plan." In *Cybernetics, Art and Ideas*, edited by J. Reichardt. London: Studio Vista: London.

———. 1975. *Conversation, Cognition and Learning: A Cybernetic Theory and Methodology*. Amsterdam: Elsevier.

———. 1982b. "SAKI: Twenty-five Years of Adaptive Training into the Microprocessor Era." *International Journal of Man-Machine Studies* 17: 69–74.

Pask, Gordon, and S. Curran. 1982a. *Microman: Computers and the Evolution of Consciousness*. New York: Macmillan.

Pask, Gordon, Robin McKinnon-Wood, and E. Pask. 1961. "Patent Specification (866279) for Apparatus for Assisting an Operator in a Skill." Available at http://v3.espacenet.com/textdoc?DB=EPODOC&IDX=CA624585&F=0&QPN=CA624585.

Price, C. 2001. Gordon Pask, *Kybernetes* 30, no. 5–6: 819–20.

Scott, B. 1980. "The Cybernetics of Gordon Pask, Part 1." *International Cybernetics Newsletter* 17: 327–36.

Thompson, Adrian. 1997. "An Evolved Circuit, Intrinsic in Silicon, Entwined with Physics." *International Conference on Evolvable Systems (ICES96)*, LNCS 1259, Berlin: Springer.

Tinbergen, Nikolaas. 1951. *The Study of Instinct.* Oxford: Oxford University Press.

Zeeuw, G. de. 2001. "Interaction of Actors Theory." *Kybernetes* 30, no. 5–6: 971–83.

9 Santiago Dreaming

Andy Beckett

When Pinochet's military overthrew the Chilean government more than thirty years ago, they discovered a revolutionary communication system, a "Socialist Internet" connecting the whole country. Its creator? An eccentric scientist from Surrey. This article recounts some of the forgotten story of Stafford Beer.

During the early seventies, in the wealthy commuter backwater of West Byfleet in Surrey, a small but rather remarkable experiment took place. In the potting shed of a house called Firkins, a teenager named Simon Beer, using bits of radios and pieces of pink and green cardboard, built a series of electrical meters for measuring public opinion. His concept—users of his meters would turn a dial to indicate how happy or unhappy they were with any political proposal—was strange and ambitious enough. And it worked. Yet what was even more jolting was his intended market: not Britain, but Chile.

Unlike West Byfleet, Chile was in revolutionary ferment. In the capital, Santiago, the beleaguered but radical Marxist government of Salvador Allende, hungry for innovations of all kinds, was employing Simon Beer's father, Stafford, to conduct a much larger technological experiment of which the meters were only a part. This was known as Project Cybersyn, and nothing like it had been tried before, or has been tried since.

Stafford Beer attempted, in his words, to "implant" an electronic "nervous system" in Chilean society. Voters, workplaces, and the government were to be linked together by a new, interactive national communications network, which would transform their relationship into something profoundly more equal and responsive than before—a sort of Socialist Internet, decades ahead of its time.

When the Allende administration was deposed in a military coup, the thirtieth anniversary of which falls this Thursday (11 September, 2003),[1] exactly how far Beer and his British and Chilean collaborators had got in constructing their high-tech utopia was soon forgotten. In the many histories of the endlessly debated, frequently mythologized Allende period, Project Cybersyn hardly gets a footnote. Yet the personalities involved, the amount they achieved, the scheme's optimism and ambition, and perhaps, in the end, its impracticality contain important truths about the most tantalizing left-wing government of the late twentieth century.

Stafford Beer, who died in 2002, was a restless and idealistic British adventurer who had long been drawn to Chile. Part scientist, part management guru, part social and political theorist, he had grown rich but increasingly frustrated in Britain during the fifties and sixties. His ideas about the similarities between biological and man-made systems, most famously expressed in his later book, *The Brain of the Firm* (1981), made him an in-demand consultant with British businesses and politicians. Yet these clients did not adopt the solutions he recommended as often as he would have liked, so Beer began taking more contracts abroad.

In the early sixties, his company did some work for the Chilean railways. Beer did not go there himself, but one of the Chileans involved, an engineering student named Fernando Flores, began reading Beer's books and was captivated by their originality and energy. By the time the Allende government was elected in 1970, a group of Beer disciples had formed in Chile. Flores became a minister in the new administration, with responsibility for nationalizing great swathes of industry. As in many areas, the Allende government wanted to do things differently from traditional Marxist regimes. "I was very much against the Soviet model of centralization," says Raul Espejo, one of Flores's senior advisers and another Beer disciple. "My gut feeling was that it was unviable."

But how should the Chilean economy be run instead? By 1971, the initial euphoria of Allende's democratic, nonauthoritarian revolution was beginning to fade; Flores and Espejo realized that their ministry had acquired a disorganized empire of mines and factories, some occupied by their employees, others still controlled by their original managers, few of them operating with complete efficiency. In July, they wrote to Beer for help.

They knew that he had left-wing sympathies, but also that he was very busy. "Our expectation was to hire someone from his team," says Espejo. But after getting the letter, Beer quickly grew fascinated by the Chilean situation. He decided to drop his other contracts and fly there. In West Byfleet, the reaction was mixed: "We thought, 'Stafford's going mad again,'" says Simon Beer.

When Beer arrived in Santiago, the Chileans were more impressed. "He was huge," Espejo remembers, "and extraordinarily exuberant. From every pore of his skin you knew he was thinking big." Beer asked for a daily fee of $500—less than he usually charged, but an enormous sum for a government being starved of U.S. dollars by its enemies in Washington—and a constant supply of chocolate, wine, and cigars.

For the next two years, as subordinates searched for these amid the food shortages, and the local press compared him to Orson Welles and Socrates, Beer worked in Chile in frenetic bursts, returning every few months to England, where a British team was also laboring over Cybersyn. What this collaboration produced was startling: a new communications system reaching the whole spindly length of Chile, from the deserts of the north to the icy grasslands of the south, carrying daily information about the output of individual factories, about the flow of important raw materials, about rates of absenteeism and other economic problems.

Until now, obtaining and processing such valuable information—even in richer, more stable countries—had taken governments at least six months. But Project Cybersyn found ways round the technical obstacles. In a forgotten warehouse, five hundred telex machines were discovered which had been bought by the previous Chilean government but left unused because nobody knew what to do with them. These were distributed to factories, and linked to two control rooms in Santiago. There a small staff gathered the economic statistics as they arrived, officially at five o'clock every afternoon, and boiled them down using a single precious computer into a briefing that was dropped off daily at La Moneda, the presidential palace.

Allende himself was enthusiastic about the scheme. Beer explained it to him on scraps of paper. Allende had once been a doctor and, Beer felt, instinctively understood his notions about the biological characteristics of networks and institutions. Just as significantly, the two men shared a belief that Cybersyn was not about the government spying on and controlling people. On the contrary, it was hoped that the system would allow workers to manage, or at least take part in the management of, their workplaces and that the daily exchange of information between the shop floor and Santiago would create trust and genuine cooperation—and the combination of individual freedom and collective achievement that had always been the political holy grail for many left-wing thinkers.

It did not always work out like that. "Some people I've talked to," says Eden Miller, an American who is writing her Ph.D. thesis partly about Cybersyn, "said it was like pulling teeth getting the factories to send these statistics."[2] In the feverish Chile of 1972 and 1973, with its shortages and strikes and jostling government initiatives, there were often other priorities. And often the workers were not willing or able to run their plants: "The people Beer's scientists dealt with," says Miller, "were primarily management."

But there were successes. In many factories, Espejo says, "Workers started to allocate a space on their own shop floor to have the same kind of graphics that we had in Santiago." Factories used their telexes to send requests and complaints back to the government, as well as vice versa. And in October 1972, when Allende faced his biggest crisis so far, Beer's invention became vital.

Across Chile, with secret support from the CIA, conservative small businessmen went on strike. Food and fuel supplies threatened to run out. Then the government realized that Cybersyn offered a way of outflanking the strikers. The telexes could be used to obtain intelligence about where scarcities were worst, and where people were still working who could alleviate them. The control rooms in Santiago were staffed day and night. People slept in them—even government ministers. "The rooms came alive in the most extraordinary way," says Espejo. "We felt that we were in the center of the universe." The strike failed to bring down Allende.

In some ways, this was the high point for Cybersyn. The following year, like the government in general, it began to encounter insoluble problems. By 1973, the sheer size of the project, involving somewhere between a quarter and half of the entire nationalized economy, meant that Beer's original band of disciples had been diluted by other, less idealistic scientists. There was constant friction between the two groups. Meanwhile, Beer himself started to focus on other schemes: using painters and folk singers to publicize the principles of high-tech socialism; testing his son's electrical public-opinion meters, which never actually saw service; and even organizing anchovy-fishing expeditions to earn the government some desperately needed foreign currency.

All the while, the right-wing plotting against Allende grew more blatant and the economy began to suffocate as other countries, encouraged by the Americans, cut off aid and investment. Beer was accused in parts of the international press of creating a Big Brother–style system of administration in South America. "There was plenty of stress in Chile," he wrote afterward. "I could have pulled out at any time, and often considered doing so."

In June 1973, after being advised to leave Santiago, he rented an anonymous house on the coast from a relative of Espejo. For a few weeks, he wrote and stared at the sea and traveled to government meetings under cover of darkness. On September 10, a room was measured in La Moneda for the installation of an updated Cybersyn control center, complete with futuristic control panels in the arms of chairs and walls of winking screens. The next day, the palace was bombed by the coup's plotters. Beer was in London, lobbying for the Chilean government, when he left his final meeting before intending to fly back to Santiago and saw a newspaper billboard that read, "Allende assassinated."

The Chilean military found the Cybersyn network intact, and called in Espejo and others to explain it to them. But they found the open, egalitarian aspects of the system unattractive and destroyed it. Espejo fled. Some of his colleagues were not so lucky. Soon after the coup, Beer left West Byfleet, his wife, and most of his possessions to live in a cottage in Wales. "He had survivor guilt, unquestionably," says Simon.

Cybersyn and Stafford's subsequent, more esoteric inventions live on in obscure Socialist websites and, more surprisingly, modern business school teachings about the importance of economic information and informal working practices. David Bowie, Brian Eno, and Tony Blair's new head of policy, Geoff Mulgan, have all cited Beer as an influence.

But perhaps more importantly, his work in Chile affected those who participated. Espejo has made a good career since as an international management consultant. He has been settled in Britain for decades. He chuckles urbanely at the mention of Pinochet's arrest in London five years ago. Yet when, after a long lunch in a pub near his home in Lincoln, I ask whether Cybersyn changed him, his playful, slightly professorial gaze turns quite serious. "Oh yes," he says. "Completely."

Notes

1. This article first appeared in *The Guardian* newspaper, London, on September 8, 2003. It is reproduced with permission. Copyright Guardian News and Media Ltd.

2. *"The State Machine: Politics, Ideology, and Computation in Chile, 1964–1973,"* Ph.D. diss., MIT, 2005. Since this article was written, Miller has finished her Ph.D. and changed her surname: she is now Eden Medina.

References

Beckett, A. 2003. *Pinochet in Piccadilly*. London: Faber.

Beer, S. 1981. *The Brain of the Firm: The Managerial Cybernetics of Organization*. New York: Wiley.

10 Steps Toward the Synthetic Method: Symbolic Information Processing and Self-Organizing Systems in Early Artificial Intelligence Modeling

Roberto Cordeschi

Marvin Minsky (1966) defined as a "turning point" the year that witnessed the simultaneous publication of three works, by Kenneth Craik (1943), Warren McCulloch and Walter Pitts (1943) and Arturo Rosenblueth, Norbert Wiener, and Julian Bigelow (1943). The year 1943 is customarily considered as the birth of cybernetics. Artificial Intelligence (AI) was officially born thirteen years later, in 1956. This chapter is about two theories of human cognitive processes developed in the context of cybernetics and early AI. The first theory is that of the cyberneticist Donald MacKay, in the framework of an original version of self-organizing systems; the second is that of Allen Newell and Herbert Simon (initially with the decisive support of Clifford Shaw) and is known as information-processing psychology (IPP). The latter represents the human-oriented tendency of early AI, in which the three authors were pioneers.

Elsewhere I have shown how IPP is situated in the context of the birth of computer science, of cybernetics and AI (see Cordeschi 2002, chapter 5). There are also popular reconstructions of the history of AI for purposes different from mine (see McCorduck 1979; Crevier 1993). Here the aim is to analyze epistemological topics of IPP in greater detail, above all during what I call its "classical" phase. This phase runs from about the mid-1950s (the years of the building of the first simulation programs) to the mid-1970s (Newell and Simon's Turing Lecture [Newell and Simon 1976] dates from 1975, and contains the formulation of the Physical Symbol System Hypothesis). Subsequently, the interests of Newell and Simon diverged, even as IPP's influence spread into cognitive science.

My interest in MacKay's theory is due to the fact that, among the cyberneticists, he was the one most sensitive to the epistemological problems raised by cybernetics, and one of the most interested in higher cognitive processes. These are also the subject matter of Newell, Shaw, and Simon's researches. In essence, MacKay's self-organizing system theory and Newell,

Shaw and Simon's symbolic information-processing theory are *process theories*. In both cases they are processes postulated to explain higher human cognitive activities, such as decision making and choice, attention, planning, complex problem solving, and, in the case of MacKay, consciousness (a topic I shall not deal with in this chapter).

MacKay introduced the study of these processes by extending the original behaviorist definition of adaptiveness and purposefulness given by cyberneticists (starting with the 1943 article by Roseblueth, Wiener, and Bigelow). In IPP, the study of these processes found a basis in the revision undertaken by Simon in the 1940s of the theory of choice, a topic shared by disciplines such as the theory of games and operations research (OR). Both MacKay and Simon introduced the analysis of processes and mechanisms underlying human choice, in particular when information is uncertain and incomplete. Both theories make use of artifacts as models of these processes and mechanisms, and thus represent steps toward the "synthetic method"—actually, *different* steps, as the two theories use very different artifacts for cognitive modeling: self-organizing systems in the case of MacKay, and computer programs in the case of IPP.

Simon's shift in interest to the context of decision making occurred in the period following World War II, and in particular in the 1950s and 1960s. It was a consequence of his awareness that only in this way could models of the processes of choice, planning, and problem solving be successful in practical applications in industry and government and military agencies. It has been shown elsewhere (Cordeschi 2002; Cordeschi and Tamburrini 2005) how, starting in the years preceding World War II, and indeed during World War I, the synthetic method developed as a mixture of epistemological issues (a modeling methodology with the aim of explaining human behavior and that of living organisms in general) and of practical applications with possible military implications (a supporting tool for human decision making and in some cases a tool for "usurping" it, to use Norbert Wiener's term). It is no coincidence that during the cybernetics era the predictor of an automatic anti-aircraft system is the most frequently mentioned example of a self-controlling and purposive device. These were the war years, and "goal-seeking missiles were literally much in the air," as W. Grey Walter recalled in his account of a meeting in the early 1940s with Kenneth Craik, who was then engaged in investigating the aiming errors of air gunners on behalf of the British government.

AI and IPP shared this context. By the 1950s and 1960s the world war was over, but the cold war had begun. Digital computers and programming science now allowed the new sciences to tackle choice as it is actually made by human beings, who are not usually fully informed decision makers when they deal with real-life problems, or operate in complex contexts. Simon's shift of interest may be viewed as lying at the intersection between OR and AI. Some exhaustive analyses insist on this point (see, for example, Mirowski 2002), and on several occasions Simon himself returned to discuss the relationship between OR and AI in these same terms. He was particularly explicit both in recognizing the potential users and funders of applications of the new decision-making theories (industry, government, and military agencies) and in emphasizing how these theories enabled applications that could not be dealt with by their predecessors (Simon and Associates 1986, p. 33):

The study of decision making and problem solving has attracted much attention through most of this century. By the end of World War II, a powerful prescriptive theory of rationality, the theory of subjective expected utility (SEU), had taken form; it was followed by the theory of games. The past forty years have seen widespread applications of these theories in economics, operations research, and statistics, and, through these disciplines, to decision making in business and government. The main limitations of SEU theory and the developments based on it are its relative neglect of the limits of human (and computer) problem-solving capabilities in the face of real-world complexity. Recognition of these limitations has produced an increasing volume of empirical research aimed at discovering how humans cope with complexity and reconcile it with their bounded computational powers.

Following this recognition, the issues raised by IPP remain to be examined in the historical context of the birth of the new theories of decision making and of computer science, as well as of the further development of more traditional disciplines, such as psychology and neurology (or neuroscience, as we call it today). This analysis of IPP occupies more of the present chapter than does MacKay's theory; this is because IPP, partly for the reasons mentioned, has enjoyed a greater degree of development and dissemination. First, IPP had less difficulty in addressing the problems raised by the synthetic method as far as human problem-solving processes were concerned. IPP also played a leading role in the field in which it was possible to obtain the most promising results at the time, namely heuristic programming. Second, IPP promptly entered into the discussions among psychologists in those years regarding the epistemological issues of their research.

MacKay's theory suffered the same fate as various cybernetic research programs, which were superseded by the success of early AI heuristic programming. However, the case of MacKay is particularly interesting because his theory, compared with other cybernetics research programs, was concerned with higher cognitive processes and not only with perceptual and low-level aspects of cognition, as I show in the following section.

In the section after that (p. 230) I examine the synthetic method as it was seen by MacKay and the founders of IPP. I shall dwell in particular on the latter, in order to show how, once the use of the computer as metaphor is rejected, the computer becomes a tool for the building and testing of theories of the mind. However, the thesis that computers merely simulate minds is rather weak. There is a stronger thesis, according to which the computer grasps the essence of mind. Here cognition is not simply *simulated* by computation—it *is* computation (see Simon 1995a for a particularly explicit statement). In the classical phase of IPP, which represents the main subject of this chapter, this strong thesis remained in the background, and I shall not discuss its implications further here.

The section following that (pp. 237–44) is entirely about IPP. In recent times, Newell and Simon's research has often been identified with "good old-fashioned AI," or GOFAI (Haugeland 1985). In general, I agree with Aaron Sloman (2002, p. 126): "This term . . . is used by many people who have read only incomplete and biased accounts of the history of AI." In particular, I suggest that IPP, in the specific historical context I am attempting to reconstruct, was the first intellectual enterprise to tackle the *entire* range of methodological and epistemological issues, which were then inherited by cognitive modeling and have continued to be used right up to the present day.

The topics introduced in these three sections lead into the topic of the final section. Here, in addition to mentioning certain developments subsequent to the classical phase of IPP, I attempt to situate both MacKay's theory and IPP within the frameworks of classical cognitive science and also of the new cognitive science that followed the readoption of neural nets in the 1980s. Both the limits of MacKay's original position in the context of early AI and its renewed vitality in the context of the new cognitive science will then become clear. At the same time I shall examine the original position of IPP and some of its limitations in relation to the new cognitive science.

I shall not discuss this topic with the intention, common in this type of reconstruction, of seeking contrasts between opposing paradigms (symbolic vs. subsymbolic, symbolic vs. situated, and so forth). The moral of this story is somewhat different. As things stand, different aspects of cognition are captured with varying degrees of success and at different levels by modeling approaches that differ greatly among themselves. A comparison of these approaches cannot be viewed as a battle on opposing fronts. This would make it impossible to objectively evaluate the strength and the limits of the synthetic method in these different approaches. Randall Beer (1998), referring to one of these battles, the "battle between computational and dynamical ideologies," decried the fact that the subjects usually examined are not "experimentally testable predictions, but rather competing intuitions about the sort of theoretical framework that will ultimately be successful in explaining cognition." He concluded, "The careful study of concrete examples is more likely to clarify the key issues than abstract debate over formal definitions" (p. 630). I believe this is a conclusion that should be endorsed.

Process Theories

If we have a rat in a very small maze, with cheese at one branch point, and if we give the rat plenty time to explore, we can predict where he will finally go without any very deep knowledge of rat psychology. We simply assume that he likes cheese (a given utility function) and that he chooses the path that leads to cheese (objective rationality). If we now transfer the rat to a maze having a number of pieces of cheese in it, but a maze that is several orders of magnitude larger than the largest maze he could possibly explore in a rat's lifetime, then the prediction is more difficult. We must now know how a rat solves problems in order to determine where he will go. We must understand what determines the paths he will try and what clues will make him continue along a path or go back.
—Herbert A. Simon, 1963

It all began in 1943, when Wiener, Bigelow, and Rosenblueth published their seminal article stating the equivalence between the teleological behavior of organisms and the behavior of negative feedback machines. As we have seen, the predictor of an automatic anti-aircraft system became the most frequently cited example of this kind of machine. It sums up the fundamental features underlying the comparison between organisms and machines: both the organism and the machine change their behavior as the conditions of the external environment change, and in doing this they exhibit goal-directedness or purposefulness.

This definition of purposefulness immediately gave rise to numerous discussions (some of which, including their premises, are mentioned in Cordeschi 2002, chapter 4). According to this definition, and that proposed some years previously by William Ross Ashby, purposefulness is defined solely in terms of the pairing of the observable behavior with the external environment. Each "subjective" and therefore "vague" element (Ashby 1940) is eliminated by this "narrow" definition of purposefulness. The three authors defined as "behavioristic" their notion of teleological behavior, as it involves only an observer who studies the relationships that a system maintains with the environment in which it is located, and its responses or "outputs" as a function of certain stimuli or "inputs" (Rosenblueth, Wiener, and Bigelow 1943, p. 24). As Ashby was later to conclude, cybernetics merely sheds some light on the *objective* aspects of behavior, namely those considered "from the point of view of an observer *outside* the system," without telling us anything about the *subjective* aspects of the system itself. By way of exception, Ashby mentioned Donald MacKay's speculations concerning a "system that 'observes' itself internally," in which the notion of negative feedback plays a central role (Ashby 1962, p. 310). Let us consider this point in some detail.

MacKay rejected the behaviorist conception of the organism as a collection of black boxes, the behavior of which may be mimicked in purely functional (or input-output) terms by an "artefact," as he called it (the latter also being likened to a collection of black boxes). Instead, organism and artifact are viewed as IFSs (information flow systems): their actual specific physical composition is irrelevant, but the internal organization and structure they share are crucial to explain adaptive and purposive forms of behavior. In this case the language of information and control is a "useful tool" in the process of "erecting, testing and . . . demolishing the psychological theory." It is also a "common language" for psychology and neurophysiology as it may be used in either field (MacKay 1956, pp. 30–31). I shall return to this point in the next section.

MacKay's assumption was that this common language allows a *neutral* level of description of purposefulness to be identified, one that is a common level for both natural and artificial IFS's. In its interaction with the environment, the IFS exhibits an adaptive and purposive behavior through negative feedback. The latter is able to eliminate the discrepancy between the "symbolic representation" of the perceived state and the "symbolic representation" of the goal state. The definition is deliberately couched in terms of representing states so that it can be applied to different systems, those that pursue a physical object, and also to systems that pursue an abstract object, in other words, "to a self-guided missile chasing an aircraft, and a man chasing the solution to a crossword puzzle" (MacKay 1956, p. 34). To acknowledge this possibility does not, however, mean neglecting the different functional organizations and structures of the two systems.

A self-guided missile is one of the simpler instances of a servomechanism as identified by Wiener. In this case, the IFS is a system—let us call it system A—that is "fully informed" in the sense that the discrepancy, as indicated by the error signal, automatically prescribes "the optimal corrective response," as MacKay put it. In this case the only degrees of freedom the IFS possesses are those determined by the error signal.

The crossword puzzle solver in MacKay's example is a much more complex instance. In this case, the IFS is a system—let us call it system B—that is normally not fully informed about the environment, and the discrepancy is not able to prescribe the optimal corrective response. In the course of the activity involving the (gradual) reduction of the discrepancy, the IFS is assisted by the memory of its past activity. The system *selects* the input patterns that are closest to the desired one (the goal), and eliminates the others. The selected patterns represent the "internal symbolic vocabulary" of the IFS, which thus behaves "as if it believes what it has perceived." In the case of system B, statistical predictions may be made concerning its future activity, that is, concerning the probability of evoking certain subsequent patterns. The system's beliefs are defined on the whole by a "matrix of transition-probability," which characterizes the IFS as a "statistical 'probabilistic' self-organizing system." The "imitative" mechanism underlying pattern selection also underlies the system's self-observational ability, which is probably what Ashby had in mind. (As for MacKay, he went on to point out that this ability of the system could serve as the basis for several forms of consciousness of increasing complexity: see Cordeschi, Tamburrini and Trautteur 1999).

In the case of the problem solver or system B, therefore, perception is an active process that, insofar as it is selective as stated, involves attention. Perception is not a kind of passive "filter" based on a template-matching method, as in the simple case of system A. Furthermore, the complexity of the problem-solving task is such that, in the case of system B, it is necessary to take into account that the IFS uses not only logical reasoning processes to attain its goal but also procedures that help it in "crossing logical gaps." In other words, system B can choose from among alternative courses of action to try and reduce the discrepancy, and does so through "statistically-controlled trial and error" (see MacKay 1951, 1952, 1959).

A man or artefact seeking to prove a geometrical theorem...is kept in activity by recurrent evidence from his test-procedure that his latest method does not work: his response to the feedback of this information (as to the discrepancy between "the outcome of my present method" and "the required conclusion") is to try new methods—to adopt new subsidiary purposes. (MacKay 1965, p. 169)

This kind of problem solver, capable of making alternative choices, was at the focus of the analysis of purposive behavior carried out by R. L. Ackoff and C. W. Churchman, two pioneers of management science and OR in the United States. Like MacKay's, their starting point is the 1943 analysis by Rosenblueth, Wiener, and Bigelow, although the conclusions they reach are quite different from MacKay's (Churchman and Ackoff 1950). Ackoff and Churchman independently distinguished the two systems described by MacKay, although they held that *genuinely* purposive behavior cannot be identified with that of a system with a zero degree of freedom, such as MacKay's system A. The latter was nevertheless *the only one in which they recognized the presence of negative feedback*. Such a system displays a single type of behavior in a specified environment, and, owing to the presence of feedback, modifies its goal-directed behavior, but only if the external environment *changes*; if it does not change, its behavior remains the same (as, for instance, in the self-guided missile).

Conversely, a system such as B is able to choose between alternative courses of action, and so can display many different behavior sequences in the same environment in order to attain the goal. In this case, as Ackoff and Churchman point out, *from the point of view of the observer*, the environment in which the system is located *does not change*. On the one hand, genuinely purposive behavior is characterized by the relative unpredictability of the system due to the system's ability to make alternative choices with the same goal, regardless of the change in the environment. On the other hand, the presence of feedback is not required in the analysis of such a behavior: the environment that does not change is precisely the one the observer (the experimenter) is interested in, whether he is a psychologist or a social scientist—think of a rat that is trying to find its way through a maze, or MacKay's IFS solving a crossword puzzle, or the mechanical chess player to which Ackoff and Churchman refer. (However, in later work they seem to offer a different judgment as to the presence of feedback in such systems: see Churchman, Ackoff, and Arnoff 1957).

To clarify matters, it is worth mentioning the possibility of distinguishing between *two* kinds of feedback, a distinction implicit in MacKay's claims for the IFS. The first of these is the one linked to the ongoing activity in the system. This feedback normally occurs when the system triggers a response, and it remains throughout the response: it is the feedback in system A. The second is a delayed feedback based on the effect of the response, the one that allows the generation of imitative patterns: it is the feedback in system B. Now, while the environment does not change *from the* (objective) *point of view of the observer* ("in the social scientist sense of 'sameness,'" in the words of Ackoff and Churchman), *from the* (subjective) *point of view of the problem solver* the external environment changes constantly. It is precisely the feedback from the effect of the response that affects the progressive reorganization of the internal representation of the problem by the problem solver. This process is made possible by the capacity of the problem solver to apply test procedures, to create subgoals and so on, as MacKay put it when he described the IFS as system B. Therefore, to limit oneself to considering the point of view of the observer, but without taking into consideration the problem solver's point of view, amounts to a failure to explain the choice mechanisms or processes used by the problem solver during its interaction with the environment, in particular with a complex environment.

It was precisely the processes of choice that, during these years, attracted the interest of Herbert Simon, who worked in the field of management science and OR. His position is the opposite of Ackoff and Churchman's: he shares with MacKay the interest in the structure and the functional organization of the problem solver, and thus in its processes and resources interacting with complex environments. The problem solver is usually not fully informed about such environments, nor can it be: take the example of chess and the combinatorial explosion of legal moves, or any real-life problem, which is always tackled by the problem solver on the basis of incomplete information. What counts for Simon is precisely the *subjective* point of view of the problem solver, that is, the environment as represented *by the problem solver*, and the perceptual and cognitive processes involved in this.

[This] requires a distinction between the objective environment in which the economic actor "really" lives and the subjective environment that he perceives and to which he responds. When this distinction is made, we can no longer predict his behavior ... from the characteristic of the objective environment; we also need to know something about his perceptual and cognitive processes. (Simon 1963, p. 710)

It is the emphasis on these processes that justifies the introduction of psychology into management science and economics, as Simon put it (hence his reference to the "economic actor"). This emphasis entails a shift of interest within the same family of disciplines in relation to their concern with decision-making behavior. This shift is based on Simon's renunciation of the normative approach of game theory in studying choice, the approach that had been adopted by authors such as Ackoff and Churchman in the context of OR. Briefly, the normative approach consists in studying the choice (or the strategy) that the agent *ought to* use objectively in order to maximize the likelihood of finding an optimal solution (Ackoff 1962). Simon shifted the attention to the study of the choice (or the strategy) that the agent normally uses insofar as this choice is conditioned by his own view of the environment in which he is operating, and about which he customarily has only incomplete information.

The agent as decision maker or problem solver was not studied by Simon from the standpoint of objective rationality, which is the essential feature of the *Homo oeconomicus* of classical economics, who is interested in the calculus of a utility function. Instead of this "ideal type" accepted by game theory and OR, Simon took a "real" agent endowed with *bounded rationality* (Simon 1947). The chess-player metaphor, frequently used in the decision-making theories of the time, remained in the foreground, but was not based on the "entirely mythical being" (as Simon was to say later) of game theory. The internal limits of the real agent and the complexity of the environment, which are clearly exemplified in the game of chess, release him from the constraint of having to find and use the best strategy in his choice of moves, and allow him to use suboptimal and incompletely informed strategies which are more or less "satisficing," to use Simon's term.

The agent's limits also involve some perceptual aspects. As in the case of MacKay's IFS, perception is not viewed as a passive activity, or as a filter:

Perception is sometimes referred to as a "filter." This term is ... misleading ...: it implies that what comes through into the central nervous system is really quite a bit like what is "out there." In fact, the filtering is not merely a passive selection of some part of a presented whole, but an active process involving attention to a very small part of the whole and exclusion, from the outset, of almost all that is not within the scope of attention. (Simon 1963, p. 711)

Unlike MacKay's process theory, the system that employs these perceptual and cognitive processes is not a statistical (analogue) self-organizing system, an IFS of which men and artifacts are instantiations (at least as far as the aspects considered are concerned). The system is viewed here as an information-processing system (IPS), a genus of which men and digital computer programs are species (Newell and Simon 1972, p. 870).

The structure of an IPS is now familiar, as it shares with early AI the notion of a computer program. The IPS, in its "psychological" version, which made it the principal player in IPP, includes several constraints. Briefly, an IPS possesses a sensory-motor apparatus through which it communicates with the external environment, or "task environment." It displays an adaptive and goal-directed behavior conditioned by the complexity of the environment and by its internal limits as an IPS. These limits are due to several of its structural features, which are variable but which recur in each individual case. These features are as follows: the IPS essentially operates in a serial mode; the small number of elementary information processes are very quickly executed; it has a rapidly accessible but limited capacity short-term memory and a slowly accessible but potentially infinite-capacity long-term memory.

In this formulation, the task environment is the objective environment of the observer or experimenter (presumably the one Ackoff and Churchman were referring to), whereas the subjective environment of the IPS, that is, the task environment as represented by it, is the "problem space." The latter includes, as Newell and Simon (1972, p. 59) put it, "the initial situation presented to him, the desired goal situation, various intermediate states, imagined or experienced, as well as any concepts he uses to describe these situations to himself." In a sense, the problem space is the machine's idiosyncratic model of the environment so insightfully described by Margaret Boden (1978, pp. 128–31) in her comment on the seminal paper by Minsky (1968, pp. 425–32) on models. The problem space suggests to the IPS the satisfactory problem-solving strategies previously mentioned—in a word, the problem-solving heuristics. An example will clarify matters and allow me to make a few final remarks.

The principal heuristic embodied in the first IPP program, the now-legendary Logic Theorist (LT), designed to prove sentence-logic theorems, was the difference-elimination heuristic. It makes it possible to eliminate the discrepancy between the initial state (the starting expression) and the final or goal state (the expression to be proved). The intermediate expressions are generated by the program, selecting the rules of logic that produce expressions progressively more similar to the final, or goal state, expression; similarity is used as the cue for the solution. "At each step a feedback of the result [of the application of a rule is obtained] that can be used to guide the next step" (Newell and Simon 1972, p. 122). It is precisely this feedback from the effect of the response that was introduced by MacKay and rejected by Churchman and Ackoff. As suggested, this kind of feedback is crucial for a cognitive-process theory, because it underlies the continuous reorganization of both the problem representation of the IFS and the problem space of the IPS. We could not predict the problem solver's behavior in the large maze or task environment of logic without postulating this kind of feedback.

The activity of a program like LT was described by Simon as a true process of selecting winning patterns, those more similar to the final state, which are like "intermediate stable forms" similar to the species in biological evolution (Simon 1996, chapter 8). In general, just as it was for MacKay, this is a selective trial-and-error goal-guided procedure, which generates a set of subgoals, that is, a number of subproblems, the solution of which might lead to the solution of the original problem.

The above comparison between two process theories that are convergent in their subject matter (the *human* processes), but very different in their premises, should not sound odd. The hierarchical organization, the limits of the resources of the purposive agent vis-à-vis complex environments, and the emphasis on the active and subjective aspects of perception and cognition are all constraints stated by theories placing the emphasis on processes *really* used by human beings. Gordon Pask (1964) gave an enlightening exposition of the close analogy between the organization of an IPP program such as the General Problem Solver (GPS) and that of a self-organizing system as conceived by MacKay. It would seem that the complex hierarchical organization of feedback and problem-solving subroutines as described by MacKay found an actual realization in the field of computer modeling in the 1950s and 1960s.

As stressed, MacKay was mainly concerned with higher cognitive processes, including forms of self-consciousness, but he did not have suggestions as to how, on the basis of his theory of self-organizing systems, it might be possible to implement effective models of these processes. As a result, on the one hand he could not, and indeed did not, go much further in examining simple or general artifacts; on the other hand, he ended up by underestimating computer-programming techniques at a time when they were beginning to show their actual power, although in a rather primitive form. Comparing his system with the computer programs of the time, he always concluded that "digital computers are deliberately designed to show as few as possible of the more human characteristics," or that they "are not designed to resemble the brain" (MacKay 1951, p. 105; 1954, p. 402). He always saw the computer above all as a logical and deterministic machine, lacking the degrees of freedom that were guaranteed by the probabilistic and self-organizing features of his IFS. It was Minsky (1959) who, taking part in the discussion that followed MacKay's (1959) talk at the Teddington Symposium in 1958, pointed out how the incompletely informed, nonlogical, and nondeterministic aspects (as MacKay seemed to define them) of the human problem solver could be handled by the newborn technique of heuristic programming.

Computer Metaphor and Symbolic Model

Any kind of working model of a process is, in a sense, an analogy. Being different it is bound somewhere to break down by showing properties not found in the process it imitates or by not possessing properties possessed by the process it imitates.
—Kenneth Craik, 1943

The concepts of analogy and metaphor are often confused with the concept of model in the study of cognition. It is, however, possible to make a distinction. For Newell and Simon (1972, p. 5) this is a crucial step:

Something ceases to be metaphor when detailed calculations can be made from it; it remains metaphor when it is rich in features in its own right, whose relevance to the object of comparison is problematic.

For MacKay, too, such a distinction is crucial, but he states it with the rather different aim of showing that the computer is the *wrong metaphor* for the brain. On the contrary, according to Newell and Simon, the computer ceases to be used as a mere metaphor *the brain is a computer* when one is not interested in its possible capacity "to exhibit humanoid behavior," but is interested in "specifying with complete rigor the system of processes that make the computer exhibit behavior," that is, in building a model of such behavior in the form of a computer program—a necessarily symbolic model (Newell and Simon 1959, p. 4). However, a different and less frequent use in IPP of these terms and of the term "theory" is contained in Simon and Newell (1956).

Through this choice, IPP enters the history of modeling methodology in the behavioral sciences, which stretches through the entire twentieth century under the label "synthetic method" (Cordeschi 2002). Briefly, this methodology starts from a theory of phenomena regarding certain features of living organisms, such as adaptation, learning, problem-solving ability, or the possession of a nervous system. The theory is based essentially on observations of overt behavior and, whenever available, on verbal reports or on biological and neurophysiological data, and is used as a basis for building a functioning artifact—a "working model," as in Craik's comment. The artifact therefore embodies the explanatory principles (the hypotheses) of the theory and is considered a test of the theory's plausibility.

Of the many issues at stake, two are of principal importance. The first is the nature of the *constraints*, suggested by the theory, that the artifact must satisfy because it is a psychologically or a neurologically realistic model—in other words, because it can help explain the phenomenon under investigation. The second issue is the following: in the course of testing the theory through the model, it may be necessary to initiate a revision process that affects the model directly, but may also involve the theory. In this case the model takes on an important heuristic function: it may actually suggest new hypotheses to explore, and may have significant implications for the theory itself.

I have deliberately given a general statement of the synthetic method in order to avoid linking it to the type of artifact, the subject matter of the theory, or the choice of the constraints. In the preceding section I stressed that MacKay's IFS and Newell, Shaw and Simon's IPS share certain structural features: both are hierarchically organized, adaptive, purposive, not fully informed, and so on. In both cases, the subject matter is human processes, in particular, choice and problem-solving processes. Both systems are to be seen as abstract systems: their structural features must occur, to different extents, in all their physical realizations. Now the IFS is a self-organizing analogical machine and the IPS is a digital machine. What is the relationship between these different abstract systems and their physical realizations?

MacKay left us no concrete examples of artifacts as realizations of the IFS, except for schemata of very simple analogue machines embodying the general features of the IFS as constraints (the examples are always the same over the years: see MacKay 1969). MacKay is not always clear on this point, although it seems that the constraints must be viewed above all as referring to the nervous system of organisms at a very general level. Indeed, he distinguishes different levels of detail: there are neural network models, which start "from an idealized model of the nerve cell" (originating with McCulloch and Pitts 1943), but his preference is for "a statistical model of the whole [nervous] system" (MacKay 1954, p. 403). Therefore his interest is focused in the first instance on the abstract system, the IFS, which may be used to guide and test the statement of hypotheses regarding the functioning of both the nervous and the humoral system. The IFS has another advantage: through it, it is possible to apply the language of information and control not only to neurological but also to psychological and even psychiatric phenomena, and so a language exists that allows "conceptual bridge-building," which can suggest "testable physical hypotheses" (p. 405).

In the case of the IPP, the "synthetic" approach (Newell, Shaw, and Simon 1960) is stated more fully. Artifacts considered as realizations of the IPS are computer programs, or, rather, *certain* computer programs: the invariant features of the IPS are constraints that must be satisfied wholly or partly by the programs, in order to be at least candidates for symbolic models. As previously seen, these features regard the IPS's structure, the problem space and the task environment, as having been identified as invariant, or common to all human IPSs, by experimental research on problem solving in the psychology of thinking. The features concern a level different from the nervous system level (as I show later). They set the IPP generalizations, defined as laws of qualitative structure, that is, qualitative statements about the fundamental structure of the IPS—more similar to the laws of Darwinian evolution than to Newtonian laws of motion, and as such supporting predictions weaker than those that can be made from more quantitative theories (see Simon 1978; these laws had already been formulated by Newell and Simon 1976, with reference to Physical Symbol Systems instead of to IPSs, and were later extended by Simon 1995b). MacKay too believed that the data of his theory were "qualitative abstractions from function" (MacKay 1954, p. 405). However, in IPP the emphasis is on the individuals, on the way in which the *individual* IPSs display their specific abilities (in terms of errors, reaction times and so forth) with regard to specific tasks. Here is a new issue for the synthetic method: How can models of *individuals* be realized?

In order to gather specific data on individual problem solvers, Newell, Shaw, and Simon from the outset used recorded protocols of subjects "thinking aloud"—reporting the procedures they followed while solving a given problem (a logic problem, for instance). This is a method used by the psychologists of the Würzburg school in Europe, and rejected as introspective by behaviorist psychologists. For the founders of IPP, who based their work on that of Jerome Bruner (see Bruner et al. 1956), De Groot (1946), and Selz (1922) among others, this judgment is unjustified: the protocol method is "as truly behavior as is circling the correct answer on a paper-and-pencil test" (Newell, Shaw, and Simon 1958, p. 156; see also Simon 1981). In this case, programs must embody the heuristic procedures as these are inferred from the individual thinking-aloud protocols, and may thus be considered as candidate models of individuals.

The foregoing suggests that the constraints on symbolic models may vary as to their generality. They may refer to (a) the structural invariants shared by programs describing the behavior of the same subject over a range of problem-solving tasks, or the behavior of different subjects in a specific task environment, or (b) the specific features describing a single subject in a single task situation. According to the degree of generality of the constraints, it is possible to imagine a hierarchy of increasingly detailed symbolic models aimed at capturing increasingly microscopic aspects of cognition. How can these idiosyncratic models ultimately be validated?

Let us begin by recalling that not every program is a simulation model. For instance, certain complex-problem solving programs using OR algorithms rely on "brute-force" capacities and inhuman computing power: as such they do not satisfy the general constraints envisaged by IPP as laws of qualitative structures, and cannot be taken into consideration as models. Conversely, a program that does *not* use these problem solving procedures is, albeit at a low level of detail, at least interesting as a demonstration of the *sufficiency* of such processes to perform a given task. Such a program embodies at least the invariant features of the IPS, and may thus be taken "as a first model" (Newell 1970, p. 367).

Therefore, already at this level the Turing test is not a useful validation criterion for a program to be a candidate as a symbolic model, as it is a test of mere functional (input-output) equivalence. It is thus a "weak test"— one that is limited to comparing the final performance of a human being with that of a program. In IPP, the emphasis is instead placed on the *internal structure* of the IPS and on the *processes* underlying this performance (Newell and Simon 1959, p. 14).

> Comparison of the move chosen by a chess program with the moves chosen by human players in the same position would be a weak test. The program might have chosen its move by quite a different process from that used by humans.... *Similarity of function does not guarantee similarity of structure, or of process.* (emphasis added)

Then, on going to the level of specific processes simulated by idiosyncratic models, a much stronger test than that of Turing is required, which Newell and Simon described as follows. Imagine you mix together in an urn some traces of programs (referring to the solution of a logic problem, for instance) with the same number of human-generated protocols, both written in some standard form. Would a qualified observer be able to distinguish one from the others? The answer is dependent on the details of the problem-solving processes effectively simulated by the traces (Newell and Simon 1959, p. 13).

For example, LT is merely a sufficiency proof: "it could certainly not pass Turing's test [in its much stronger version] if compared with thinking-aloud protocols, although many of its quantitative features match those of human problem solvers" (Newell and Simon 1959, p. 17). These features consist essentially in the selectivity of the theorem proving strategies of LT (its heuristics), which exclude the use of brute force. LT therefore embodies parts of the IPS theory, but only at a fairly coarse level. To improve the level of approximation of LT to the explanandum (*how* a human being solves logic problems), it is necessary to initiate a procedure of matching between traces and protocols, with the aim of achieving greater adherence to the protocol in an increasing number of details, modifying the original program where necessary. The General Problem Solver, as a "revised version" of LT, is a case in point: unlike LT, "it was derived largely from the analysis of human protocols, and probably can pass Turing's test [in its much stronger version] for a limited range of tasks" (p. 19).

Apart from this optimistic evaluation of the old GPS, it is necessary to examine two things: (a) to what level of detail the model succeeds in reproducing processes that may be inferred from a protocol, and (b) to what extent the model may be revised in order to match the protocol ever more closely. The building and validation of the model follow a "forward motion": "its path will be helical rather than circular, producing successively more adequate explanations of behavior" (Paige and Simon 1966/1979, p. 203). Successes and gaps in the model suggest how it might be improved in the light of the data.

This helical path had already been described by John Holland, Nathaniel Rochester, and their coworkers in their computer simulation of Hebb's theory (Cordeschi 2002, pp. 176–77). MacKay (1954, p. 404) also described it very clearly:

> We can think of [the model] as a kind of template which we construct on some hypothetical principle, and then hold up against the real thing in order that the discrepancies between the two may yield us fresh information. This in turn should enable us to modify the template in some respect, after which a fresh comparison may yield further information, and so on.

The "hypothetical principle" leading to the choice of the constraints is also essential here to characterize a model above and beyond mere functional (or input-output) equivalence. The model must not be limited to *mimicking* the brain's behavior but must "*work internally on the same principles* as the brain." For MacKay, as already seen, it is the self-organizing probabilistic system that "could handle and respond to information in the same way" as the brain (MacKay 1954, pp. 402–3).

If the statement of general processes and mechanisms seems to be the distinctive flavor of MacKay's model making, the "idiographic" method of thinking-aloud protocols of individual subjects surely represents the "hallmark" of IPP (Newell and Simon 1972, p. 12). Thus, on the one hand IPP refuses to use the methods of nomothetic experimental psychology, as IPP models do not aim to imitate relatively uniform behaviors (the individual as an intersection of a statistically defined population); on the other, IPP is not concerned with ideal abstractions that may be isolated from the multiplicity of empirical data, but with individuals with their own specificity.

Different questions have been raised regarding the reliability of thinking-aloud protocols, and the actual testability of the correspondence between protocols and traces. As for the latter issue, there is no definitive technique for comparing a protocol with a trace in order to decide (a) which processes have actually been simulated in the individual model, and (b) how faithfully they have been simulated, given possible errors of omission (when the model does not possess "properties possessed by the process it imitates," as Craik's put it) or errors of commission (when the model possesses "properties not found in the process it imitates"). On the one hand, the "partial," "highly incomplete" nature of the protocols is on several occasions acknowledged by Newell and Simon; on the other hand, the reproducibility in the program of as many details of the protocol as possible remains to guarantee, as Newell emphasized, that one is not up against the usual metaphor, in which it is no longer clear what "breaks down," as Craik put it, or where. Indeed, some reflections on the analysis of sufficiency suggested to Newell (1970) a conclusion that appears critical of his own approach, and also of the previous neural network models (p. 388):

This same point [presently reached with respect to cognitive modeling and psychology] was reached some years ago with respect to neural modeling and physiology. No neural modeling is of much interest anymore, unless it faces the detail of real physiological data. The novelty and difficulty of the task undertaken by heuristic programming has tended to push the corresponding day of reckoning off by a few years. The development of symbolic systems that would behave in any way intelligently produced sufficiency analyses that were in fact relevant to the psychology of thinking. But the automatic relevance of such efforts seems to me about past.

Without doubt, the emphasis on sufficiency sometimes proved misleading. For example, the theorem-proving methods of J. Alan Robinson (1965) and the rather different methods of Hao Wang (1960) were also considered by Newell and Simon as proofs of sufficiency, simply because they were *selective* methods (for further details, see Cordeschi 1996). An additional factor is that the term "heuristic," when used by Newell and Simon to refer to problem-solving processes, qualified the human processes, but in the early AI community the term was more generally used to designate the efficiency of the programs' performance, regardless of the conditions of human-process simulation. AI researchers also referred to the sufficiency of the processes. Nevertheless, sufficiency, as the basis for increasingly strong successive approximations, has remained a crucial feature of cognitive models ever since the classical phase of IPP.

The vast amount of evidence available on protocol-based cognitive modeling throws into relief all of the above-mentioned difficulties. (This applies not only to Newell and Simon [1972] but also to subsequent research undertaken by Simon with different coworkers until his death). I shall deal with some other difficulties later, but in the meantime it should be emphasized that, as Zenon Pylyshyn (1979) pointed out, there is a "grain problem": it is not always clear at what level the trace and the protocol are to be compared. How can it be recognized that a strong proof of sufficiency has been reached in order to speak, or to begin to speak, of cognitive modeling and not, for instance, of the mere implementation of an AI program?

As I pointed out earlier, consideration of the constraints related to cognitive limits do not render the symbolic model a simplification of the "ideal case" type. Nevertheless, the model is clearly simplified vis-à-vis the biological or neurological constraints. We shall see in the next section how refraining from considering these constraints is a choice prompted by a particular concept of scientific research, which tries to guarantee a relative autonomy for mind science when viewed in terms of IPP. However, it must be stressed here that constraints considered "biological" by Newell and Simon also include those concerning emotional states. They are included among the limitations of the human problem solver that affect his performance; in fact, they appear to be deeply integrated into the problem solver's activity. Simon (1967) gave some hints for the simulation of those states of mind defined as emotional, but he simply seemed to define certain computer routines in mental terms. The conclusion was that these and other mental states acknowledged to be dependent on biological constraints (after-images, illusions, and so forth) may be neglected if it is allowed that IPP refers to a "normal situation ... in which these biological limits are not exceeded" (Newell and Simon 1972, p. 866). But how is a human IPS to be placed in a *normal* problem-solving situation? After all, in IPP any problem posed by the external environment is not "normal"; any given problem always tends to be ill-structured for an individual IPS.

Neglecting these kinds of biological constraints means that human problem-solving theory remains a "first approximation," because the theory would consider an IPS as a "whole person," not just as a "thinking person" (Simon 1996, p. 53). Of course, a model is always an approximation to the full phenomenon, and it provides scientific insight because it is an approximation. When stated in these terms (the "whole person"), the issue of constraints in cognitive models is prone to take on a paradoxical aspect, one that may be summarized in Pylyshyn's (1979) words: "If we apply minimal constraints we will have a Turing machine. If we apply all the constraints there may be no place to stop short of producing a human" (p. 49).

Psychology from Natural Science to a "Science of the Artificial"

The history of psychology ... is very much a history of changing views, doctrines, images about *what to emulate in the natural sciences*—especially physics.
—Sigmund Koch, 1959

As far as the great debates about the empty organism, behaviorism, intervening variables, and hypothetical constructs are concerned, we take these simply as a phase in the historical development of psychology.
—Allen Newell and Herbert Simon, 1972

Newell, Shaw, and Simon (1958), in an article published in the *Psychological Review* that may be considered their manifesto, proposed a comparison between their theory of human behavior simulation and other problem-solving theories that was actually a particular interpretation of the state of psychological research at the time. The three authors held that psychology was deadlocked by the "polarization," as they called it, of opposing aims proposed by Gestalt and behaviorist psychologists. The former supported a "question-oriented" psychology, that is, one committed to finding answers to the difficult questions raised by the theory (the problem of meaning, insight, imagination, creative problem solving); the others supported a rigorously operational, or "method-oriented," psychology, based on the observation of quantitatively assessable experimental data (overt behavior, trial-and-error learning, or psychophysiological data). On the one hand, important problems were addressed using methods lacking rigor; on the other, more easily testable problems were addressed using operational methods, but deeper and more difficult issues were often neglected.

Although believing it possible to accept the mechanistic legacy of behaviorism, Newell, Shaw, and Simon felt that, since an IPS consists of a complex and goal-guided hierarchy of mechanisms, IPP "resembles much more closely" several problem-solving theories proposed by the Gestalt and Würzburg psychologists (Newell, Shaw, and Simon 1958, p. 164).

I would like to show that two issues are involved in superseding polarization. On the one hand, it was necessary to deal with the cluster of methodological issues marking the development of scientific psychology—the entities to be postulated between stimulus and response, the level of psychological explanation, the relationship between psychology and neurology, and the (possible) reduction of the former to the latter. On the other hand, the outcome of this task was the reexamination of certain popular methodological claims in psychology, with the aim of giving psychology a new epistemological status with respect to that of natural science. In 1959 they wrote (Newell and Simon 1959, pp. 2–3):

Until a decade ago the only instruments we had for building theories about human behavior were the tools we borrowed and adapted from the natural sciences: operationalism and classical mathematics. And so inadequate are these tools to the task that a highly respected psychologist offered in earnest the doctrine that we must build a science without a theory—surely a doctrine of desperation.

Reading between the lines of this criticism, we see that it is directed against several of the main proponents of the various behaviorist tendencies in psychology, with all the methodological discussions of the period in the background.

The "doctrine of desperation" can apparently be identified with radical behaviorism a la Skinner. Elsewhere Newell and Simon are more explicit, and clearly characterize Skinner's position as one of radical skepticism concerning the unobserved entities that may be used in explaining mind, whether they consist of intervening variables or hypothetical constructs. (We shall not dwell on the distinction here, although it was much discussed by psychologists at the time: see Zuriff 1985). The description given by Skinner in terms of the "empty organism" ("empty of neurons and almost of intervening variables," as Boring had said, in a remark referred to in Newell and Simon 1972, p. 875) was deemed by them to be an oversimplification. The complexity and the dynamic features of human cognitive processes call for the introduction into the explanation of "hypothetical entities," such as intermediate constructs and processes between stimulus and response. This amounts to asking how to fill Skinner's empty organism with the right constructs and processes.

It was precisely operationism and the methods of logic and classical mathematics that, to varying degrees, served as inspiration to the American behaviorist psychologists, as a result of the growing prestige of European neopositivism that had been transplanted into the United States in the 1930s by the European "emigrants" (see Smith 1986). One philosopher who followed the teachings of Rudolf Carnap was Herbert Feigl, the Vienna "emissary," as Edwin Boring (1964) called him. He patiently explained, as Boring put it, the new operational positivism to the American psychologists who in the 1930s had enthusiastically embraced this epistemological proposal. Sigmund Koch, a former pupil of Feigl's, introduced an image of science and of scientific explanation like that set out in table 10.1, which soon became popular among American psychologists (Feigl 1948).

Without going into too much detail it can be said that this image may be broken down into a hierarchy of levels that, starting from the bottom, eventually attains the most fundamental and unifying level for each field of research, namely the level of the higher theoretical constructs regarding the microstructure of the phenomena located at a lower level. It is at the higher level that the causes of the lower-level phenomena may be identified. In this view, research proceeds by intertheoretic reduction—the reduction of one theory to another at a higher level—the consequence of which is explanatory unification. This has always been considered an advantage for science: if a theory is reduced to a "higher" theory, two apparently

Table 10.1
Feigl's Hierarchy of Scientific Explanation

Theories, second order	Still more penetrating interpretation (still higher constructs)
Theories, first order	Sets of assumptions using higher-order constructs that are results of abstraction and inferences; deeper interpretation of the facts than that rendered on the level of empirical law
Empirical laws	Functional relationships between relatively directly observable or measurable magnitudes
Descriptions	Simple account of individual facts or events (data) as more or less immediately observable

different orders of phenomena may be described by the latter, and the two phenomena are identical. One classical example is the reduction of optics to electromagnetism (a case, as we shall see, of complete reduction), which gives rise to the unification of concepts and laws from the two respective levels.

Nevertheless, according to the "liberalized" version of this hierarchy proposed by Carnap, scientific research is not exhausted by the reduction: in order to progress, macrolevel explanations do not have to wait for the microlevel ones to be developed. For example, chemistry developed at the level of the theory of valence before the latter was explained by atomic theory. By analogy with the case of atomic theory in chemistry, the (future) neurophysiological microexplanation in the sciences of behavior will play a similar role as regards its unifying power—at least this is to be hoped for.

Indeed, the unitary science neopositivist project took this evolutionary tendency of the individual "special" sciences to an extreme, pointing to the possibility, at least in principle, of the final step, as Carnap (1956) put it: the reduction of the concepts and laws of the individual sciences, and in particular neurophysiology, to the level of microphysics.

It was Edwin Tolman ("the farthest from the dominant S-R position" among American psychologists, as he was described by Newell and Simon 1972, p. 874) who distinguished "molar" behaviorism from "molecular" behaviorism. Both make use of intervening variables between stimulus and response, and thus fill the empty organism, although they do so at different levels: the first does it at the macrolevel of overt behavior, and the second at the microlevel of neurophysiology. In Tolman's view, the different approach to intervening variables followed by psychology on the one hand and physiology on the other does not raise any problems of competition between the two disciplines. Following the neopositivist approach described, Tolman describes a hierarchy of levels that are equally legitimate as far as explanatory power is concerned. Carnap's reductionist claim involving the most fundamental molecular level also seems to be valid for Tolman, too; this level provides the final explanation of the facts and laws of the molar level (Tolman, 1935/1958).

As Clark Hull put it in the language of his nomological-deductive approach, the postulates laid down by those working at the molar level will ultimately appear as theorems to those working at the molecular level (Hull 1943). For Hull, the existing gap between current neurological research and that required for an adequately grounded molar theory of behavior is presently insurmountable. However, to make the progress of psychology contingent upon the progress made in neurophysiology would be a paradox comparable to that of imagining the pioneers of the mechanics of macrophenomena having to delay the development of their discipline until the development of microphysics. Therefore, the molar psychologist, on the strength of a "division of labor" with the neurologist, "can still properly demand his own place in the sun" (Tolman 1935/1958, p. 102).

However, Tolman ended up accepting the idea of a certain utility of neurological hypotheses in psychology, in the form of a kind of "pseudo-brain model," while at the same time confirming his skepticism concerning the "premature" neurology of the time, "even if it be called 'Cybernetics'" (Tolman 1949, p. 48). Hull, long before cybernetics, had instead identified in the building of mechanical models of behavior a "kind of experimental shortcut" to the study of behavior in view of the future reduction of psychology to neurology (Cordeschi 2002, chapter 3). That left Skinner, suspicious as he was toward practically every kind of theoretical construct (the "doctrine of desperation"), to continue to reject modeling and neurocybernetic approaches as speculation on what happens "inside the head" (Skinner 1974, p. 217).

At the time of IPP's entry onto the scene in the mid-1950s, the image of science as shown in table 10.1 was still influential in its liberalized version, and the echo of the discussions among behaviorists in their interactions with the Gestalt psychologists was still strong. It was above all Simon, who was a pupil of Carnap's at the University of Chicago in the late 1930s, who attempted to situate IPP in relation to the traditional mind sciences, psychology and neurology, in a framework typical of the neopositivists' liberalized view of the science.

In Simon's (1961) modified diagram in table 10.2, as in Feigl's in table 10.1, the scientific explanation concerns theoretical constructs or entities located in a hierarchy of levels of variable generality. The main difference

Table 10.2
Simon's Diagram of Scientific Explanation

	Chemistry	Psychology of Thinking
Genetics		
Level 3 Biochemistry of genes	Nuclear physics	Biochemistry of brain processes
Level 2 Genes and chromosomes	Atomic theory	Neurophysiology
Level 1 Mendel: statistics of plant variety	Molecular reactions	Information processing

is that, in addition to the exemplification rendered canonical by the neo-positivist tradition and taken over by the behaviorist psychologists (genetics, chemistry), space is given to an information-processing level, the level of the newborn IPP.

Compared with radical behaviorism, the mind and mental processes again become the objects of psychological research, but at the particular level of symbolic information processing. As seen earlier (pp. 230–37), psychology is given a novel task: the building of detailed models of cognition based on information processes. The founders of IPP, therefore, propose an intermediate level of explanation, between overt behavior (a kind of level 0) and neurophysiology, with the aim of finding a new role for psychological research. It is at this level that the psychologist satisfies what Hebb (1951–52) called the "need of theory," that is, the need for explanatory entities and processes at a level other than that of neurophysiology. Without this, one ends up by merely inducing psychologists to "couch their theories in physiological language" (Simon and Newell 1964, p. 283). Karl Lashley is given as an example of this, whereas Hebb's "general methodological position is not inconsistent" with IPP, as Hebb does not insist on an exclusively physiological basis for psychological theory (p. 299).

It should be noted that the entities and symbolic processes of IPP (that is, elementary information entities and processes, heuristic processes, and so on) are viewed as genuinely hypothetical: they are molar vis-à-vis the neurophysiological processes, although as such they fill the empty organism, explaining to a first approximation the complexity of mental processes.

Shifting the role of the hypotheses from the neurophysiological level to the IPP level makes it possible to recover the important Gestalt psychology issues related to phenomena that are not immediately observable in terms of overt behavior. This is achieved by building simulation programs as working models, thus superseding the above-mentioned "polarization" between Gestalt psychology and behaviorism (Newell and Simon 1965, p. 146).

Program...is a hypothetical explanatory entity in exactly the same way that genes, when first proposed in biology, were hypothetical entities with no specific or specifiable physiological substrate. Subsequent evidence has allowed biologists...to treat them...as "real."

The constraints on models thus refer to the hypothetical entities at the explanatory level at which psychology is located as IPP. This explains, as seen earlier (pp. 223–30), why cognitive models do not include biological constraints. The future will tell us whether and to what extent the information processes postulated by IPP can be reduced to the physical basis of the mind, to the brain, according to the strategy described of gradual unification of scientific theories. This strategy does not entail any reference to an absolutely privileged level of hierarchy, essentially that of microphysics, as postulated by an absurd ideal of extreme reductionism, labeled "Laplacian" by Simon (1973). Conversely, Simon's idea of near decomposability, which refers to a hierarchical organization of semi-independent units (Simon 1996), can be applied to this hierarchy. In this case, near decomposability means that it is possible to have an approximate knowledge of a system at a given level without knowing what happens at a higher level. In other words, an abstraction may be made from the details of the latter, and every level of the hierarchy has its own reality, which derives from its near independence of the others. In actual scientific practice, as the evolution of various theories has shown, all levels may legitimately have their own autonomous functions within the hierarchy. At present, not enough is known about the relationship between information processes, as they may be inferred from the raw data of the protocols, and brain processes—for instance, little is known about how elementary information processes could be stored and executed by specific neural structures. Thus, psychology, in the form of IPP, can develop autonomously out of neurophysiology, in a similar fashion to the emergence of other disciplines out of genetics or chemistry (and, it is hoped, with the same chances of success).

We thus have confirmation of the thesis proposed by molar behaviorism of the "division of labor," as endorsed by Newell and Simon. The scientist of the mind can claim to have his own "place in the sun," but this time in the domain of IPP. In view of the "gulf of ignorance" that exists between IPP and neurophysiology (Simon 1966, p. 6), the open problem, as Newell and Simon repeatedly claimed, is that of the "bridge" between the two levels of explanation.

The original ambition of experimental psychology of becoming a "full-blown natural science," to use Hull's expression, was contradicted by IPP even before it was acknowledged by authoritative neopositivists that "the contention of behaviorism that psychology is a natural science...must now be more carefully scrutinized" (Feigl 1963, p. 251). Newell and Simon's skepticism concerning the "tools...borrowed and adapted from the natural sciences" has ended up by involving the neopositivists' physicalist claim (Newell 1968, p. 272):

The emergence of a science of symbol processing has particularly encouraged psychology to return to the study of higher mental behavior, after its long sojourn with a behaviorism that viewed as legitimate only descriptions in the language of physics.

In conclusion, the notion of IPS as a complex goal-directed hierarchy, the rejection of the empty organism, the use of models including psychological constraints (not "black boxes," but "artifacts...we can open them up and look inside" [Newell and Simon 1976, p. 114]), the criteria for the validation of models—all of these justify the original aim of the founders of IPP: "Methodology requires a re-examination today, both because of the novel substantive problems that the behavioral sciences face and because of the novel devices that are now available to help us solve these problems" (Simon and Newell 1956, p. 83).

Psychology, in other words, no longer has the status of a natural science: it appears as an empirical discipline and, given the "novel devices" from symbolic information processing, as a "science of the artificial" (Simon 1996).

To Conclude, and to Continue

To what degree is the Rock of Gibraltar a model of the brain?—It persists; so does the brain; they are isomorphic at the lowest level.
—W. Ross Ashby, 1956

I have reconstructed several concepts of IPP in what I called its classical phase (circa 1955–1975), and defined IPP as a human-process theory, like that of MacKay. On the one hand, the history of IPP reaches into our times, as it effectively became part of the research program of cognitive science. On the other, MacKay's theory of self-organizing systems, after being eclipsed for a period, may be said to have returned to the limelight within the framework of the so-called "new AI." It is of interest to dwell on both phenomena in order to attempt to make a final assessment.

IPP, like cognitive science after it, aimed at the psychological plausibility of models, claiming that they should satisfy constraints imposed by a theory of cognitive process, in particular of higher cognitive processes. Starting in the 1980s new models began to be proposed that were sometimes associated with research projects from the cybernetics era: neural networks a la Rosenblatt, robotics a la Grey Walter, and self-organizing systems a la Ashby or MacKay. The principal aim of these new models is neurological and also biological plausibility, albeit at a highly variable degree of generality. They included connectionist models based on neural networks (and on networks from artificial life), dynamical systems, and "situated" robotics, both behavior-based and evolutionary, the latter enabled by the development of genetic algorithms (Clark 1997; Pfeifer and Scheier 1999). Some developments along the lines of Gerald Edelman's "synthetic neural modeling" converged on a number of situated-robotics topics (Verschure et al. 1995). This set of research programs, referring mainly but not exclusively to low-level processes such as perception-action and forms of adaptation and simple learning, represents the new AI; in what follows I prefer to refer to it as "new cognitive science."

I would like to emphasize that both old ("symbolic") and new cognitive science share the modeling, or synthetic, method (as it is now called, for example, by Steels 1995). I have exemplified different approaches to this method through the comparison between MacKay's IFS and Newell, Shaw, and Simon's IPS. I have emphasized how this method was not directly dependent on the kind of artifact (digital or analogue), the subject matter (high or low level processes), or the choice of constraints (psychological or neurological-biological) imposed on the models, and that IPP is a particular view of the relationships between the different levels of explanation involved in mind science. My claim is that what distinguishes old and new cognitive science is the choice of the level of explanation at which the right constraints for the models are to be introduced. This is not without repercussions on the assessment of the successes and failures of research programs. As in every scientific undertaking, the choice of level of explanation is related to the researcher's interests, but it is conditioned above all by the state of knowledge and of the relevant technology. I will touch upon this issue in some detail here as regards the view of IPP as a trend in the history of modeling methodology, before concluding with a brief reference to the nature of modeling in new cognitive science.

At the outset of AI and IPP, and at least throughout the following two decades dominated by symbolic cognitive science, there are two different factors that encouraged and made plausible a definite choice regarding the constraints to be taken as right. These factors were the state of computer technology and the state of knowledge regarding neuroscience. They had a considerable influence on the choice made by the IPP approach of building models at a well-defined level—models embodying psychologically plausible hypotheses rather than neurologically or biologically plausible ones.

The first factor, that of computer technology, may be said to have counted much more than any external factors, some of which carried some weight; these included the well-known DARPA research funding exercises, which were directed more toward AI than other sectors. As James McClelland concluded, "The world wasn't ready for neural networks.... The computing power available in the early sixties was totally insufficient for [simulating neural networks on the computers of the time]" (see Crevier 1993, p. 309). The same may be said of MacKay's and other cyberneticists' self-organizing systems approaches in the 1950s and 1960s. For MacKay, the problem of the computer simulation of these systems was not taken for granted, and like many other cyberneticists he underestimated the digital computer. It is a fact, however, that the entire new cognitive science—including neural networks, dynamical systems, and genetic algorithms—would not have been the same without the development of digital machines with increasing computing power, although the same could be said also for much of AI, starting with expert systems (see Cordeschi 2006 on this point). As for cybernetic robotics, its specific limiting factors were appropriately identified by Rodney Brooks (1995, p. 38) in two points:

(a) The technology of building small self-contained robots when the computational elements were miniature (a relative term) vacuum tubes, and (b) the lack of mechanisms for abstractly describing behavior at a level below the complete behavior, so that an implementation could reflect those simpler components. Thus, in the first instance the models of thought were limited by technological barriers to implementing those models, and in the second instance, the lack of certain critical components of a model (organization into submodules) restricted the ability to build better technological implementations.

The internal limits of these research programs have always been assessed in the light of the comparison with the successes of heuristic programming in early AI, whose subject matter consisted of problems requiring little knowledge, rightly considered at that time to be the *Drosophila* of AI. As far as early IPP is concerned, it is no coincidence that its *Drosophila* was not chess but logic, the field in which the first successful simulation program, LT, was deployed. As Simon (1991) tells us, he and Newell gave up chess in favor of logic because they realized the importance in chess of eye movements, which are difficult to simulate.

As for the second factor, the state of neuroscience, this was known to be particularly backward in the 1950s and 1960s. Whether or not to consider neurological hypotheses originating out of particularly weak or speculative theories was a topic discussed by psychologists before IPP, as we saw earlier. The "division of labor" between psychology and neurology under the auspices of IPP may be viewed as an attempt to resolve the conflict between the two disciplines at a time of the extreme weakness of the second.

In this undertaking, IPP brought about a reexamination of the methodology followed in the study of behavior at the time: the epistemological standard of psychology was not considered to be that of physics. The "Newtonian" style (Pylyshyn 1978, p. 95) of building and validating theories had no significant impact on cognitive modeling in general. John Haugeland (1978, p. 215) stigmatized the "misguided effort [of psychology] to emulate physics and chemistry," and reserved the "systematic explanation" for cognitivism and IPP, based on the notion of a system as a hierarchy of interacting components.

We saw that this notion lies at the basis of very different information-processing systems, such as IPS and MacKay's IFS, and that the hierarchical organization of the former was described by Simon in terms of near decomposability. Near decomposability enjoyed considerable success in cognitive science, for in addition to proposing a mechanistic explanation paradigm, it could be viewed as a further step forward with regard to the neopositivists' original image of science. In the context of the evolution of theories, near decomposability is that "dynamic criterion" emphasized by William Wimsatt (1976, p. 242) which accounts for the evolution of theories and which was missing, at least according to critics such as Wimsatt himself and Churchland (1986), from the neopositivists' image of science (however, see Wimsatt's critique, 1972). Moreover, the reduction of one theory to another is not always (indeed is almost never) complete to the point of giving rise to explanatory unification nor, in the final analysis, to the identification of two theories (as in the case of optics and electromagnetism mentioned earlier), although with time it may be resolved into one or more approximations to this ideal goal.

Simon's (1976) example is that of the reduction of chemistry to quantum mechanics. This is a reduction that can be successful only in simple cases even now, let alone in the year in which Simon was writing. The reduction of psychology to neuroscience could be seen as a similar case. From this point of view, reductionism in the form of the elimination of a lower level (psychology, however defined) in favor of a higher one (neuroscience)—that is, the ideal level of the alleged genuine explanation—is not plausible. Even if it achieved a high degree of success, as in the case of chemistry, a "division of labor" among different experts would remain (Simon 1976, p. 64).

The idea of near decomposability suggested a way of overcoming the claim that it was possible to make a complete reduction of the laws and concepts of a given approach to those of a privileged or "more fundamental" science, in which the ultimate explanatory causes of the phenomena could be sought. In place of this program, which is the one suggested by the explanatory unification ideal attributed to neopositivists, it is possible to propose another that is closer to the practice and the evolution of science: that of unification as the identification of explanatory principles shared by different biological or physical systems, without the concepts and laws of one of them being taken to be "more fundamental" (see, for example, Glennan 2002).

However, this "division of labor" had ended up by introducing over time a kind of rigidity in the relations between psychology, viewed in IPP terms, and neuroscience. To some extent, this rigidity has been the consequence of the relative lack of "co-evolution" (to use Wimsatt's term) of theories at the two respective levels, or of the scarce feedback, always supported by Newell and Simon, that occurred in the long run between the two levels. (McCauley and Bechtel also spoke of co-evolution but pointed out its bi-directionality; see McCauley and Bechtel 2001). But above all, in view of the progress made by neuroscience over the past two decades (and of the dissemination of increasingly more advanced information technologies, which has allowed experimentation with innovative architectures with sophisticated abilities in the perceptual-motor sphere), the independence of the study of cognitive processes from neuroscience proposed by early IPP proved more difficult to sustain as the only possible choice for the study of cognition. Here is Newell's (1990, p. 483) opinion:

Throughout the history of IPP, an important element of its rhetoric has been that it is not anchored to the structure of the brain.... It was, I think, an important stance and the right one for its time. [Nevertheless] information processing can no longer be divorced from brain structure.... The great gain ... is that many additional constraints can be brought to bear as we examine information processing more deeply. But that puts it too abstractly. The great gain is that we finally bridge the chasm between biology and psychology with a ten-lane freeway.

And yet, Newell's claim proved in one sense to be overemphatic and in another rather reductive. In the computational cognitive architecture SOAR, to which Newell refers in this passage, "additional constraints" come into play from what, referring to the neurobiological-evolutionary sciences as a whole, Newell called the "biological band"; an example of this is the real-time constraint brought to the fore by the new robotics. However, the real comparison attempted by Newell is not between SOAR and effective neuroscience research but between SOAR and the connectionist models of the 1980s. The latter, as he was fully aware, conformed only loosely to the requirement of biological plausibility, forcing him to the slightly disconsolate conclusion that, even allowing that progress is being made in our understanding of the computational mechanisms involved in the brain, "connectionism is what to do until functioning neurophysiological system architectures arrive" (Newell 1992, p. 486). How to "bridge the chasm" has remained an open question for the cognitive theories concerned with higher cognitive processes since the days of early IPP (see pp. 237–44). Indeed, the biological constraints related to the emotional aspects of problem solving, or even of the consciousness that interested MacKay above everything, found no room in SOAR and analogous symbolic systems.

The foregoing leads us again to the issue of the constraints to be imposed on models. In the first place, without constraints of some kind, models are completely underdetermined, bringing us once again to the practically unlimited generation of functionally isomorphic machines (Ashby 1956), or of artifacts that merely mimic and do not explain anything (for details, see Cordeschi 2002, pp. 250ff.). In the second place, Newell and Simon always had clearly in mind the so-called "degree-of-freedom problem," which arises when, in the helical path of model revision, action is taken to modify the program so that it better fits the data. The risk was that of allowing too much freedom to the theoretician in this process, since "if we allow a parameter change or a new mechanism to be introduced [in a program] for each bit of behavior to be explained, we will have explained nothing" (pp. 33–34). Lastly, this problem is linked to the "irrelevant specification problem," which is closely connected to the use of models expressed in terms of simulation programs: how can we distinguish in such models the parts embodying the hypotheses, that is, the theoretically significant ones, from the merely implementational parts, which might be the result of choices of convenience?

These problems have been raised successively with reference both to classical models (Pylyshyn 1984) and to connectionist models (Green 1998).

Pylyshyn, for example, claims that the underdetermination problem and the degree-of-freedom problem can be redimensioned by identifying constraints at the level of cognitive architecture. Newell claimed that there is a possible solution for the irrelevant-specification problem: instead of treating cognition by using an unordered collection of programs and data structures, reliance can be placed on a unified theory of cognition. (This is an implicit critique of the idiosyncratic models of the early IPP; see also Newell 1973). A system like SOAR establishes a single theoretical core (as regards the mechanisms of memory, learning, and so forth), and this increases the reliability of the details regarding the simulation of different subjects and in different tasks. Moreover, unification also serves the purpose of reducing the degrees of freedom of the parameters requiring modification to fit the data (Newell 1990, pp. 22–23). As a unified theory, SOAR is thus an attempt to reduce the freedom of the theoretician—and thus reduce the risk of ad hoc simulations—by imposing as many "additional constraints" as possible, far beyond those envisaged in early IPP.

Simon was skeptical about a single unified theory a la Newell. His preference was for "middle-level" theories. The various versions of the Elementary Perceiver and Recognizer (EPAM), first programmed for a computer in 1959 (see Feigenbaum and Simon 1962), represent his favorite example (in GPS he already saw an early attempt in this direction). Simon remained faithful to the verbal-protocol method and retained the thesis of the independence of levels, expressing the hope that there would be collaborations using IPP, first with connectionism and then later with the new robotics, rather than a confrontation between "paradigms." In the "forward motion" procedure, the next step for him would be the testing of the model with reference to the gradual expandability of the postulated mechanisms. From his point of view, one way of reducing the risk of the degree-of-freedom problem would be to shift the ratio of data to modifiable parameters in favor of data, for example by cross-linking the data on verbal protocols with those on eye movements, whenever this could be done (Simon 1992). Nevertheless, the basic idea remains the same as that of Newell: to increase the number of constraints, since less-constrained programs (those with a larger number of parameters) can be modified more easily and perhaps arbitrarily using ad hoc procedures than more-constrained programs (those with a smaller number of modifiable parameters). This increases the validity of taking such programs seriously as suitable models (Simon and Wallach 1999).

I will conclude my argument with a brief reference to what I have characterized as new cognitive science. As noted, this, too, uses the synthetic method, but the constraints considered right are above all those referring to the neurological or biological level. New cognitive science has been able to profit from recent progress both in information technology and in neuroscience, and has attempted to "bridge the chasm," as Newell put it, with different degrees of generality, and in different fields. In this context one could mention the modeling of neural brain functions through neural nets (of a kind very different from the early connectionist varieties), and the modeling of the behavior of simple organisms using an ethological approach (Holland and McFarland 2001).

The underdetermination problem and the related problems mentioned are believed to afflict the models of classical AI and cognitive science, as also is the "symbol-grounding problem" (Harnad 1990). In fact, in this context, symbol grounding is not a recent issue, having been raised by MacKay when he proposed his human-process models. He believed that the latter leave less scope for the theoretician or system designer because they develop their symbols by means of self-organization processes. This is an ability that he did not see embodied in the computer programs of the time. Dealing with this issue later, he identified as a weakness the "practice in artificial intelligence of taking a non-biological approach to the internal representation of the external world, using intrinsically meaningless symbols" (MacKay 1986, p. 149). This is a judgment that new cognitive science would endorse.

Elsewhere I have argued that the symbol-grounding problem, as well as the underdetermination problem, cannot be automatically solved by appealing to constraints at the neurological or biological level (Cordeschi 2002, chapter 7). This argument can be extended much further; for an insightful discussion, see Webb (2001). It seems however that the issue of establishing the right constraints by reference to some privileged level (or levels) of explanation is still an open question. While there might one day be a univocal response to this question, perhaps through achieving an effective unification of the relevant knowledge, I believe it certainly does not exist at the moment. Of course, it might be possible to state the question with less radical aims, as is suggested by some current proposals for "hybrid" models (from symbolic-connectionist to reactive-reasoning systems). However, when this is not the case, the situation in which we find ourselves could be simplified as follows:

On the one hand, if the aim is to build a model of an ability such as syllogistic reasoning, the best way to proceed is still that suggested by IPP. The constraints should be chosen with reference to a psychological theory, that is, to hypothetical constructs at the psychological level. For example, Philip

Johnson-Laird's mental models, or the analogous constructs implemented in SOAR as a syllogistic demonstrator, are constructs of this kind, and could be considered psychologically plausible. In this case we have a "classical" example of a computer simulation as a model. On the other hand, if the aim is to build a model of an ability such as discrimination, it might be possible to choose the constraints with reference to a neurological theory, or to hypothetical constructs at the neurological level, and possibly to avoid the risk of a "non-biological approach" as identified by MacKay. For example, the hypotheses of Edelman's neural Darwinism are constructs of this kind, and could be considered as neurologically plausible. In this case the model is expressed as a computer simulation with particular neural networks, such as DARWIN III, and also as a situated robot, such as NOMAD (see Verschure et al. 1995).

We do not yet know whether mental models or other similar theoretical constructs (for example, those of SOAR) are reducible—wholly or in part, or through co-evolution of the theories concerned—to effective neurological structures; still less do we know how this might be done. In the two cases discussed above (mental models and neural Darwinism) we have hypothetical constructs at different levels. Nevertheless, in both cases the shared epistemological assumption is the same, that of the validity of the synthetic method. It might be not a coincidence that a behaviorist following the "doctrine of desperation" would refute both such theoretical constructs: the mental models (and similar symbolic structures) are for him a mere mentalist speculation, while those deriving from neural Darwinism are a kind of "conceptual nervous system," to use Skinner's expressions.

References

Ackoff, R. L. 1962. *Scientific Method*. New York: Wiley.

Ashby W. Ross. 1940. "Adaptiveness and Equilibrium." *Journal of Mental Science* 86: 478–83.

———. 1956. *An Introduction to Cybernetics*. London: Chapman & Hall.

———. 1962. "What Is Mind? Objective and Subjective Aspects in Cybernetics." In *Theories of Mind*, edited by J. M. Sher. New York: Free Press.

Beer, Randall D. 1998. "Framing the Debate Between Computational and Dynamical Approaches to Cognitive Science." *Behavioral and Brain Sciences* 21: 630.

Boden, Margaret A. 1978. *Purposive Explanation in Psychology*. Hassocks, UK: Harvester.

Boring, E. G. 1964. "The Trend Toward Mechanism." *Proceedings of the American Philosophical Society* 108: 451–54.

Brooks, Rodney A. 1995. "Intelligence Without Reason." In *The Artificial Life Route to Artificial Intelligence: Building Embodied, Situated Agents*, edited by L. Steels and R. Brooks. Hillsdale, N.J.: Lawrence Erlbaum.

Bruner, Jerome S., J. J. Goodnow, and G. A. Austin. 1956. *A Study of Thinking*. New York: Wiley.

Carnap, Rudolf. 1956. "The Methodological Character of Theoretical Concepts." In *Minnesota Studies in the Philosophy of Science*. Volume 1, edited by H. Feigl and M. Scriven. Minneapolis: University of Minnesota Press.

Churchland, P. S. 1986. *Neurophilosophy: Toward a Unified Science of the Mind-Brain*. Cambridge, Mass.: MIT Press.

Churchman, C. W., and R. L. Ackoff. 1950. "Purposive Behavior and Cybernetics." *Social Forces* 29: 32–39.

Churchman, C. W., R. L. Ackoff, and E. L. Arnoff. 1957. *Introduction to Operation Research*. New York: Wiley.

Clark, A. 1997. *Being There: Putting Brain, Body, and World Together Again*. Cambridge, Mass.: MIT Press.

Cordeschi, Roberto. 1996. "The Role of Heuristics in Automated Theorem Proving: J. A. Robinson's Resolution Principle." *Mathware and Soft Computing* 3: 281–93.

———. 2002. *The Discovery of the Artificial: Behavior, Mind and Machines Before and Beyond Cybernetics*. Dordrecht: Kluwer Academic Publishers.

———. 2006. "Searching in a Maze, in Search of Knowledge: Issues in Early Artificial Intelligence." In *Reasoning, Action, and Interaction in AI Theories and Systems*, edited by O. Stock and M. Schaerf. Berlin: Springer.

Cordeschi, Roberto, and G. Tamburrini. 2005. "Intelligent Machines and Warfare: Historical Debates and Epistemologically Motivated Concerns." *Proceedings of the European Computing and Philosophy Conference ECAP 2004*. London: College Publications.

Cordeschi, Roberto, G. Tamburrini, and G. Trautteur. 1999. "The Notion of Loop in the Study of Consciousness." In *Neuronal Bases and Psychological Aspects of Consciousness*, edited by C. Taddei-Ferretti and C. Musio. Singapore: World Scientific.

Craik, Kenneth J. W. 1943. *The Nature of Explanation*. Cambridge: Cambridge University Press.

Crevier, D. 1993. *AI: The Tumultuous History of the Search for Artificial Intelligence*. New York: Basic Books.

De Groot, A. 1946. *Het Denken van den Schaker.* [Thought and Choice in Chess.] Amsterdam: N. H. Utig. Mij.

Feigenbaum, E. A., and H. A. Simon. 1962. "A Theory of the Serial Position Effect." *British Journal of Psychology* 53: 307–20.

Feigl, H. 1948. "Some Remarks on the Meaning of Scientific Explanation." *Psychological Review* 52: 250–59.

———. 1963. "Physicalism, Unity of Science, and the Foundation of Psychology." In *The Philosophy of Rudolf Carnap*, edited by P. A. Schilpp. La Salle, Illinois: Open Court.

Glennan, S. 2002. "Rethinking Mechanistic Explanation." *Philosophy of Science* 69: 342–53.

Green, Christopher. 1998. "The Degrees of Freedom Would Be Tolerable If Nodes Were Neural." *Psycoloquy* 9, no. 26. On-line journal. Available at http://www.cogsci .ecs.soton.ac.uk/cgi/psyc/newpsy?9.26.

Harnad, S. 1990. "The Symbol Grounding Problem." *Physica* (series D) 42: 335–46.

Haugeland, J. 1978. "The Nature and Plausibility of Cognitivism." *Behavioral and Brain Sciences* 1: 93–127.

———. 1985. *Artificial Intelligence: The Very Idea.* Cambridge, Mass: Bradford/MIT Press.

Hebb, D. O. 1951–52. "The Role of Neurological Ideas in Psychology." *Journal of Personality* 20: 39–55.

Holland, Owen, and D. McFarland. 2001. *Artificial Ethology.* Oxford: Oxford University Press.

Hull, C. L. 1943. *The Principles of Behavior.* New York: Appleton-Century.

MacKay, Donald M. 1951. "Mindlike Behaviour in Artefacts." *British Journal for the Philosophy of Science* 2: 105–21.

———. 1952. "Mentality in Machines." *Proceedings of the Aristotelian Society* (supplements) 26: 61–86.

———. 1954. "On Comparing the Brain with Machines." *Advancement of Science* 10: 402–6.

———. 1956. "Towards an Information-Flow Model of Human Behaviour." *British Journal of Psychology* 47: 30–43.

———. 1959. "Operational Aspects of Intellect." *Proceedings of the Teddington Symposium on Mechanisation of Thought Processes.* Volume 1. London: Her Majesty's Stationery Office.

——. 1965. "From Mechanism to Mind." In *Brain and Mind*, edited by J. R. Smythies. London: Routledge & Kegan Paul.

——. 1969. *Information, Mechanism and Mind*. Cambridge, Mass.: MIT Press.

——. 1986. Intrinsic Versus Contrived Intentionality. *Behavioral and Brain Sciences* 9: 149–50.

McCauley, R. N., and W. Bechtel. 2001. "Explanatory Pluralism and the Heuristic Identity Theory." *Theory and Psychology* 11: 737–60.

McCorduck, P. 1979. *Machines Who Think*. San Francisco: Freeman.

McCulloch, W. S., and W. Pitts. 1943. "A Logical Calculus of the Ideas Immanent in Nervous Activity." *Bulletin of Mathematical Biophysics* 5: 115–37.

Minsky, Marvin L. 1959. "Discussion." *Proceedings of the Teddington Symposium on Mechanisation of Thought Processes*. volume 1. London: Her Majesty's Stationery Office.

——. 1966. Artificial Intelligence. *Scientific American* 215, no. 3: 247–61.

——, ed. 1968. *Semantic Information Processing*. Cambridge, Mass.: MIT Press.

Mirowski, P. 2002. *Machine Dreams. Economics Becomes a Cyborg Science*. Cambridge: Cambridge University Press.

Newell, Allen. 1968. "The Trip Towards Flexibility." In *Bio-engineering: An Engineering View*, edited by G. Bugliarello. San Francisco: San Francisco Press.

——. 1970. "Remarks on the Relationship Between Artificial Intelligence and Cognitive Psychology." In *Theoretical Approaches to Non-Numerical Problem Solving*, edited by R. Banerji and M. Mesarovic. Berlin: Springer.

——. 1973. "You Can't Play 20 Questions with Nature and Win." In *Visual Information Processing*, edited by W. G. Chase. New York and London: Academic Press.

——. 1990. *Unified Theories of Cognition*. Cambridge, Mass.: Harvard University Press.

——. 1992. Précis of *Unified Theories of Cognition*. *Behavioral and Brain Sciences* 15: 425–92.

Newell, Allen, J. C. Shaw, and Herbert A. Simon. 1958. "Elements of a Theory of Human Problem-Solving." *Psychological Review* 65: 151–66.

——. 1960. "Report on a General Problem-Solving Program for a Computer." In *Proceedings of the International Conference on Information Processing*. Paris: UNESCO.

Newell, Allen, and Herbert A. Simon. 1959. "The Simulation of Human Thought." Paper P-1734. Santa Monica: RAND Corporation, Mathematics Division.

————. 1965. "Programs as Theories of Higher Mental Processes." In *Computers in Biomedical Research*, edited by R. W. Stacey and B. Waxman. Volume 2. New York: Academic Press.

————. 1972. *Human Problem Solving*. Englewood Cliffs, N.J.: Prentice-Hall.

————. 1976. "Computer Science as Empirical Inquiry: Symbols and Search." *Communications of the ACM* 19: 113–26.

Paige, J. M., and Herbert A. Simon. 1966/1979. "Cognitive Processes in Solving Algebra Word Problems." In *Problem Solving*, edited by B. Kleinmuntz. New York: Wiley. Also in *Models of Thought*, edited by Herbert A. Simon. New Haven and London: Yale University Press.

Pask Gordon. 1964. "A Discussion on Artificial Intelligence and Self-Organization." *Advances in Computers* 5: 109–226.

Pfeifer, R., and C. Scheier. 1999. *Understanding Intelligence*. Cambridge, Mass.: MIT Press.

Pylyshyn, Z. W. 1978. "Computational Models and Empirical Constraints." *Behavioral and Brain Sciences* 1: 93–127.

————. 1979. "Complexity and the Study of Artificial and Human Intelligence." In *Philosophical Perspectives in Artificial Intelligence*, edited by M. Ringle. Brighton, UK: Harvester.

————. 1984. *Computation and Cognition: Toward a Foundation for Cognitive Science*. Cambridge, Mass.: MIT Press.

Robinson, J. A. 1965. "A Machine Oriented Logic Based on the Resolution Principle." *Journal of the Association for Computing Machinery* 12: 23–41.

Rosenblueth, Arturo, Norbert Wiener, and Julian Bigelow. 1943. "Behavior, Purpose and Teleology." *Philosophy of Science* 10: 18–24.

Selz, O. 1922. *Zur Psychologie des produktiven Denkens und des Irrtums*. Bonn: Friedrich Cohen.

Simon Herbert A. 1947. *Administrative Behavior*. Macmillan: New York.

————. 1961. "The Control of Mind by Reality: Human Cognition and Problem Solving." In *Man and Civilization*, edited by S. M. Farber and R. H. L. Wilson. New York: McGraw-Hill.

————. 1963. "Economics and Psychology." In *Psychology: A Study of a Science*, edited by S. Kock. Volume 4. New York: McGraw-Hill.

————. 1966. Thinking by Computers. In *Mind and Cosmos: Essays in Contemporary Science and Philosophy*, edited by R. G. Colodny. Pittsburgh: University of Pittsburgh Press.

————. 1967. "Motivational and Emotional Controls of Cognition." *Psychological Review* 74: 29–39.

————. 1973. "The Organization of Complex Systems." In *Hierarchy Theory*, edited by H. H. Pattee. New York: Braziller.

————. 1976. "The Information Storage System Called 'Human Memory.'" In *Neural Mechanism of Learning and Memory*, edited by M. R. Rosenzweig and E. L. Bennett. Cambridge, Mass.: MIT Press. Also in *Models of Thought*, edited by Herbert A. Simon. New Haven and London: Yale University Press.

————. 1978. "Information-Processing Theory of Human Problem Solving." In *Handbook of Learning and Cognitive Processes*, edited by W. K. Estes. Volume 5. Hillsdale, N.J.: Lawrence Erlbaum.

————. 1981. "Otto Selz and Information-Processing Psychology." In *Otto Selz: His Contribution to Psychology*, edited by N. H. Frijda and A. D. de Groot. The Hague: Mouton.

————. 1991. *Models of My Life*. New York: Basic Books.

————. 1992. "What Is an 'Explanation' of Behavior?" *Psychological Science* 3: 150–61.

————. 1995a. "Machine as Mind." In *Android Epistemology*, edited by K. M. Ford, C. Glymour, and P. J. Hayes. Menlo Park, Calif.: AAAI/MIT Press.

————. 1995b. "Artificial Intelligence: An Empirical Science." *Artificial Intelligence* 77: 95–127.

————. 1996. *The Sciences of the Artificial*. Cambridge, Mass.: MIT Press.

Simon, Herbert A., and associates 1986. "Decision Making and Problem Solving." *Research Briefings 1986: Report of the Research Briefing Panel on Decision Making and Problem Solving*. Washington, D.C.: National Academy of Sciences.

Simon, Herbert A., and Allen Newell. 1956. "Models: Their Uses and Limitations." In *The State of the Social Sciences*, edited by L. D. White. Chicago: Chicago University Press.

————. 1964. "Information Processing in Computer and Man." *American Scientist* 53: 281–300.

Simon, Herbert A., and D. Wallach. 1999. "Cognitive Modeling in Perspective." *Kognitionwissenschaft* 8: 1–4.

Skinner, B. F. 1974. *About Behaviorism*. Knopf: New York.

Sloman, A. 2002. "The Irrelevance of Turing Machines to Artificial Intelligence." In *Computationalism: New Directions*, edited by M. Scheutz. Cambridge, Mass.: MIT Press.

Smith, L. D. 1986. *Behaviorism and Logical Positivism: A Reassessment of the Alliance.* Palo Alto: Stanford University Press.

Steels, L. 1995. "Building Agents out of Autonomous Systems." In *The Artificial Life Route to Artificial Intelligence: Building Embodied, Situated Agents,* edited by L. Steels and Rodney Brooks. Hillsdale, N.J.: Erlbaum.

Tolman E. C. 1935/1958. "Psychology Versus Immediate Experience." *Philosophy of Science* 2: 356–80. Also in E. C. Tolman, *Behavior and Psychological Man.* Berkeley and Los Angeles: University of California Press.

———. 1949. "Discussion." *Journal of Personality* 18: 48–50.

Verschure, P. F. M. J., J. Wray, O. Sporns, T. Tononi, and G. M. Edelman. 1995. "Multilevel Analysis of Classical Conditioning in a Behaving Real World Artifact." *Robotics and Autonomous Systems* 16: 247–65.

Wang, H. 1960. "Toward Mechanical Mathematics." *IBM Journal for Research and Development* 4: 2–22.

Webb, B. 2001. "Can Robots Make Good Models of Biological Behaviour?" *Behavioral and Brain Sciences* 24: 1033–50.

Wimsatt, W. C. 1972. "Complexity and Organization." In *Boston Studies in the Philosophy of Science,* edited by K. F. Schaffner and R. S. Cohen. Volume 20. Dordrecht: Reidel.

———. 1976. "Reductionism, Levels of Organizations, and the Mind-Body Problem." In *Consciousness and the Brain: Scientific and Philosophical Enquiry,* edited by G. G. Globus, G. Maxwell, and I. Savodnik. New York: Plenum.

Zuriff, G. E. 1985. *Behaviorism: a Conceptual Reconstruction.* New York: Columbia University Press.

11 The Mechanization of Art

Paul Brown

Sorry miss, I was giving myself an oil-job.
—Robby the Robot, in *Forbidden Planet*

I'm sorry Dave, I can't do that.
—HAL 9000, in *2001: A Space Odyssey*

This chapter is an idiosyncratic account of the development of "the mechanization of art." I am an artisan, a maker of art, and neither an historian nor a scholar, and so it describes only those parts of the narrative with which I am familiar. As the German Dadaist Kurt Schwitters, the architect of Merz (a movement embracing dance, theater, visual art, and poetry), once claimed, "I am the meaning of the coincidence." I have also chosen to end my account in the late 1970s. By then the personal computer had arrived and the world was changed forever. The ensuing proliferation of artworks and ideas are still difficult, for me at least, to record and contextualize.

A comprehensive overview of the historical developments that led to the flowering of the mechanization of art in the twentieth century is beyond the scope of this chapter. However, a few examples are worthy of note, since they give a context and demonstrate that this pursuit of knowledge has a long and intriguing pedigree that stretches back even into prehistory. The Chinese *I Ching*, or *Book of Changes*, is believed to have first taken form in about 1800 B.C.E. and is attributed to the legendary "founder" of China, Fu Hsi. The book was restructured and derived its modern format in the early Chou dynasty, following revisions attributed to King Wen and his son Tan, the Duke of Chou, around 1100 B.C.E. Further commentaries were added by Confucius (511–479 B.C.E.) and his school and are known as the Ten Wings. Although the book has been perceived in the West as a divination system or oracle, Joseph Needham and later scholars emphasize its importance in the history of Chinese scientific thought and philosophy and describe its method as "coordinative" or "associative," in contrast to the European "subordinate" form of inquiry.[1] The book may be interpreted as a cosmology where the unitary "one" first divides into the binary principles—the yin and the yang, represented by a broken or whole line, respectively—which are then permutated to form the eight trigrams. These, as the name suggests, are three-line structures that may also be interpreted as the vertices of a unit cube—the three dimensions of the material world. The trigrams are then permutated with each other to form the sixty-four hexagrams (or archetypes) and then each (any and all) of the six lines that make up the hexagram can flip into its opposite (yin to yang, broken to whole, and vice versa), which enables any hexagram to change to any other and so give the final 4,096 changes to which the title refers. The book may be "consulted" by a process of chance operations, flipping coins or dividing groups of yarrow stalks, a process that identifies the unique "time" of the consultation. Jesuit missionaries sent a copy of the book to Gottfried Leibniz, who introduced the binary mathematical notation system to Europe, and the *I Ching* has had an ongoing effect on Western scientific and artistic thought ever since. This gained momentum after a scholarly translation by Richard Wilhelm and Cary F. Baynes, with an introduction by Carl Jung, was published in 1968, coinciding with the cognitive experimentation of the psychedelic movement.[2]

During the first century C.E. the Greek engineer Hero of Alexandria designed and constructed sophisticated automata that were powered by water, gravity, air, and steam. As the Christian Dark Ages closed in over Europe, the ancient Greek and Egyptian knowledge was preserved and developed in the Arab world. Al Jaziri's *Al Jami' Bain Al 'Ilm Wal 'Amal Al Nafi Fi Sina'at Al Hiyal*, or *The Book of Knowledge of Ingenious Mechanical Devices* (about 1206) describes many of al Jaziri's automata and has been recently placed in the context of art and science history by Gunalan Nadarajan.[3] Among the devices that al Jaziri describes is an automatic wine server that was used at royal parties at the Urtuq court of Diyar-Bakir, who were his patrons. It randomly selected guests to serve so some got very intoxicated while others remained completely sober, to the great amusement of all.

Not long after this, Ramon Lull (1235–1315) was born in Palma, Majorca. He was a Christian writer and philosopher living in Spain when it was part of the Islamic Moorish empire, which included Portugal and parts of North Africa. Unlike his Northern European contemporaries, who were still living under the repressive Catholic rule appropriately named the Dark Ages, Lull had access to Arab knowledge dating back to Greece and culled from around the rapidly expanding Islamic sphere of influence. Although his contribution to knowledge was broad, of particular interest here are his Lullian Circles. Described in his *Ars Generalis Ultima*, or *Ars Magna*, published in 1305, these consist of a number of concentric disks that can be rotated independently on a common spindle. Each disk is scribed with symbols representing attributes, or archetypes, that can be permutated together to form compound expressions. The system forms a combinatorial logic that is remarkably similar in concept (though not in implementation) to the generative method employed by the much earlier *I Ching*. Two centuries later Leibniz (who, as mentioned, knew about the *I Ching*) developed Lull's idea for his investigations into the philosophy of science. Leibniz named the method *Ars Combinatoria*. Machines like Lull's appear in literature: in *Gulliver's Travels* (1721) Jonathan Swift describes a system that creates knowledge by combining words at random, a passage that is believed to be a parody of Lull's work. More recent fictional combinatorial knowledge machines appear in books such as Hermann Hesse's *The Glass Bead Game* and Umberto Eco's *The Island of the Day Before*.[4]

The Christian reconquest of Spain during the fifteenth century enabled the European rediscovery of the long-suppressed knowledge preserved by Islam, and this was a major cause of the flowering of the Renaissance (literally "rebirth"). The polymath Leonardo da Vinci (1452–1519) is known for his lateral and experimental approach to both art and science. Among his prolific output, around 1495 he recorded in a sketchbook a design for an anatomically correct humanoid automaton; there is no record that Leonardo's Robot, as it is now known, was ever built. The German artist Albrecht Dürer (1471–1528) was another polymath who made significant contributions to both mathematics and the visual arts. In his *Treatise on Measurement* (1525) he included several woodcut prints of perspective-drawing systems that can be retrospectively acknowledged as early precursors of analogue computing machines.

By the seventeenth century the French mathematician and philosopher René Descartes (1596–1650) proposed that animals were nothing more than complex machines. By suggesting a correspondence between the mechanical and the organic, Descartes laid the groundwork for a more formal study of autonomy. The production of automata flourished with ever more complex and sophisticated examples. The Jesuit alchemist Athanasius Kircher (1602–1680) is reputed to have made a statue that could carry on a conversation via a speaking tube (he's also credited with building a perpetuum mobile!). However, it was in 1737 that the French engineer and inventor Jacques de Vaucanson (1709–1782) made what is considered the first

major automaton of the modern age. His Flute Player was not only intended to entertain but was also a serious investigation into human respiration. As such it stands as an early precursor of the art-science collaborations
that developed in the twentieth century. Vaucanson's automated loom of
1744 did not use the punch cards proposed by Falcon in 1728, but instead
used the paper-tape control system invented by Basile Bouchon in 1725. By
1801 Joseph Marie Jacquard had created a robust card-driven loom, a design that was still in use in the late twentieth century. Jacquard's card
system had another major and arguably more influential outcome when
Charles Babbage (1791–1871) selected it as the control and storage mechanism for his Analytical Engine. Later, Herman Hollerith (1860–1929) took
up the idea and went on to found the company known today as IBM. It's
an early and excellent example of how research in the arts can have a profound effect on science and technology and demonstrates how the modern
science of computing has clearly defined roots in the art of weaving, which
is, after all, an ancient system for the codification, manipulation, storage,
and reproduction of pattern.

Religious warnings about human intervention in the work of God
accompanied many of these developments and emerged in literature. The
archetypical text is Mary Shelley's wonderful *Frankenstein* (1818).[5] Similar
concerns continue to this day in many of the detractors and critics of artificial intelligence and artificial life, as well as many other aspects of science
and technology such as evolution, nanotechnology, and stem-cell research.

Developments continued throughout the nineteenth century. The paper-
tape and punch-card control systems developed for weaving were adapted
for use in other applications. Orchestral machines such as steam organs
toured the fairs, and pianolas and music boxes were mass-produced. Paper
pianola scrolls enabled people to hear performances by contemporary vir-
tuosi, and also formed a valuable historical record. They created a demand
for pre-programmed music that would later be satisfied by shellac and vinyl
gramophone recordings and contemporary compact disks and MP3 players.
In the visual arts and sciences the invention of photographic recording by
Joseph Niépce in 1827 was improved by Louis Daguerre. In 1835, William
Henry Fox Talbot devised a method to duplicate images by printing multiple positives from one negative. The Renaissance experiments into perspective, Dürer's drawing systems, and other devices such as the camera obscura
were automated—image making was now a mechanical process. By 1888
Kodak's founder, George Eastman, could coin the slogan "You press the
button, we do the rest." During the same decades French Postimpressionist
artists such as Paul Cézanne (1839–1906) and Georges Seurat (1859–1891)

challenged the role of painting as representation, a function that had in any case been usurped by photography, and emphasised instead its analytical role. Both artists were concerned with a proto-semiological exploration of the relationship between the flat plane of the canvas, the representation, and the three-dimensional world, the represented. Neither would break completely with the figurative. That would happen early in the twentieth century, when the Russian artist Wassily Kandinsky (1866–1944), a theosophist, recalled some illustrations he had seen in a book called *Thought Forms*, by Annie Besant and C. W. Leadbeater (1888) and painted what he (amazingly, in retrospect) titled *First Abstract Watercolour* in 1910.[6] The visual arts had been freed from their anchor in "the real" and a colossal explosion in creativity ensued, causing ripples throughout the art world.

A decade later Karel Čapek (1890–1938) wrote the play *Rossum's Universal Robots*, or *R.U.R.* It was first performed in Prague in 1921, then in New York City, in 1922. Karel's brother, Josef, had coined the term *robot*: *robota* is Czech for "drudgery" or "servitude," and a *robotnik* is a peasant or serf. The play is either a utopia or dystopia, depending on your point of view. Robots are created as cheap labor who ultimately revolt and kill all the humans except one. The robots learn to replicate themselves and the play closes when two of them, Helena and Primus, fall in love and are dubbed Adam and Eve by Alquist, the last human (see chapter 12 for a detailed discussion of the play). Responding to criticism by George B. Shaw and G. K. Chesterton, among others, Čapek stated that he was much more interested in men than in robots. He predicted the sentiments of William Gibson who, over sixty years later, would express his concern when he discovered that computer graphics enthusiasts at the annual SIGGRAPH Conference were busy implementing the dystopian virtual reality he created for his Orwellian-style Cyberspace Trilogy: *Neuromancer*, *Count Zero*, and *Mona Lisa Overdrive*.[7] In 1927, five years after *R.U.R.*, Fritz Lang (1890–1976) wrote and directed his legendary film *Metropolis* (restored in 2002). Based on the novel by his wife, Thea von Harbou, it's a parable of socialist class struggle where the Lord of Metropolis, Johann Fredersen, wants to replace his human workers with robots. Their leader, Maria, is cloned by the evil scientist Rotwang into a robot "femme fatale" as part of a plot to incite a revolution that Johann hopes will give him the excuse to eliminate the workers and replace them with Rotwang's machines. A decade later, in 1936, the German Marxist historian and cultural theorist Walter Benjamin (1892–1940) published his essay "The Work of Art in the Age of Mechanical Reproduction," in which he argued that the artwork is democratized by mass-production technology but the result is that its unique intrinsic value is threatened.[8] The essay was influential, particularly in the latter half of the twentieth century, when the concept of the art object gave way to art as process.

The French artist Marcel Duchamp (1887–1968) is recognized as one of the major intellects of twentieth-century art. As a key member of the Dada movement he questioned the entire nature of the artwork when he introduced his ready-mades with *Roue de Bicyclette* (*Bicycle Wheel*) in 1913. During the 1920s Duchamp worked on a number of "Rotoreliefs," and some were recorded in his film *Anémic Cinéma* (1925–1926). The rotating disks produced 3-D illusions and progressed Duchamp's interest in both art-as-machine and as cognitive process. László Moholy-Nagy (1895–1946) created his light-space modulator in 1930 after some years of experimentation. It's a kinetic sculpture that he described as an "apparatus for the demonstration of the effects of light and movement." These effects are recorded in his film *Lichtspiel, schwarz-weiss-grau*, (Light-play, black-white-gray), made the same year. The original light-space modulator is preserved in the collection of the Busch-Reisinger Museum in Cambridge, Massachusetts, and a number of working reconstructions have been made. Alexander Calder (1898–1976) was a Paris-based American sculptor best known for the kinetic sculptures, dubbed "mobiles" by Duchamp, that he started constructing in 1931. Though his early experiments were motor-driven, he soon developed the graceful wind- and gravity-powered mobiles for which he is now best known.

The Swiss artist Jean Tinguely (1925–1991) belonged to a later generation of artists who were influenced by both Dada and these early kinetic experiments. In 1944 he began making his Metamechanics, or Metamatics, eccentric machines that often expended high energy doing nothing. Although his early work is playful and entertaining, there is always a dark undercurrent. By the 1960s the early whimsy had evaporated, to be replaced by a more somber mood reflective of the times. Among Tinguely's best-known work of this period is *Homage to New York* (1960), an ambitious autodestructive installation in the courtyard of New York's Museum of Modern Art, which was documented in Robert Breer's film *Homage to Jean Tinguely's "Homage to New York."* It is further notable because it was the first collaboration with an artist of the Bell Telephone Lab engineer Billy Klüver (1927–2004), who went on to cofound the influential EAT, Experiments in Art and Technology.

Takis (1925–) was born in Athens but, like Calder and Tinguely, based himself in Paris and began making his illuminated Signaux—Signals—in 1955. They become kinetic in 1956 and in 1958 Takis integrated electromagnetic elements that gave his works chaotic dynamics. Frank Malina (1912–1981) was an American aerospace engineer who did pioneering work on rocketry and was a cofounder and the first director of Caltech-NASA's Jet Propulsion Lab in Pasadena. Disillusioned with the increasing military application of his research, he left in 1947 to join UNESCO before committing himself full-time to his art practice in 1953. He based himself in Paris, where many of the European kinetic artists were congregated. His son, Roger, has recently commented that he "was amazed that artists created so little artwork depicting the new landscapes we now see, thanks to telescopes, microscopes and robots that explore the ocean and space."[9] In 1954 Malina introduced electric lights into his work and in 1955 began his kinetic paintings. In 1968 he founded the influential publication *Leonardo*, the journal of the International Society for Arts, Science and Technology (ISAST).[10]

It was in Paris in the 1950s that the artist Nicolas Schöffer (1912–1992) formulated his idea of a kinetic art that was not only active and reactive, like the work of his contemporaries, but also autonomous and proactive. He developed sculptural concepts he called Spatiodynamism (1948), Luminodynamism (1957), and Chronodynamism (1959) and was influenced by the new ideas that had been popularized by Norbert Wiener and Ross Ashby.[11] His *CYSP 1* (1956, figure 1) is accepted as the first autonomous cybernetic sculpture. Its name is formed from CYbernetic SPatiodynamism. It was controlled by an "electronic brain" (almost certainly an analogue circuit) that was provided by the Dutch electronics company Philips. In addition to its internal movement, *CYSP 1* was mounted on a mobile base that contained the actuators and control system. Photosensitive cells and a microphone sampled variations in color, light, and sound (see figure 11.1). It was

...excited by the colour blue, which means that it moves forward, retreats or makes a quick turn, and makes its plates turn fast; it becomes calm with red, but at the same time it is excited by silence and calmed by noise. It is also excited in the dark and becomes calm in intense light.

On its second outing *CYSP 1* performed with Maurice Béjart's ballet company on the roof of Le Corbusier's Cité Radieuse, as part of the Avant-Garde Art Festival held in Marseilles in 1956. Schöffer said of his work: "Spatiodynamic sculpture, for the first time, makes it possible to replace man with a work of abstract art, acting on its own initiative, which introduces into the show world a new being whose behaviour and career are capable of ample developments."[12]

Figure 11.1
Nicolas Schöffer, CYSP 1, 1956. © ADAGP, Banque d'Images, Paris 2007. Printed with permission.

Schöffer worked closely with composers and choreographers, including Pierre Henry and Alwin Nikolais. These three together created *KYLDEX*, the first experimental cybernetic show, at the Hamburg Opera House in 1973. Schöffer is also credited with making the first video production in the history of television, *Variations Luminodynamiques 1*, for Télévision Française in 1960 and so in addition to his considerable contribution to the world of kinetics and autonomous arts he is also remembered as the "father" of video art.

The same year that *CYSP 1* danced in Marseilles, 1956, across the Channel in the United Kingdom the Independent Group—consisting of artists, architects, designers, and critics who challenged prevailing approaches to culture—put together a show at London's Whitechapel Gallery, "This Is Tomorrow," which became an influential landmark in the history of the contemporary arts in the UK. Charlie Gere has pointed out that the catalogue contains what is possibly the first reference to punch cards and paper tape as artistic media.[13] Robby the Robot, star of Fred Wilcox's then recently released (1956) film *Forbidden Planet*, attended the opening and the show received a high popular profile in the British press. *Forbidden Planet* bucked the trend of most American sci-fi movies of the time—where Communists disguised as aliens are taught that freedom and democracy come out of the barrel of a gun—with a thoughtful script that was loosely based on Shakespeare's *The Tempest*. But in the film the spirit world is a product of cybernetic amplification of the human subconscious. The film was influenced by the popular science and psychology of the day and also contains echoes of Shelley's *Frankenstein*.

The mood of the time was strongly pro-science—the public action of the Campaign for Nuclear Disarmament (founded 1958) and televized atrocities of the Vietnam War, which would alienate people from science's perceived military agenda, were still a decade in the future. *Eagle* was a popular comic book of the day geared toward middle-class boys, one issue of which featured a car powered by a small nuclear power pack that would never need refueling and was expected on Britain's roads before the turn of the century! In 1963 the Labour prime minster Harold Wilson promised that the "white heat of technology" would solve the country's problems, and a golden age of plenty, delivered by science and its machines, seemed imminent.

In Germany Herbert Franke produced his first Oszillogramms in 1956. The mathematician, physicist, and philosopher Max Bense (1910–1990) proposed his concept of Information Aesthetics the next year, when he

Figure 11.2
Galerie Wendelin Niedlich, Stuttgart. Screenshot of virtual reconstruction of the gallery room with exhibition of computer art by Frieder Nake and Georg Nees, Nov. 1965. Courtesy Yan Lin-Olthoff.

brought together aspects of information theory, cybernetics, and aesthetics.[14] At about the same time the French theorist Abraham Moles (1920–1992) published his work in the area.[15] A decade later, in 1965, Bense curated what is believed to be the first public exhibition of computer art in the world when he invited the computer-graphics artist Georg Nees to show his work at the Studiengalerie der Technischen Hochschule (Technical University) in Stuttgart. The exhibition ran February 5 to 19. This encouraged the artist Frieder Nake to show his work, along with Nees later that year from November 5 to 26 at Stuttgart's Galerie Wendelin Niedlich (figure 11.2). Many of the European artists working in the new field congregated in Zagreb in August 1968 for a colloquy, "Computers and Visual Research," that was part of the New Tendencies Movement; it led to a major exhibition called "Tendencies 4," which ran May 5 to August 30, 1969. Rainer Usselmann has suggested that these meetings confronted sociopolitical issues associated with the new technologies (and especially the military agendas) that were absent from the more playful British debate—especially the signal event that has come to epitomize the period.[16]

A suggestion from Max Bense in 1965 inspired writer and curator Jasia Reichardt to organize the exhibition that now stands as a defining moment in the history of the computational arts. The show "Cybernetic Serendipity" opened at London's Institute of Contemporary Art on August 2, 1968 and ran until 20 October 1968.[17] Reichardt recently described it as

...the first exhibition to attempt to demonstrate all aspects of computer-aided creative activity: art, music, poetry, dance, sculpture, animation. The principal idea was to examine the role of cybernetics in contemporary arts. The exhibition included robots, poetry, music and painting machines, as well as all sorts of works where chance was an important ingredient.

The show coincided with and complemented the release of one of the major cultural artifacts of the period, Stanley Kubrick's enigmatic film *2001: A Space Odyssey*. It features a self-aware artificial intelligence—HAL 9000—that has a psychotic breakdown when it is unable to resolve conflicting data.

Among work by over three hundred scientists and artists at "Cybernetic Serendipity" was a piece by the British cybernetician Gordon Pask (1928–1996). The *Colloquy of Mobiles* (figure 11.3) consisted of five ceiling-mounted kinetic systems—two "males" and three "females." Using light and sound they could communicate with each other in order to achieve "mutual satisfaction." The system could learn, and the mobiles optimized their behavior so that their goal could be achieved with the least expenditure of energy. Members of the public, using flashlights and mirrors, could also interact with the mobiles and influence the process.[18]

Pask also worked with the architect John Frazer, the artist Roy Ascott, and others as an adviser to the Fun Palace Project, conceived by Archigram's Cedric Price and the socialist theatrical entrepreneur Joan Littlewood.[19] Although the Fun Palace, a dynamically reconfigurable interactive building, was never built, it had a wide influence; for example, it inspired Richard Rogers and Renzo Piano's Centre Georges Pompidou in Paris. In the seventies Frazer worked closely with Pask at the Architectural Association and is notable for his concept of the Intelligent Building.[20]

"Cybernetic Serendipity" also included Edward Ihnatowicz's (1926–1988) sound-activated mobile, or SAM. Ihnatowicz would later describe himself as a Cybernetic Sculptor.[21] SAM consisted of four parabolic reflectors shaped like the petals of a flower, on an articulating neck. Each reflector focused ambient sound on its own microphone; an analogue

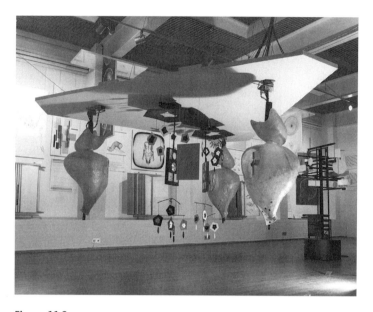

Figure 11.3
Gordon Pask, Colloquy of Mobiles, installation shot from Cybernetic Serendipity (1968). Courtesy Jasia Reichardt.

circuit could then compare inputs and operate hydraulics that positioned the flower so it pointed toward the dominant sound. SAM could track moving sounds, and this gave visitors the eerie feeling that they were being observed. Not long after, Ihnatowicz was commissioned by Philips to create the Senster (figure 11.4) for the company's Evoluon science center in Eindhoven. The Senster was a twelve-foot ambitious minicomputer-controlled interactive sculpture that responded to sound and movement in a way that was exceptionally "life like" (it was exhibited from 1970 to 1974, when it was dismantled because of high maintenance costs).[22] Ihnatowicz was an early proponent of a "bottom-up" approach to artificial intelligence—what we would now call artificial life. His reading of the work of the developmental psychologist Jean Piaget inspired him to suggest that machines would never attain intelligence until they learned to interact with their environments.[23]

The socialist techno-utopian vision that played a major role in European politics and culture of the period was less influential in the Communist-phobic United States. In consequence, developments there were less centralized, more sporadic, and often linked to artists' initiatives or the

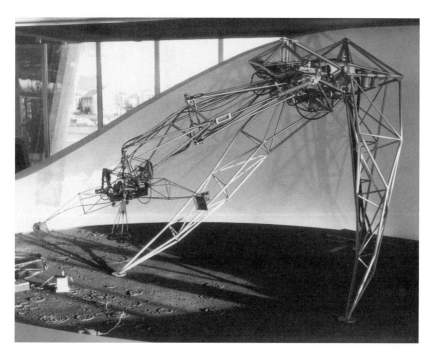

Figure 11.4
Edward Ihnatowicz, The Senster, 1970. Courtesy Olga Ihnatowicz.

commercial art world rather than state-patronized social agendas. Ben Laposky (1914–2000) began to make his analogue Oscillons in 1950, the same year the composer John Cage (1912–1992) discovered the *I Ching*. This profoundly influenced Cage's career, which increasingly involved technology and chance elements. He used coin tosses to determine pitch, rhythm, dynamics, and duration of his "Music of Changes," written in 1951, and he created the masterpiece "4′ 33″" the next year. In this work the performer stands still on the stage and the audience listens to the ambient sounds and silence. In 1952 Cage began working with electronic music, and in 1967, with Lejaren Hiller, he produced the ambitious computer-assisted "HPSCHD." The name reflects the contemporary use of a "high level" programming language, FORTRAN (FORmula TRANslation), that allowed only six-character names, in uppercase, and that often omitted vowels. The year before, in 1966, Cage was one of many artists who contributed to the defining event of art-technology collaborations in the United States. "9 Evenings: Theater and Engineering" was produced by

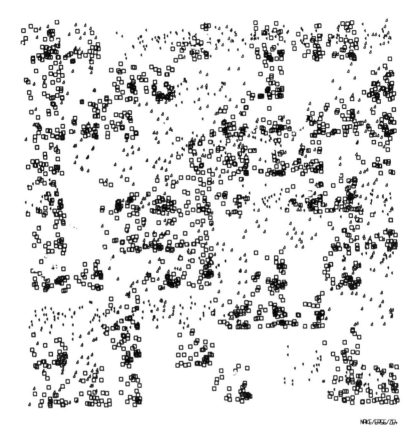

Figure 11.5
Frieder Nake: 13/9/65 Nr. 5, "Random distributions of elementary signs," China ink on paper, 51 × 51 cm, 1965. Possession of Sammlung Etzold, Museum Abteiberg, Mönchengladbach. First prize Computer Art Contest 1966, Computers and Automation. Printed with permission.

the Experiments in Art and Technology (EAT) group, and was set up by Billy Klüver and Fred Waldhauer with the artists Robert Rauschenberg and Robert Whitman.[24]

Starting in 1963, the journal *Computers and Automation* sponsored a computer art competition; in 1963 and 1964 the winning entries were visualizations from the U.S. Ballistics Research Lab at the Aberdeen Proving Ground in Maryland. Michael Noll won in 1965 and Frieder Nake in 1966 (see figure 11.5). Noll had produced the first computer graphics artwork in 1962. The United States' first computer art exhibition, "Computer Generated Pictures," was held April 6 to 24, 1965 at the Howard Wise Gallery in

New York (just three months after the pioneering Stuttgart show) and featured work by Noll and Bela Julesz (1928–2003). Charies "Chuck" Csuri, a sculptor, established a pioneering computer arts lab at Ohio State University, where Tom Defanti completed his Ph.D. before collaborating with the artist-engineer and video art pioneer Dan Sandin. In 1974 together Defanti and Sandin established the Electronic Visualization Lab at the University of Illinois, Chicago Circle, and later the world's first M.F.A. program in computer arts. It's believed that Copper Gilloth was the first graduate. A year earlier, Myron Kruger, who had collaborated with Sandin, coined the term "artificial reality" to describe his interactive immersive computer-based art installations.

London in the 1960s was "swinging" and the art world was fertile anarchistic ground for any and all new ideas. Jim Haynes set up the London Arts Lab on Drury Lane and the London Filmmakers Coop was established. Later the Arts Lab moved to Camden as an artist-run space called the Institute for Research into Art and Technology; from 1969 it included the Electronics and Cybernetics Workshop (possibly a single mechanical teletype and a 300-baud modem) that was organized by John Lifton and offered free and exclusive computer access to artists for the first time. At Ealing College in 1961 the recently graduated Roy Ascott was appointed head of Foundation Studies, where he developed the influential Groundcourse.[25] He recruited an impressive team of young artists as teachers, and visitors included Pask and the linguist Basil Bernstein. Ascott and others believed that it was the process, rather than the product, that provided the essential content of the artwork. This became a dominant aesthetic of the arts in the latter part of the twentieth century, influencing the formation of several movements including Art & Language, Conceptual Art, and Systems Art.[26] Stephen Willats was a student of Ascott's who went on to produce some major works linking art and technology with a social agenda; his contribution has recently been reassessed.[27] Stroud Cornock, a colleague of Ascott's, moved to the City of Leicester Polytechnic, where he met the artist and mathematician Ernest Edmonds. They coauthored the influential paper "The Creative Process Where the Artist is Amplified or Superseded by the Computer," and Edmonds went on to establish the Creativity and Cognition Lab (originally at Leicester, then at Loughborough, and now at the University of Technology, Sydney), as well as found the ACM Creativity and Cognition conference series.[28] Ascott later pioneered the use of communication networks in the arts and more recently has established the Planetary Collegium as a global initiative intended to encourage scholarly research in the field of art, technology, and consciousness.[29]

Figure 11.6
George Mallen, The Ecogame, 1969. Courtesy the Computer Arts Society.

In 1969 the Computer Arts Society was cofounded by Alan Sutcliffe, John Lansdown, and George Mallen.[30] Mallen had worked closely with Gordon Pask at his company Systems Research, and for the CAS launch—Event One, an exhibition at the Royal College of Art—he produced a remarkably sophisticated (especially considering the rudimentary technology of the time) interactive computer artwork called *The Ecogame* (figure 11.6).[31] The CAS bulletin, *PAGE*, originally edited by Gustav Metzger, is still in print and forms a valuable historical record.[32] The same year that CAS was formed, Penguin published a book called *Systems Thinking*, edited by an Australian, Fred Emery, as an inexpensive paperback special.[33] It contained chapters by W. Ross Ashby and Geoff Summerhoff, among others, and because of its accessibility it was widely influential throughout the art world in the UK, being on the recommended book list for many foundation and undergraduate fine arts courses in the UK. Two books by the left-wing cybernetician Stafford Beer, *Designing Freedom* and *Platform for Change*, were also influential as the 1970s progressed.[34] Although the systems art movement was pan-European, the Systems Group was primarily based in the UK. Malcolm Hughes, a member, was also head of postgraduate studies at

Figure 11.7
Paul Brown, CBI North West Export Award, 1976. An early alife work by the author that was driven by a dedicated digital circuit.

the Slade School of Fine Art, University College, London. He set up what became the Experimental and Computing Department, or EXP, in 1973 under Chris Briscoe, where the systems ethos was transferred into the computer domain. The emerging ideas of deterministic chaos, fractals, and cellular automata were influences and the output of EXP forms a root of both the computational and generative arts and the scientific pursuit of A-life (see figures 11.7 and 11.8).[35] Edward Ihnatowicz, who was then based in the Mechanical Engineering School at University College London, was a regular visitor, as was Harold Cohen who was working on an early version of his expert drawing system, AARON, at the University of California, San Diego.[36] From 1974 to 1982, when it closed, EXP was a major focus for artists from around Europe who were working in the computational domain.

In 1970 two important exhibitions took place in New York. Kynaston McShine's "Information" show at the Museum of Modern Art was an eclectic, idiosyncratic mix of conceptual formalism, linguistic and information

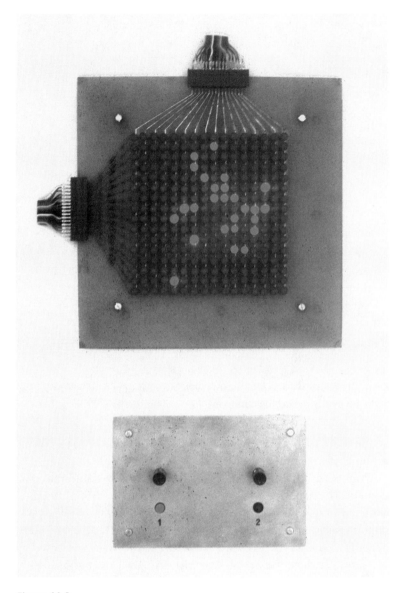

Figure 11.8
Paul Brown, Life/Builder Eater, 1978. An alife work by the author produced at EXP.
Believed to be the first artwork to have an embedded microprocessor.

theories, and sociopolitical activism.[37] Jack Burnham's software show, "Software—Information Technology: Its New Meaning for Art," at the Jewish Museum, was intended to draw parallels between conceptual art and theories of information such as cybernetics. The show included work by a young architect named Nicholas Negroponte, who would later found MIT's Media Lab. Burnham, in his earlier influential book, *Beyond Modern Sculpture* (1968) had suggested that art's future lay in the production of "life-simulation systems."[38] Many artists of the time agreed and believed that the world of art would be radically transformed by an imminent revolution and undergo what the philosopher of science Thomas Kuhn had recently described as a "paradigm shift."[39] The art world did change, but not in the radical way these artists and theorists expected; by the 1980s it was being driven by humanities-educated graduates who identified more with the eclecticism of McShine than with the focused analytical vision of Burnham and the systems and conceptual artists. They adopted the emerging theories of postmodernism and tended to be unfamiliar with, and deeply suspicious of, computing and information technology, which they identified with the growth in power of what later became known as the military-industrial-entertainment complex. In my opinion they made a singular mistake: by identifying the kind of developments I have described with the absolute narratives of utopian modernism (which, to be fair, is not an altogether unreasonable association) they ignored aspects such as emergence, nonlinearity, hypermediation, interaction, networking, self-similarity, self-regulation that should have been seen—and more recently have been acknowledged—as central to the postmodern debate. It was a classic case of throwing out the baby with the bathwater.

The ongoing lack of support for computer art from the arts mainstream throughout the latter decades of the twentieth century led to the formation of an international "salon des refusés." The Computer Arts Society ran several exhibitions in the unused shells of computer trade shows in the late 1970s and early '80s in the UK and in 1981 in the United States the first SIGGRAPH Art Show was curated by Darcy Gerbarg and Ray Lauzzana; the latter also established the influential bulletin board "fineArt forum" in 1987.[40] The Austrian Ars Electronica convention and Prix was launched in 1979, and in 1988 the International Symposium on Electronic Arts series began in Utrecht, The Netherlands.[41] These international opportunities were, and most of them remain, important venues for debate and exhibition of work that until recently rarely found its way into the established gallery system. Thanks in major part to this "patronage," a younger generation of computational and generative artists emerged in the 1980s and early '90s, whose ranks include Stelarc, Karl Sims, Yoichiro Kawaguchi, William Latham, the Algorists, Michael Tolson, Simon Penny, Jon McCormack, Troy Innocent, Ken Rinaldo, Richard Brown, and many others.

Acknowledgments

My apologies to those whom I have left out because of space limitations. This essay would not have been possible without my participation in the CACHe project (Computer Arts, Contexts, Histories, etc.), which took place from 2002 to 2005 in the History of Art, Film, and Visual Media Department at Birkbeck College, University of London. I would like to thank my colleagues on that project: Charlie Gere, Nick Lambert, and Catherine Mason for their ongoing contributions and support, especially for their informed comments about early revisions of this chapter. I am indebted to Frieder Nake and Margit Rosen for helpful comments on an earlier draft. I must also thank the UK's Arts and Humanities Research Council, who funded the CACHe project, and Phil Husbands, the coeditor of this volume, for his valuable feedback.

Notes

1. J. Needham, *Science and Civilisation in China*, volume 2: *History of Scientific Thought* (Cambridge: Cambridge University Press, 1956).

2. R. Wilhelm, trans., *The I Ching or Book of Changes*, edited by Cary F. Baynes (London: Routledge & Kegan Paul, 1968).

3. Gunalan Nadarajan, "Islamic Automation: A Reading of Al-Jazari's *Book of Ingenious Mechanical Devices* (1206)," in *MediaArtHistories*, edited by Oliver Grau (Cambridge, Mass.: MIT Press/Leonardo, 2007).

4. Hermann Hesse, Das Glasperlenspiel [*The Glass Bead Game*; also *Magister Ludi*] (Zurich: Fretz & Wasmuth Verlag, 1943), and Umberto Eco, *The Island of the Day Before*, translated by William Weaver (Harcourt Brace, 1984).

5. Mary Shelley, *Frankenstein* (London: Lackington, Hughes, Harding, Mavor & Jones, 1818).

6. The illustrations as part of the e-book are available at "Thought Forms, by Annie Besant and C. W. Leadbetter," http://www.anandgholap.net/Thought_Forms-AB_CWL.htm.

7. William Gibson, The Cyberspace Trilogy: *Neuromancer* (New York: Ace Books, 1984); *Count Zero* (Arbor House, 1986); *Mona Lisa Overdrive* (Bantam Spectra, 1988).

8. Walter Benjamin, "The Work of Art in the Age of Mechanical Reproduction," in *Illuminations*, translated by H. Zohn, edited and with an introduction by Hannah Arendt (New York: Schocken Books, 1969).

9. Roger Malina, Fluid Arts website, "Educational Projects in Art, Science and Technology," at http://www.fluidarts.org/projects/space.html.

10. An obituary of Frank Malina, the founder of *Leonardo*, is available at the Fondation Daniel Langlois website, at http://www.fondation-langlois.org/html/e/page .php?NumPage=233.

11. Norbert Wiener, *Cybernetics or Control and Communication in the Animal and the Machine* (Cambridge, Mass.: MIT Press, 1948), and W. Ross Ashby, *Introduction to Cybernetics* (London: Chapman & Hall, 1956). The complete book can be downloaded from the Principia Cybernetica website, http://pespmcl.vub.ac.e/ASHBROOK.html.

12. Schöffer quoted in "CYSP 1. (1956), The First Cybernetic Sculpture of Art's History," at http://www.olats.org/schoffer/cyspe.htm.

13. Charlie Gere, "Introduction," in *White Heat Cold Logic: British Computer Art 1960–1980*, edited by Paul Brown, Charles Gere, Nicholas Lambert, and Catherine Mason (Cambridge, Mass.: MIT Press/Leonardo, forthcoming).

14. Max Bense, *Aesthetica* (Baden-Baden: agis Verlag, 1965). This volume contains four separate publications, *Aesthetica* I to IV, published by agis between 1954 and 1960.

15. A. A. Moles, *Théorie de l'information et perception esthétique* (Paris: Flammarion, 1958); published in English as *Information Theory and Esthetic Perception* (University of Illinois Press, 1966).

16. Christoph Klütsch, "The Summer of 1968 in London and Zagreb: Starting or End Point for Computer Art?," in *Proceedings of the Fifth Conference on Creativity and Cognition* (London, April 12–15, 2005), *C&C '05* (New York: ACM Press, 2005); and Rainer Usselmann, "The Dilemma of Media Art: Cybernetic Serendipity at the ICA," *Leonardo* 36, no. 5: 389–96 (MIT Press, 2003).

17. Jasia Reichardt, "In the Beginning," and B. MacGregor, "Cybernetic Serendipity Revisited," both in Brown, Gere, Lambert, and Mason, *White Heat Cold Logic*; Reichardt, "Cybernetic Serendity," press release, available at http://www.medienkunstnetz/de/ exhibitions/serendipity.

18. Margit Rosen, "Gordon Pask—The Colloquy of Mobiles," available at http:// medienkunstnetz.de/works/colloquy-of-mobiles.

19. Design Museum, "Cedric Price, Architect (1934 to 2003)," available at http:// www.designmuseum.org/design/cedric-price.

20. James Frazer, "Interactive Architecture," in Brown, Gere, Lambert, and Mason, *White Heat Cold Logic*.

21. See Alex Zivanovic's comprehensive website on Ihnatowicz's work, "Senster—A website devoted to Edward Ihnatowicz, cybernetic sculptor," at http://www.senster .com.

22. Alex Zivanovic, "The Technologies of Edward Ihnatowicz," in Brown, Gere, Lambert, and Mason, *White Heat Cold Logic*.

23. Edward Ihnatowicz, personal communication with Paul Brown, mid-1970s.

24. Fondation Daniel Langlois, "Billy Klüver—Experiments in Art and Technology (E.A.T)," available at http://www.fondation-langlois.org/html/e/page.php? NumPage=306.

25. Roy Ascott, "Creative Cybernetics: The Emergence of an Art Based on Interaction, Process and System," in Brown, Gere, Lambert, and Mason, *White Heat Cold Logic*.

26. Lucy Lippard, *Six Years: The Dematerialization of the Art Object from 1966 to 1972* (London: Studio Vista, 1973); G. Howard, "Conceptual Art, Language, Diagrams and Indexes," in Brown, Gere, Lambert, and Mason, *White Heat Cold Logic*.

27. George Mallen, "Stephen Willats: An Interview on Art, Cybernetics and Social Intervention," *PAGE 60, Bulletin of the Computer Arts Society*, Spring 2005.

28. S. Cornock and E. A. Edmonds, "The Creative Process Where the Artist Is Amplified or Superseded by the Computer," *Leonardo* 6, no. 1: 11–15 (MIT Press, 1973); E. A. Edmonds, "Constructive Computation," in Brown, Gere, Lambert, and Mason, *White Heat Cold Logic*.

29. Roy Ascott and C. E. Loeffler, eds., "Connectivity—Art and Interactive Telecommunications," *Leonardo* (special issue) 24, no. 2 (MIT Press, 1991).

30. The Computer Arts Society is a specialist group of the British Computer Society. For more information on CAS, see http://www.computer-arts-society.org; Alan Sutcliffe, "Patterns in Context," in Brown, Gere, Lambert, and Mason, *White Heat Cold Logic*.

31. George Mallen, "Bridging Computing in the Arts and Software Development," in Brown, Gere, Lambert, and Mason, *White Heat Cold Logic*.

32. Issues of *PAGE* are accessible on-line at http://www.computer-arts-society.org/ page/index.html.

33. Fred Emery, ed., *Systems Thinking* (Harmondsworth, UK: Penguin Books, 1969).

34. Stafford Beer, *Designing Freedom* (Toronto: CBC Learning Systems, 1974; London and New York: Wiley, 1975); Beer, *Platform for Change* (London and New York: Wiley, 1975; rev. ed., 1978).

35. Paul Brown, "From Systems Art to Artificial Life: Early Generative Art at the Slade School of Fine Art," in Brown, Gere, Lambert, and Mason, *White Heat Cold Logic*; Brown, "The CBI North West Award," *PAGE* 62 (Autumn 2005), pp. 12–14.

36. Harold Cohen, "Reconfiguring," in Brown, Gere, Lambert, and Mason, *White Heat Cold Logic*.

37. E. Meltzer, "The Dream of the Information World," *Oxford Art Journal* 29, no. 1: 115–35(2006).

38. Jack Burnham, *Beyond Modern Sculpture: The Effects of Science and Technology on the Sculpture of This Century* (New York: Braziller, 1968).

39. Thomas S. Kuhn, *The Structure of Scientific Revolutions* (Chicago: University of Chicago Press, 1962).

40. For more information on ACM SIGGRAPH (the Association for Computing Machinery's Special Interest Group on Graphics and Interactive Techniques) see http://www.siggraph.org; for more information on the fineArt forum, see their website, "fineArt forum: art + technology net news," at http://www.fineartforum.org.

41. For further information about Ars Electronica, see their website at http://www.aec.at/en/index.asp; for further information on the Inter-Society for Electronic Arts, see their website at http://www.isea-web.org/eng/index.html.

12 The Robot Story: Why Robots Were Born and How They Grew Up

Jana Horáková and Jozef Kelemen

This is a story of why, how, and where the word *robot* was born, how it changed its meaning, how it grew up, and how it spread, becoming a part of languages all over the world. Today we talk of robots as of one of the myths of the second half of the past century and at least the first decade of the twenty-first century. These myths live in the form of jokes, fairy tales, and legends, often without our knowing their authors.[1] But we encounter robots also in science fiction novels and movies, and we bump into them in art exhibitions and in the press as well as in research laboratories. How did it all start?

In order to make our robot story not only as objective as possible but also provocative and inspiring, we have felt free to include, from time to time, our own impressions, views, and hypotheses on the robot's history and destiny. We are going to describe the concept of a robot in the context of its inventors' work as well as in the wider cultural context of the period when the famous play *R.U.R.* (*Rossum's Universal Robots*), in which the word *robot* first appeared, was written, and of the contexts in which it spread. We have decided to proceed in this way because we believe that a broader knowledge of the cultural background to present-day research in the fields of advanced robotics, artificial intelligence, artificial life, and other related fields will help specialists and maybe many others to understand these fields in a wider context. This will also aid the understanding of positive public expectations, as well as misgivings, connected with the possible results of these fields of enquiry.

The Robot's Parents

It is a relatively commonly known fact that the word *robot* first appeared in 1920 in the play *R.U.R.* (*Rossum's Universal Robots*) by the Czechoslovak journalist and writer Karel Čapek (1890–1938). The play was at least

Figure 12.1
The Spatown Trenčianske Teplice in Slovakia in 1986. The sanatorium Pax replaced the Čapek brothers' parents' house, where Karel Čapek wrote at least some parts of the *R.U.R.* and where Josef Čapek invented the neologism *robot*. Photo by Jozef Kelemen.

partially written during the summer of 1920, when he, his brother, Josef (1887–1945), and their sister, Helena (1886–1961), were on vacation in their parents' house in the spa town of Trenčianske Teplice, Slovakia, where their father worked as a physician (see figure 12.1).

There are a number of stories about how the idea and the word emerged. For example, Karel Čapek published following version of the story in the December 24, 1933, issue of the Prague newspaper *Lidové noviny* (People's News):[2]

A reference by Professor Chudoba to the *Oxford Dictionary* account of the word *Robot*'s origin and its entry into the English language reminds me of an old debt. The author of the play *R.U.R.* did not, in fact, invent that word; he merely ushered it into existence. It was like this: The idea for the play came to said author in a single, unguarded moment. And while it was still warm he rushed immediately to his brother, Josef, the painter, who was standing before an easel and painting away at a canvas till it rustled.

"Listen, Josef," the author began, "I think I have an idea for a play."

"What kind," the painter mumbled (he really did mumble, because at the moment he was holding a brush in his mouth).

The author told him as briefly as he could.

"Then write it," the painter remarked, without taking the brush from his mouth or halting work on the canvas. The indifference was quite insulting.

"But," the author said, "I don't know what to call these artificial workers. I could call them Labori, but that strikes me as a bit bookish."

"Then call them Robots," the painter muttered, brush in mouth, and went on painting. And that's how it was. Thus was the word *Robot* born; let this acknowledge its true creator.

The word *robot* is a neologism derived etymologically from the archaic Czech word *robota*, a word rooted in the ancient Slavonic protolanguage from which today's Slavonic languages (Czech, Polish, Russian, Slovak, Ukrainian, and so forth) have developed. In present-day Czech and Slovak, *robota* means something like a serf's obligatory work.

However, it is (at least internationally) not so commonly known that the true coiner of the word *robot*, Josef Čapek, is recognized as a representative of Czech cubism (influenced by the naive style), and that his work as a painter is an important contribution to twentieth-century Czech art. But in the history of Czech culture Josef Čapek is also highly regarded as a writer and as the author of numerous short stories. Moreover, the Čapek brothers collaborated on many works, especially at the beginning of their careers, and they remained important sources of inspiration for each other until the ends of their lives (see figure 12.2). Especially at the beginning of their careers, both were influenced by the time they spent together in Paris, where they were exposed to all the latest modern styles and the various *-isms* that emerged during the first third of the twentieth century.

The Pedigree

We have cited Karel Čapek's own description of the birth of the word *robot*. He mentioned more about the birth of the *idea* in an article in the British newspaper *The Evening Standard*, published on July 2, 1924:

Robots were a result of my traveling by tram. One day I had to go to Prague by a suburban tram and it was uncomfortably full. I was astonished with how modern conditions made people unobservant of the common comforts of life. They were stuffed inside as well as on stairs, not as sheep but as machines. I started to think about humans not as individuals but as machines and on my way home I was thinking about an expression that would refer to humans capable of work but not of thinking. This idea is expressed by a Czech word, *robot*.

More generally, the idea of Čapek's robots might be viewed as a twentieth-century reincarnation of an old idea present in European culture—the idea of a man created by man. In Genesis 2:7 we read, "The Lord God formed

Figure 12.2
Josef and Karel Čapek, about 1922. Reproduced from Čapkova (1986) with permission.

man of the dust of the ground and breathed into his nostrils the breath of life; and man become a living soul." From this perspective, Adam, a product of the technology of pottery, is the oldest predecessor of robots. Moreover, stating that "God said: Let us make man in our image" (Genesis 1:26), the Bible also gave to Western civilization the ideological assurance that not only God but also man is able to perform creative acts.

The role of machines that interact with each other and cooperate with human beings is also present in Aristotle's contemplations on the possibility of changes to the social structure of human society. In Book 2 of his fundamental work *Politics* he wrote (Aristotle 1941, pp. 33–39):

For if every instrument could accomplish its own work, obeying or anticipating the will of others, like the statue of Daedalus, or the tripods of Hephaestus, which, says the poet, "of their own accord entered the assembly of the Gods"; if, in like manner, the shuttle would weave and the plectrum touch the lyre without a hand to guide them, chief workmen would not want servants, nor master slaves.

In *The Iliad*, Homer (1998) expressed this dream of artificial humanlike creatures, created by means of the technology of metalworking, as follows: "There were golden handmaids also who worked for him, and were like real young women, with sense and reason, voice also and strength, and all the learning of the immortals" (pp. 415–20).

In the September 23, 1935, issue of the German-language Prague newspaper *Prager Tagblatt*, Karel Čapek himself mentioned the relationship of his robots to one of the most famous artificial servants of man, the Golem, a medieval legend that combined material technology with the mysterious power of symbols: "*R.U.R.* is in fact a transformation of the Golem legend into a modern form," Čapek wrote. "However, I realized this only when the piece was done. 'To hell, it is in fact Golem,' I said to myself. 'Robots are Golem made with factory mass production.'"

The legend of the Golem lives on in Prague up to the present day (see Petiška 1991). According to the legend, a famous Prague rabbi at the end of the sixteenth century and the beginning of the seventeenth century, Judah Loew ben Bezalel (a real person who is buried in the Jewish cemetery, in Prague's Old Town, constructed a creature of human form, the Prague Golem (figure 12.3). He proceeded in two main stages: First, he and his collaborators constructed the earthen sculpture of a manlike figure. Second, he found the appropriate text, wrote it down on a slip of paper, and pushed it into the Golem's mouth. So long as this seal remained in the Golem's mouth, he had to work and do the bidding of his master, performing all kinds of chores for him, helping him and the Jews of Prague in many ways.

Another root of the concept behind robots can be traced to *androids*, a term that first appeared in about 1727, referring to humanoid automata, and an idea still current in the Čapeks' time. The historical age in which such androids were particularly popular, the eighteenth century, is usually called the Age of Reason. "Reason" in the Czech language is *rozum*, pronounced "rossum." We can see why the name of the first constructor of robots in Čapek's play *R.U.R.* is Rossum, and why *Rossum's Universal Robots* is the title of the play.[3] Further conceptual forefathers of robots that are only occasionally mentioned appeared in the writings of the Čapek brothers before 1920 in the specific political and cultural context of that period in Europe.

In addition to the dreams expressed in such influential books as the Old Testament, Homer's *Iliad*, or in the legends of the Golem, we can also find real artifacts from at least the beginning of the eighteenth century that are evidence of engineers' efforts to design and produce human-like machines.

Figure 12.3

The Prague Old-New Synagogue (Altneuschul) connected with the legend of the Golem, and the usual shape of Golems in present-day Prague gift shops. Photos by Jozef Kelemen.

First, let us look to the Age of Reason both for ideas concerning robots and for technical activities. The famous mechanical duck developed by Jacques de Vaucanson from 1738 or the mechanical puppets constructed by the Jaquet-Droz family in the period 1772 to 1774, and now exhibited in the Historical Museum in Neuchâtel, Switzerland (see Capuis and Droz 1956, for more details), are good examples of inspiration of this latter type. However, the well-known chess-playing mechanical Turk developed in this period is more closely related to robots.

During the eighteenth century an Austro-Hungarian nobleman, Johannes Wolfgang von Kempelen (1734–1804), who lived in the Central European city of Pressburg (now Bratislava), the capital of the Slovak Republic, was full of ideas.[4] Ideas about how to organize the production and safe transport of salt in the Austro-Hungarian Empire, about how to build a bridge over the Danube River in Pressburg, how to construct a speaking machine for the dumb, and—last but not least—how to construct a mechanical chess-playing machine. This machine, constructed in 1770—in a certain sense something like today's autonomous embodied agents—would behave not in the traditional way of acting according to the intentions of its user, but in the opposite way. It would act against its user's intentions: a mechanical human-like machine, camouflaged to look like a human-size moving puppet in the form of a pipe-smoking Turk, it would sit at a chess board and move the pieces in the right way to win games.[5]

Artificial humanoid beings also played a significant part in the modernist view of humanity, which affirms the power of people to improve quality of life with the aid of science and technology. We can find them for instance in the symbolist theater conventions of the beginning of the twentieth century. In Expressionist plays we often meet schematized characters; in fact, two Expressionist plays by Georg Kaiser, *Gass I* and *Gass II*, are generally regarded as having influenced Karel Čapek's work). In art, the cubist image of a human as a union of squares and triangles is reminiscent of deconstructed human-like machines (or machine-like humans?). Two significant dimensions of futurism are its yearning for the mechanization of humans and the adulation of the "cold beauty" of machines made of steel and tubes as depicted in many futurist works.[6]

The Conception

Not only can we trace a hypothetical line between androids and robots, but we can also meet androids in the Čapek brothers' early works. That being the case, it becomes particularly interesting to try and find out why Karel Čapek decided to seek out a new word for his artificial characters and thus, in a certain way, to partly disconnect robots from their forefathers, giving them more contemporary connotations.

The Čapek brothers had already started to deal with the subject of human-like creatures in the form of androids as well as with ideal workers in a few of their works written before 1920.

In 1908 Josef and Karel wrote a short story entitled "System" that was included in the brothers' collection of short stories *The Krakonoš Garden* (in Czech *Krakonošova zahrada*), first published in 1911. In this story the authors expressed their misgivings concerning the reduction of human beings to easily manageable and controllable uniformed workers, as implied by Taylorism and Fordism. The Čapeks understood that the creation of such workers would lead toward the mechanization of humans, who would then become merely a pieces of equipment or tools. The story satirizes the organization of human work within industrial mass production while also critically reflecting the social and political situation at the beginning of the twentieth century in Europe. In the story, in order to produce "ideal" workers, workers are brought together and then aesthetically and emotionally deprived; the ideal worker is called "a kind of construction of *Operarius utilis Ripratoni*" (Mr. Ripraton is the owner of the factory where the so-called "cultural reform" is carried out). Later, one of the workers discovers the existence of individual beauty (in the form of a naked woman), and organizes a revolt in the region in which the *Operarius utilis* is located. During the rebellion the factory owners as well as their families are killed. The similarity of this story to the plot of *R.U.R.* is striking, as we shall see, and the story is often referred to as a conceptual draft of the play.

In 1910 Karel and Josef wrote the short story "L'Eventaille," which featured both a mechanical lady with a fan and also a historical person—Droz with his androids.[7] The story, set in the atmosphere of a rococo carnival ball in a garden, is based on the motif of the interchangeability of masked ladies and gentlemen, androids, and even sleeping and dead people.

In Josef Čapek's "The Drunkard" (in Czech, "Opilec"), published in *Lelio*, a 1917 collection of his short stories, we again come across the idea of a mechanical alter ego of man that predates *R.U.R.* It takes the form of a humanoid automaton that carries out the commands of its creator, an engineer, and again we can recognize a conceptual predecessor of the robot. However, this mechanical alter ego seems to be useless to its creator because it cannot be used to replace him either in his work or in spying on his sweetheart.

Finally, in 1924, after *R.U.R* had been produced, Josef published a long essay entitled "Homo Artefactus" (see Čapek 1997). This work connected the contemporary idea of man-machines with the cultural history of artificial creatures, including homunculi as well as Golems, and examined problematical human-machine relationships—for example, the soldier as being a man in armor carrying a weapon—throughout the whole of history. As noted above, Josef Čapek had been very influenced by cubism and futurism at the beginning of his artistic career, and this essay is his own way of getting over his futurist period by directly facing the consequences of the fact that the futurist movement inclined toward fascism and the adoration of war.[8] From our point of view it is remarkable that he also put into this essay a paragraph dedicated to his brother's robots, writing about them in a characteristically ironic style (Čapek 1997, p. 196):

The action of a young scholar dr. Karel Čapek was very overrated. This rather adventurous writer made his robot in American factories and then he sent this article into the word, leading all educated people abroad into the misapprehension that there is no other literature in Czech other than that for export.... According to Čapek's theories and promises this robot should replace workers, but we are claiming openly that it was not very useful in practice; it was used only in theatrical services.... For that matter, just as living automata of older times were fully constructed from machinery, so they were not in fact humans, Čapek's robots were made exclusively from an organic jelly so they are neither machines nor human. The intuition of a critical countryman was very good when he promptly recognized Čapek's trick and after a first production of robots stated that there had to be some swindle in it. (Translated from Czech by Jana Horáková)

The Plot

The play *R.U.R.* was among the first Czech science fiction texts. The drama is set (as was for example H. G. Wells's *The Island of Dr. Moreau*) on an isolated island. In this isolated and distant place (reminiscent of many others in the history of fantastic literature, particularly of those housing various kinds of utopias), we find only the factory of Rossum's Universal Robots. There is nothing else on the island, so the island is a factory and the factory is an island.

In this utopian island factory, robots, a specific kind of artificial worker— are mass produced and distributed all around the world. They are physically stronger than humans, and do not become exhausted by mechanical work. Originally, they were developed by Rossum senior, a scientist of the "age of knowledge," who wanted to make artificial people "in order to depose God through science." The robots produced by the *R.U.R.* factory are the "younger generation of the old robots"; they are not complete replicas of humans but are very effective in use, being ergonomic devices developed by Rossum junior, an engineer, as the "mind children" of the "age of industry."

As is the case in many literary works of science fiction, the invention—in this case the robots—provides the story's central drive. While H. G. Wells, with whom Čapek is often compared, usually mentions numerous scientific details to make his fantastic inventions credible, there are very few references to the origin of robots in *R.U.R.* (and those are ambiguous). There is mention of some chemical processes needed for the living jelly from which parts of robots are made; we know a little bit about the serial production of different organs, which are then collected on the assembly lines of the factory. (This reference to robots as mechanisms, in contrast to the chemical basis of the organs, relates to the more general picture of the Cartesian view of the human body as a machine, a view emphasized by La Mettrie). Čapek describes the robots' incredible powers of memory and their ability to communicate and count, along with their lack of creative thought and initiative. This is very reminiscent of computers controlling their robotic bodies: "If you were to read a twenty-volume encyclopedia to them, they'd repeat it all to you with absolute accuracy," he mentions, and adds ironically: "They could very well teach at universities" (Čapek 1983, p. 130; this last sentence appeared in the original Czech text of the play, but was not included in the first English translation; see Čapek and Čapek 1961, p. 15).

The robots are being sold on the world market as a cheap labor force. Helena, now the wife of Mr. Domin, makes the top production engineer, Dr. Gall (who is in love with her, like all the other directors of the factory), give the robots "an irritability" that causes outbursts of anger; later these emotions allow the emergence of something like individuality—an ability to make decisions and to behave humanly. When the robots realize their physical and mental superiority over humankind they want to replace them at the top of the hierarchy of living creatures. They declare war against all humans, and destroy the entire human race.

By now, people have lost the ability of reproducing naturally, so Helena destroys the recipe for producing robots because she believes that this is the way to save humanity. But her act is counterproductive and she only accelerates the conflict between robots and humans. The robots kill all the humans on the island, except the master builder, Alquist, and in the rest of the world. However, the robots had all been made without any reproductive system, and could live for only twenty years. For this reason they try to force Alquist to write down the recipe for robot (re)production again. He reproaches them for their mad plan to massacre all humans. When the robots tell him that he is in fact the last living human, he orders them from his study in a fury and falls asleep. He is woken up by two robots, Helena and Primus, who behave like a young human couple falling in love. Unlike the other robots, they have feelings, they sense love, they are willing to protect each other's "life." Alquist sends them excitedly into the world as a new Adam and Eve, as the new generation. This first couple of living robots is already indistinguishable from humans. So the end of the play is very typical for Čapek: Nobody is completely guiltless and nobody is altogether innocent.

Karel Čapek wrote some instructions about the behavior of the robots in the play. In the prologue the robots are dressed like people. Their movements and speech are laconic, their faces are expressionless, and their gaze is fixed. In the play they wear linen fatigues tightened at the waist with a belt, and have brass numbers on their chests. In contrast to the robots in the prologue, in the last act of the play the robotess Helena and the robot Primus talk and act like humans, and behave even more humanly than humans in the play. The female robot Helena even uses Helena Glory's typical articulation of the letter *R*, as if the first couple of robots really carry on a human heritage.

In the play Čapek shows his attitude toward technology and progress. The horrors of World War I, in which technology was so extensively misused, were very recent. He also points out that humans themselves have to be aware of the possibility of falling into stereotyped behaviors, and that the inclination of individuals to identify with a crowd can lead them toward robot-like behaviour. He also thinks about an important part of our life: work. According to him, work is an integral part of human life. It is not only one of our duties, but also one of our essential needs. Without work humankind will degenerate because people will not have any need to improve themselves.

Antecedents

As we have shown already, it is possible to discuss *R.U.R.* in the wider context of the Čapek brothers' stories, particularly those emphasizing the themes of androids and automata. This point of view allows us to see the play *R.U.R.*, as well as Čapek's robots, from a historical perspective. However, we can also understand the play in a context in which it often appears, that of science fiction. We can find several other pieces inspired by fictitious scientific inventions or imaginary devices in Karel Čapek's work: in the novels *Krakatit*, *Továrna na Absolutno* (*Factory of the Absolute*) and *Válka s Mloky* (*War with the Newts*), as well as in his other plays, such as *Věc Makropulos* (*The Macropulos Case*) and *Bílá Nemoc* (*White Disease*).

There are two themes underlying *R.U.R.* First, the introduction of the idea of robots was Čapek's artistic reaction to the contemporary political situation in Europe, which was dominated by problems concerning the social and political status of the proletariat in the industrial society of the time. Second, the play is also an expression of the Čapeks' ambivalent attitude to science, especially nineteenth-century science: on the one hand, science provided a hope for effective solutions to various social problems, but on the other, it evoked fears concerning its misuse or the unexpected consequences of its use.

In the American periodical *The Saturday Review of Literature*, on July 23, 1923, Karel Čapek expressed his views on the origin of his robots by clearly explaining that "the old Rossum . . . is no more and no less than the typical scientific materialist of the past century [the nineteenth]. His dream to create an artificial man—artificial in the chemical and biological sense, not in the mechanical one—is inspired by his obstinate desire to prove that god is unnecessary and meaningless." Twelve years later, in the Prague newspaper *Lidové noviny* (June 9, 1935), he set down his thoughts as to the substance from which the robots are constructed in the play:

Robots are not mechanisms. They have not been made from tin and cogwheels. They have not been built for the glory of mechanical engineering. The author intended to show admiration for the human mind; this was not the admiration of technology, but of science. I am terrified by the responsibility for the idea that machines may replace humans in the future, and that in their cogwheels may emerge something like life, love or revolt.

In the play *R.U.R.* he explained the ontology of robots very clearly through the words of Mr. Domin, the president of the R.U.R. robot factory, recollecting the beginnings of the idea of robots for Helena Glory, who is visiting the factory (Čapek and Čapek 1961, p. 6):

And then, Miss Glory, old Rossum wrote the following in his day book: "Nature has found only one method of organizing living matter. There is, however, another method more simple, flexible, and rapid, which has not yet occurred to nature at all. This second process by which life can be developed was discovered by me today." Imagine him, Miss Glory, writing those wonderful words. Imagine him sitting over a test-tube and thinking how the whole tree of life would grow from it, how all animals would proceed from it, beginning with some sort of beetle and ending with man himself. A man of different substance from ours. Miss Glory, that was a tremendous moment.

The Newborn

R.U.R.'s debut had been planned for the end of 1920 in Prague's National Theater, but it was delayed, probably because of unrest connected with the appointment of Karel Hugo Hilar, a famous Czech Expressionistic stage director, to the position of head of the Theater's actors' chorus. During the delay, an amateur troupe called Klicpera from Hradec Králové (a town about sixty miles east of Prague, where Karel Čapek briefly attended high school) mounted the first production of *R.U.R.* on January 2, 1921, in spite of an official prohibition from the National Theater. The director of this production was Bedřich Stein, an inspector of the Czechoslovak State Railway. F. Paclt performed the role of the robot Primus.[9] Unfortunately, there is no photo documentation of this first production of the play. According to a couple of reviews in local newspapers, the Hradec Králové premier was quite successful, but the troupe was punished with a not-insignificant fine.

The official first night of the play took place three weeks later, on January 25, 1921, in the National Theater (figure 12.4). The director of the production was Vojta Novák. The stage designer was Bedřich Feuerstein, a young Czech architect. Costumes were designed by Josef Čapek.[10]

Vojta Novák had directed the most recent of Karel Čapek's plays produced in the National Theater, *The Robber*, and apparently Čapek himself chose him again. Encouraged by this, Novák on the whole respected the new play's text. In the first act he did make some fairly large cuts, but only to move more quickly to the heart of the piece. However, he shortened the third act so that it became only a brief epilogue. Novák was impressed by the international nature of the cast of characters; writing in a theater booklet for a 1958 production of *R.U.R.* at the Karlovy Vary (Carlsbad) theater (Čapek 1966), he said it represented "the cream of the creative experimental science of leading European nations—the English engineer Fabry, the French physiologist Gall, the German psychologist Hallemeier, the Jewish businessman Busman, and the central director with his Latin surname Domin and first name Harry, referring probably to a U.S. citizenship. They are not just inventors but superior representatives of human progress—modern versions of heroes from old Greek dramas with abilities to achieve miracles" (p. 110).

The stage set, by Bedřich Feuerstein, whom Čapek had also recommended, was in a very contemporary style, in which sober cubist and Expressionistic shapes were used, painted in symbolically lurid colors (figure 12.5).

Josef Čapek worked as a costume designer for the first time for the production of *R.U.R.* At that time it was unusual for much attention to be paid to the costumes, but the first night of *R.U.R.* was an exception. For members of Domin's team Čapek made chef-like jackets with padding to intensify the impression of masculinity. Following Karel Čapek's recommendations in the script, for the robots he designed basically unisex gray-blue or blue fatigues with numbers on the chests for male and female robots. In summary, the Čapek brothers' robots were much more like humans behaving like machines than machines behaving like humans.

Appearing in the roles of robots were Eduard Kohout as Primus and Eva Vrchlická as both Helena Glory and the female robot Helena. Further robots were played by Eugen Wiesner, Anna Červená, Eduard Tesař, Karel

Figure 12.4

A view of the National Theater in Prague,—the venue of the official premier of *R.U.R.* (photo by Jozef Kelemen), and the first picture of a robot, in the robot costume design by Josef Čapek for the National Theater production. On the robot's shirt front is the date of Prague's first night; the face is a caricature of Karel Čapek.

Figure 12.5
The first scene (Prologue) of the National Theater production of *R.U.R.*, directed by
Bedřich Feuerstein. Photo reproduced from Černy (2000) with permission.

Kolár, Václav Zintl, Karel Váňa, Emil Focht, Hynek Lažanský and Václav
Zatíranda.

The director followed the author's idea, in the opening scenes of the
play, of having robots behave "mechanically," speak monotonously, and
cut words to syllables. Later they became more human-like, even though
they maintained a certain stiffness. Later still, Primus and Helena, the pro-
genitors of a new generation, were indistinguishable from humans in all
characteristics.

The critics wrote about the robotess Helena, as performed by Eva
Vrchlická, as a kind of poetic Eve of the new generation. It is interesting
that Eva Vrchlická persuaded Čapek that she would play Helena Glory as
well as the female robot Helena. Čapek hadn't thought of this possibility
before, but was delighted with it, for it bolstered the production's stress on
the continuity between the last human people and the new generation of
robots.

Regrettably, no photos were taken of the robots in action on the first
night; only later, after some small changes had been made in the cast,
were two actors in robot costumes photographed in a studio.

According to the records, the production of the play was very success-ful.[11] There were long queues for tickets in front of the National Theater and the performances sold out in a couple of hours. The show ran until 1927, with thirty-six re-runs. Many theater critics praised its cosmopolitan character and the originality of the theme, and predicted worldwide suc-cess. Regardless of whether or not they liked the play, they expressed their admiration for Čapek's way of thinking.

The critics were right. The play was performed in New York (1922), Lon-don (1923), Vienna (1923), Paris (1924), and Tokyo (1924) as well as in many other cities, and it was soon translated and published in book form all around the world. After the Slovenian (1921) and Hungarian (1922) translations, the German and English versions were published (1923).

After the Prague premier, Karel Čapek became internationally recognized as an Expressionist playwright and an author of science fiction. The subject matter of the play was very topical in many ways and at various levels of interpretation, and its novelty distinguished Čapek from other dramatists of the time. However, different critics saw the relevance of the play in dif-ferent ways. It is interesting to note the contradiction between the author's intentions and the audiences' interpretation of the play that emerged im-mediately after the first production. The audience usually understood the play as a warning against technology and machines, which threatened to wrest control out of human hands. But Čapek never viewed machines as enemies of humans, and according to him the fact that technology could overwhelm humankind was not the main idea of the play. This was one of the reasons why he repeatedly explained his own interpretation of the play.

Some reviewers found parallels between the robot revolt and the contem-porary struggles of the working class, even thought they didn't assert that this was the author's primary viewpoint. However, the play was written in a period, after the First World War, when several countries in Central Eu-rope were experiencing the culmination of various revolutionary workers' movements fighting for changes in their social conditions and status. Čapek expressed in *R.U.R.* something that until that time had no prece-dent. For the actors, interpreting the human-like machines of the modern age was an entirely new and challenging goal; visualizing these robots was equally challenging for stage and costume designers. *R.U.R.* was—and still is—a play that forces you to think about its content, whether you want to or not. It is a play about the similarities of two totally different worlds that mutually overlap, that live and die in each other.

R.U.R., with its futuristic and Expressionist features and cosmopolitan at-mosphere, was quickly appropriated to become a part of North American culture. The play opened on October 9, 1922, in New York, at the Garrick Theatre, where it was performed by the Theatre Guild, a company specializing in modern drama. The director was Nigel Moeller, and the play was translated from the Czech by Paul Selver and Nigel Playfair.[12] The first night was a success. The critic of the *New York Evening Sun* wrote on October 10, 1922: "Like H. G. Wells of an earlier day, the dramatist frees his imagination and lets it soar away without restraint and his audience is only too delighted to go along on a trip that exceeds even Jules Verne's wildest dreams. The Guild has put theatregoers in its debt this season. *R.U.R.* is super-melodrama—the melodrama of action plus idea, a combination that is rarely seen on our stage." The *New York Herald* theater reviewer, A. Woolcott, emphasizing the play's social dimension, wrote on October 10, 1922 that it was a "remarkable, murderous social satire done in terms of the most hair raising melodrama [with] as many social implications as the most heady of Shavian comedies."

In an article by Alan Parker in *The Independent* on November 25, 1922, we can even read an irritated reaction to Čapek. The only reason for the positive reception of Čapek's play, according to Parker, was the success of its premier at the National Theater in Prague and thus, he stated, it was "received with all the respect and reverence that is evoked nowadays by anything that comes out of 'Central Europe.' Had this piece been of American authorship, no producer on Broadway could have been induced to mount it."

In general, however, the play brought its author great fame in the United States, but in the context of American culture the play lost its social satirical edge and the theme was categorized like so many sci-fi stories in which an atavistic folk interest in human-like creatures predominated along with a fear of conflict between human beings and machines (robots) or human-like monsters. In this context it is worth remembering again the Golem, Mary Shelley's Frankenstein, and many other characters in stories from European literature from the centuries before *R.U.R.* was first performed.

It was the topicality of the play that made the greatest impression in the UK. The significant progress of industrial civilization and its social impact and associated economic theories were much discussed in England. The interest generated by *R.U.R.* can be seen in the public discussion organized at St. Martin's Theatre in London, on June 23, 1923, in response to the excited reception the play had received. Such influential personalities of London's political and cultural life as G. K. Chesterton, Lieutenant-Commander Joseph Kenworthy, and George Bernard Shaw participated in the debate (see *The Spectator*, June 30, 1923):

Mr. Chesterton...was at his most amusing when he talked about the "headlong yet casual" rise of capitalism. Mr. Kenworthy saw in the play lessons on the madness of war and the need for internationalism. Mr. Shaw, at one point, turned to the audience calling them Robots, because they read party press and its opinions are imposed on them. Man cannot be completely free, because he is the slave of nature. He recommended a division of the slavery: "If it has to be, I would like to be Robot for two hours a day in order to be Bernard Shaw for the rest of the day."

The Fates

Despite the fact that robots had been intended by their author as a metaphor for workers dehumanized by hard monotonous work, this understanding soon shifted or the robot was misinterpreted as a metaphor for high technology, which would destroy humankind because of humans' inability to prohibit its misuse. The theme of powerful machines jeopardizing humankind seems to have already been current in Čapek's day; perhaps it entered his text unknowingly, against the author's intention.

Meanwhile, another factor—one quite understandable in a theatrical context—influenced the metamorphoses of the meaning of the play: The author is never the sole owner of his work and ideas—the director, the stage designer, the costume designer, and also the actors become coauthors of the performance, which is the right form for a drama's existence. Theatrical performance is a collective work. It is more like a modern kind of ritual that happens again and again "here and now" than an expression of individual talent and ideas related to the subject matter of the author. As a theatrical performance it is possible to see the play R.U.R. as a ritual that represents our relationships, and our fears and desires, to the most significant topics of our times.

So, two fates, two goddesses of destiny stood next to the newborn robot in 1921 and determined its destiny: the first one was Culture, the second one Industry. These fates opened up R.U.R. for interpretation in terms of perhaps the two most appealing topics of twentieth-century intellectual discourse: the problem of human-machine interaction, and the problem of human-like machines. Reflecting the social and political situation of Europe immediately after the end of the World War I, the robots were interpreted first of all as a metaphor for workers dehumanized by hard repetitive work, and consequently as an easily abusable social class.

From the artistic point of view, the artificial humanoid beings used by Čapek in his play may also be understood as his humanistic reaction to the trendy concepts dominating the modernist view of human beings in the first third of the twentieth century, namely, the concept of the "new man" in symbolist theater conventions, in Expressionism, in cubism, and so forth, and most significantly in futurist manifestos full of adulation of the "cold beauty" of the machines made of steel and tubes that they often depicted in their artworks.

As mentioned earlier, in the short story "System" Karel and Josef Čapek expressed their misgivings concerning the simplification and homogenization of human beings into an easily controllable mass of depersonalized workers without human desires, emotions, aesthetics, or even dreams. The style of production of such workers was supposed to be based on the education of human children. The satirical caricature of the organization of mass production, as well as of the goals of education, are extremely clear in this short story, critically reflecting the social and political situation at the time in Europe.

To summarize Karel Čapek's position as expressed in *R.U.R.*, he thought of robots as *simplified humans*, educated in a suitable manner, or perhaps mass-produced using a suitable "biotechnology" in the form of humanoid organic, biochemically based systems in order to form the components most suitable for industry. This conviction was also clearly reflected in the first production of the play, and especially in Josef Čapek's designs for the robots—male and female human beings in simple uniform-like costumes.

R.U.R had been accepted in Prague as a sociocritical drama (another interpretation, as a comedy of confusion, has been proposed in Horáková 2005). However, immediately after its New York premier it was accepted in a rather different way, being compared with famous pieces of science fiction literature, a genre that had only recently emerged, and was interpreted not as a social commentary but in an industrial context. In fact, robots in the American tradition, now widespread all over the industrialized world, have become *complicated machines* instead of Čapek's *simplified human beings*. These complicated mechanisms resemble a twentieth-century continuation of the dreams of older European engineers such as Vaucanson, Jaquet-Droz, Kempelen, and many others, but now stuffed with electronics and microprocessors, and often programmed to replace some human workers.[13] This new understanding of robots was also accepted in certain quarters in Europe, perhaps because of its apparent continuity with the efforts of some of the modernist tendencies in European culture of that time, especially futurism.

The Presence (of Cyborgs)

In 1988, in his lecture delivered at the Ars Electronica festival in Linz, Austria, the French philosopher Jean Baudrillard asked whether he was now a human or a machine: "Bin ich nun Mensch, oder bin ich Maschine?" (Baudrillard 1989). He claimed that today we who are searching for an answer are obviously and subjectively people, but virtually, as he points out, we are approaching machines. It is a statement of the ambivalence and uncertainty created by the current form of workers' relationships to machines in industrial plants, and the postmodern approach to machine processing and the mass dissemination of information. Technology gradually eliminates the basic dichotomies of man/machine and object/subject, and also perhaps some others such as freedom/restraint.

People often have the impression that the problem of the relationship between the mind and the body is only a philosophical matter, fairly remote from something that can actually affect us. It is as if thoughts about a subject in other than an anthropomorphic context were by definition pointless, and thoughts about cyborgs—a certain kind of biotechnological fusion of humans and robots-belonged exclusively to science fiction or in the postmodern theme arsenal. But it is not so. The mind and body—in reality the mind and the body of a machine—today stand at the center of the current tangle of problems in the theory and technology of artificial intelligence and robotics, the disciplines that on the one hand evoke the greatest concern and on the other the greatest hope in connection with cyborgs. At the same time it is not possible to exclude certain philosophical implications of current research; in fact, it is actually much more realistic to expect them.

If the Čapeks' robot can be seen as a modern artificial humanoid machine (the body of a worker or a soldier as an ideal prototype of members of a modern society), then the *cyborg* is a symbol of the postmodern human being (as a metaphor for our experience of the information society). As long as we are able to free ourselves from the traditional binary mode of articulating reality, there is nothing to stop us from seeing reality as basically a "hybrid." Then reality seen in terms of binary opposition (human versus robot, man versus machine) is the product more of our thoughts than of anything else (figure 12.6). What we have in mind can be more closely explained using the metaphor of twilight. Twilight is not a hybrid of light and dark, but light and dark (human and machine) are opposite extremes of twilight. Similarly, the cyborg is perhaps not a hybrid of the organic and mechanical, but, rather, the "organic" and the "mechanical" are two extremes of the *cyborg state*. This is the basic thesis of the ontology of twilight (explained in more detail in Kelemen 1999). The two different approaches to robots demonstrated at the time of the first performances of

Figure 12.6
Honda's humanoid robot ASIMO laying a bunch of flowers at the foot of the pedestal bearing the bust of Karel Čapek in the Czech National Museum, Prague, August 22, 2003. Courtesy of *Lidové Noviny*.

R.U.R. in Prague and New York, respectively, might in fact reflect human intuition concerning this kind of ontology.

Acknowledgments

This chapter is partially based on the text of the authors' tutorial lecture delivered during ALIFE IX, the 9th International Conference on Synthesis and Simulation of Living Systems, in Boston, September 12, 2004.

Thanks to Phil Husbands and Owen Holland for help with writing this article in English.

Notes

1. Karel Čapek thought of himself as a kind of simple storyteller, and not as a real writer. He expressed his attitude to the play *R.U.R.* by saying, in a letter to H. G. Wells, "It could have been written by anybody" (Harkins 1962, p. 94).

2. The event is similarly described by Helena Čapková (1986) in her memoir (pp. 314–15).

3. The title of the play has been translated into German as *Werstands Universal Robots*, or *W.U.R.*

4. Bratislava, which straddles the Danube, has over the course of history been called *Istropolis* or *Posonium* in Latin, *Pressburg* in German, and *Pozsony* in Hungarian.

5. For more on Kempelen's Turk see T. Standage (2003) or M. Sussman (2001).

6. "A racing car whose hood is adorned with great pipes, like serpents of explosive breath—a roaring car that seems to ride … is more beautiful than the *Victory of Samothrace*," wrote Filippo Marinetti in his first "Manifesto of Futurism" (see note 8).

7. The historical Henri Jacquet-Droz had already appeared in Karel and Josef Čapek's "Instructive Story" (*Povídka poučná*), written in 1908.

8. Futurism as an aesthetic program was initiated by Filippo T. Marinetti when he published his manifesto "The Founding and Manifesto of Futurism," in the Paris newspaper *Le Figaro* on February 20, 1909. In it Marinetti emphasized the need to discard the static and unimportant art of the past and to rejoice in turbulence, originality, and creativity in art, culture, and society. He stressed the importance of new technology (automobiles, airplanes, locomotives, and so forth, all mechanical devices par excellence) because of the power and complexity it used and conveyed. Several years later, in 1915, Marinetti introduced fascist ideas in his volume of poems *War, the World's Only Hygiene.*

9. Karel Čapek spent four of his school years, 1901 to 1905, at the gymnasium—the central European high schools for university-bound pupils—in Hradec Králové, but the two events are not believed to be connected. He had another ambivalent experience connected with the town: In the 1904–5 school year he was expelled from school for belonging to an anarchist society. After this he finished his high school education in Brno, the capital of Moravia, the eastern part of the present-day Czech Republic, before attending Charles University, in Prague.

10. Please see F. Cerný (2000) for more information concerning the Czech premiers of *R.U.R.*

11. Even though the premier of the play was a great success, it is not the work most revered and performed by Czechs. Another of his plays, *The Insect Play* (*Ze života hmyzu*), an allegory of little human imperfections, first performed in February 1922, is more appealing to the Czech mentality.

12. There were some major differences between Paul Selver's American translation (published in 1923 by the Theatre Guild) and the Czech original, and Selver has often been criticized for this. Selver and Playfair also collaborated on the English production in 1923; this translation, published by Oxford University Press in 1923, differed from the American script, and was in some ways closer to the original Czech version. An illuminating account of both translations, which substantially exonerates Selver, can be found in Robert Philmus (2001).

13. A good argument illustrating the origin of robots as complicated machines in the context of North American cultural traditions can be found in Stuart Chase's book on early impressions of problems concerning the man-machine interaction (Chase 1929) in which he noted his impression from a 1927 presentation of R. J. Wensley's Westinghouse robot, called Mr. Televox, describing it as metallic and looking as if it had been shaped by a cubist sculptor.

References

Aristotle. 1941. *The Basic Works of Aristotle.* Edited by R. McKeon. New York: Random House.

Baudrillard, Jean. 1989. "Videowelt und Fraktales Subjekt." [The world of video and the fractal subject.] *Philosophie der neuen Technologie.* [The philosophy of the new technology.] Merve Diskurs, no. 146. Berlin: Merve Verlag.

Čapek, Josef. 1997. *Ledacos; Umělý člověk.* [Anything; Homo artefactus.] Prague: Dauphin.

Čapek, Josef, and Karel Čapek. 1961. *R.U.R. and The Insect Play.* Translated by Paul Selver. Oxford: Oxford University Press.

Čapek, Karel. 1966. *R.U.R. Rossum's Universal Robots*. Edited by M. Halík. Prague: Československý spisovatel.

———. 1983. *Loupežník; R.U.R.; Bílá nemoc*. [The robber; R.U.R.; White disease.] Prague: Československý spisovatel.

Čapková, Helena. 1986. *Moji milí bratři*. [My nice brothers.] Prague: Československý spisovatel, 1986.

Capuis, A., and E. Droz. 1956. *The Jacquet-Droz Mechanical Puppets*. Neuchâtel: Historical Museum.

Chase, S. 1929. *Men and Machines*. New York: Macmillan.

Černý, F. 2000. *Premiéry bratří Čapků*. [Premiers of the Čapek brothers.] Prague: Hynek, 2000.

Harkins, W. E. 1962. *Karel Čapek*. New York: Columbia University Press.

Homer. 1998. *The Iliad*. New York: Penguin Books.

Horáková, Jana. 2005. "*R.U.R.*—Comedy About Robots." *Disk, a Selection from the Czech Journal for the Study of Dramatic Arts* 1: 86–103.

Kelemen, Josef. 1999. "On the Post-Modern machine." In *Scepticism and Hope,* edited by M. Kollár. Bratislava: Kalligram.

Marinetti, F. 1915. *War, the World's Only Hygiene*. [In Italian.]

Petiška, E. 1991. *Golem*. Prague: Martin.

Philmus, Robert. 2001. "Matters of Translation: Karel Čapek and Paul Selver." *Science Fiction Studies* 83, no. 28: 7–32.

Standage, T. 2003. *The Turk*. New York: Berkley.

Sussman, M. 2001. "Performing the Intelligent Machine—Description and Enchantment in the Life of the Automaton Chess Player." In *Puppets, Masks, and Performing Objects*, edited by J. Bell. Cambridge, Mass.: MIT Press.

13 God's Machines: Descartes on the Mechanization of Mind

Michael Wheeler

Never Underestimate Descartes

In 1637 the great philosopher, mathematician, and natural scientist René Descartes (1596–1650) published one of his most important texts, *Discourse on the Method of Rightly Conducting One's Reason and Seeking the Truth in the Sciences*, commonly known simply as the *Discourse* (Cottingham, Stoothoff, and Murdoch 1985a).[1] This event happened over three hundred years before Alan Turing, W. Ross Ashby, Allen Newell, Herbert Simon, Norbert Wiener, and the other giants of cybernetics and early artificial intelligence (AI) produced their seminal work. Approximately the same time span separates the *Discourse* from the advent of the digital computer. Given these facts it will probably come as something of a surprise to at least some readers of this volume to discover that, in this text, Descartes reflects on the possibility of mechanizing mind. Not only that but, as I shall argue in this chapter, he elegantly identifies, and takes a far from anachronistic or historically discredited stand on, a key question regarding the mechanization of mind, a question that, if we're honest with ourselves, we still don't really know how to answer. Never underestimate Descartes.

Cartesian Machines

Before we turn to the key passage from the *Discourse* itself, we need to fill in some background. And to do that we need to understand what Descartes means by a machine. In fact, given the different ways in which Descartes writes of machines and mechanisms, there are three things that he might mean by that term:

A. A material system that unfolds purely according to the laws of blind physical causation

B. A material system that is a machine in the sense of A, but to which in addition certain norms of correct and incorrect functioning apply

C. A material system that is a machine in the sense of B, but that is also either (1) a special-purpose system or (2) an integrated collection of special-purpose subsystems.[2]

As we shall see, Descartes thinks that there are plenty of systems in the actual world that meet condition A alone, but that there is nothing in the actual world that meets condition B but not condition C. Nevertheless, it is conceivable that something might meet B but not C, so it is important to keep these two conditions distinct.

Let's say that conditions A, B, and C define three different types of machine: type A, type B, and type C. So what sorts of things are there that count as type A machines? Here the key observation for our purposes is that when it came to nonmental natural phenomena, Descartes was, for his time, a radical scientific reductionist. What made him so radical was his contention that (put crudely) biology was just a local branch of physics. Prior to Descartes, this was simply not a generally recognized option. The strategy had overwhelmingly been to account for biological phenomena by appealing to the presence of special vital forces, Aristotelian forms, or incorporeal powers of some kind. In stark contrast, Descartes argued that not only all the nonvital material aspects of nature, but also all the processes of organic bodily life—from reproduction, digestion, and growth to what we would now identify as the biochemical and neurobiological processes going on in human and nonhuman animal brains—would succumb to explanations of the same fundamental character as those found in physics. But what was that character? According to Descartes, the distinctive feature of explanation in physical science was its wholly *mechanistic* nature. What matters here is not the details of one's science of mechanics. In particular, nothing hangs on Descartes's own understanding of the science of mechanics as being ultimately the study of nothing other than geometric changes in modes of extension.[3] What matters here is simply a general feature of mechanistic explanation, one shared by Descartes's science of mechanics and our own, namely, the view that in a mechanistic process, one event occurs after another, in a law-like way, through the relentless operation of blind physical causation. What all this tells us is that, for Descartes, the entire physical universe is "just" one giant type A machine. And that giant type A machine consists of lots of smaller type A machines, some of which are the organic bodies of nonhuman animals and human beings.

So far, so good. But when we say of a particular material system that it is a machine, we often mean something richer than that its behavior can be explained by the fundamental laws of mechanics. We are judging, additionally, that certain norms of correct and incorrect functioning are applicable to that system. For example, a clock has the function of telling the time. A broken clock fails to meet that norm. Where such norms apply, the system in question is a type B machine. To see how the introduction of type B machines gives us explanatory leverage, we need note only that a broken type B machine—one that fails to function correctly judged against the relevant set of norms—continues to follow the fundamental laws of mechanics just the same as if it were working properly. A broken clock fails to perform its function of telling the time, but not by constituting an exception to the fundamental laws of mechanics. Thus we need the notion of a type B machine, a machine as a *norm-governed* material system, to explain what changed about the clock, as a machine, when it stopped working. (Descartes himself makes these sorts of observations; see the "Sixth Meditation," Cottingham, Stoothoff, and Murdoch 1985b.)

It is a key feature of our understanding of the organic bodies of nonhuman animals and human beings—what I shall henceforth refer to as bodily machines or, to stress their generically shared principles of operation, as *the bodily machine*—that such systems count as machines in the richer, normatively loaded, type B sense. This is essential to our understanding of health and disease. Thus, a heart that doesn't work properly is judged to be failing to perform its function of pumping blood around the body. Descartes recognizes explicitly the normatively loaded character of the bodily machine. So where does he locate the source of the all-important norms of proper functioning? As G. Hatfield (1992) notes, Descartes vacillated on this point. Sometimes he seems to argue that all normative talk about bodily machines is in truth no more than a useful fiction in the mind of the observer, what he calls an "extraneous label." Thus, in the "Sixth Meditation," he says, "When we say, then, with respect to the body suffering from dropsy, that it has a disordered nature because it has a dry throat but does not need a drink, the term 'nature' [the idea that the body is subject to norms of correct and incorrect functioning] is here used merely as an extraneous label" (Cottingham, Stoothoff, and Murdoch 1985b, p. 69). At other times, however, an alternative wellspring of normativity presents itself. Descartes is clear that the bodily machine was designed by God. As he puts it in the *Discourse*, the body is a machine that was "made by the hands of God" (Cottingham, Stoothoff, and Murdoch 1985a, p. 139). For Descartes, then, organic bodies, including those of human beings and nonhuman animals,

are God's machines. Now, it seems correct to say that the functional normativity of a human-made machine is grounded in what the human designer of that artifact intended it to do. This suggests that the functional
normativity of the bodily machine might reasonably be grounded in what
its designer, namely God, intended it to do. Either way, the key point for
our purposes is that some Cartesian machines, including all bodily
machines, are explicable as norm-governed systems. Given the surely plausible thought that useful fictions can be explanatorily powerful, that would
be true on either of Descartes's candidate views of the source of such
normativity.[4]

Time to turn to the notion of a type C machine—a machine as (additionally) a special-purpose system or as an integrated collection of special-
purpose subsystems. To make the transition from type B to type C
machines, we need to pay particular attention to the workings of the Cartesian bodily machine. A good place to start is with Descartes's account of the
body's neurophysiological mechanisms.[5] According to Descartes, the nervous system is a network of tiny tubes along which flow the "animal spirits," inner vapors whose origin is the heart. By acting in a way that (as
Descartes himself explains it in the *Treatise on Man*) is rather like the bel-
lows of a church organ pushing air into the wind-chests, the heart and
arteries push the animal spirits out through the pineal gland into pores
located in various cavities of the brain (Cottingham, Stoothoff, and Murdoch 1985a, p. 104). From these pores, the spirits flow down neural tubes
that lead to the muscles, and thus inflate or contract those muscles to
cause bodily movements. Of course, the animal spirits need to be suitably
directed so that the outcome is a bodily movement appropriate to the situation in which the agent finds herself. According to Descartes, this is
achieved in the following way: Thin nerve fibers stretch from specific locations on the sensory periphery to specific locations in the brain. When sensory stimulation occurs in a particular organ, the connecting fiber tenses
up. This action opens a linked pore in the cavities of the brain, and thus
releases a flow of animal spirits through a corresponding point on the
pineal gland. Without further modification, this flow may be sufficient to
cause an appropriate bodily movement. However, the precise pattern of
the spirit flow, and thus which behavior actually gets performed, may depend also on certain guiding psychological interventions resulting from
the effects of memory, the passions, and (crucially for what is to follow)
reason.

The fine-grained details of Descartes's neurophysiological theory are, of
course, wrong. However, if we shift to a more abstract structural level of de-

scription, what emerges from that theory is a high-level specification for a control architecture, one that might be realized just as easily by a system of electrical and biochemical transmissions—that is, by a system of the sort recognized by contemporary neuroscience—as it is by Descartes's ingenious system of hydraulics. To reveal this specification let's assume that the bodily machine is left to its own devices (that is, without the benefit of psychological interventions) and ask, "What might be expected of it?" As we have seen, Descartes describes the presence of dedicated links between specific peripheral sites at which the sensory stimulation occurs, and specific locations in the brain through which particular flows of movement-producing animal spirits are released. This makes it tempting to think that the structural organization of the unaided (by the mind) bodily machine would in effect be that of a look-up table, a finite table of stored if-this-then-do-that transitions between particular inputs and particular outputs. This interpretation, however, ignores an important feature of Descartes's neurophysiological theory, one that we have not yet mentioned. The pattern of released spirits (and thus exactly which behavior occurs) is sensitive to the physical structure of the brain. Crucially, as animal spirits flow through the neural tubes, they will sometimes modify the physical structure of the brain around those tubes, and thereby alter the precise effects of any future sensory stimulations. Thus, Descartes clearly envisages the existence of locally acting bodily processes through which the unaided machine can, in principle, continually modify itself, so that its future responses to incoming stimuli are partially determined by its past interactions with its environment. The presence of such processes suggests that the bodily machine, on its own, is potentially capable of intralifetime adaptation and, it seems, certain simple forms of learning and memory. Therefore (on some occasions at least) the bodily machine is the home of mechanisms more complex than rigid look-up tables.

What we need right now, then, is a high-level specification of the generic control architecture realized by the bodily machine, one that not only captures the intrinsic specificity of Descartes's dedicated mechanisms but also allows those mechanisms to feature internal states and intrinsic dynamics that are more complex than those of, for example, look-up tables. Here is the suggestion: The bodily machine should be conceptualized as an integrated collection of special-purpose subsystems, where the qualifier "special-purpose" indicates that each subsystem is capable of producing appropriate actions only within some restricted task domain. Look-up tables constitute limiting cases of such an architecture. More complex arrangements, involving the possibility of locally determined adaptive change within the task domain, are, however, possible. What all this tells us is that, according to Descartes, the bodily machine is a type C machine.

That concludes our brief tour of the space of Cartesian machines. Now, what about mechanizing the mind?

The Limits of the Machine

As we have seen, for Descartes, the phenomena of bodily life can be understood mechanistically. But did he think that the same mechanistic fate awaited the phenomena of mind? It might seem that the answer to this question must be a resounding no. One of the first things that anyone ever learns about Descartes is that he was a substance dualist. He conceptualized mind as a separate substance (metaphysically distinct from physical stuff) that causally interacts with the material world on an intermittent basis during perception and action. But if mind is immaterial, then (it seems) it can't be a machine in any of the three ways that Descartes recognizes, since each of those makes materiality a necessary condition of machinehood.

Game over? Not quite. Let's approach the issue from a different angle, by asking an alternative question, namely, "What sort of capacities might the bodily machine realize?" Since the bodily machine is a type C machine, this gives us a local (organism-centered) answer to the question "What sort of capacities might a type C machine realize?" One might think that the answer to this question must be autonomic responses and simple reflex actions (some of which may be modified adaptively over time), but not much else. If this is your inclination, then an answer that Descartes himself gives in the *Treatise on Man* might include the odd surprise, since he identifies not only "the digestion of food, the beating of the heart and arteries, the nourishment and growth of the limbs, respiration, waking and sleeping [and] the reception by the external sense organs of light, sounds, smells, tastes, heat and other such qualities," but also "the imprinting of the idea of these qualities in the organ of the 'common' sense and the imagination, the retention or stamping of these ideas in the memory, the internal movements of the appetites and passions, and finally the external movements of all the limbs (movements which are ... appropriate not only to the actions of objects presented to the sense, but also to the passions and impressions found in memory)" (Cottingham, Stoothoff, and Murdoch 1985a, p. 108). In the latter part of this passage, then, Descartes takes a range of capacities that many theorists, even now, would be tempted to regard as psychological in character, and judges them to be explicable by appeal to nothing more fancy than the workings of the bodily machine. And here is another example of Descartes's enormous faith in the power of a "mere" organic mechanism. According to Descartes, the first stage in the phenomenon of hunger is excitatory activity in certain nerves in the stomach. And he claims that this purely physical activity is sufficient to initiate bodily movements that are appropriate to food finding and eating. Thus, once again we learn from Descartes that the body, unaided by the mind, is already capable of realizing relatively complex adaptive abilities. (This is, of course, not the whole story about hunger. I'll fill in the rest later.)

Should we be surprised by Descartes's account of what the bodily machine can do? Not really. As we have seen, Descartes often appeals to artifacts as a way of illustrating the workings of the bodily machine. When he does this, he doesn't focus on artifacts that in his day would have been thought of as dull or mundane, examples that might reasonably lead one to suspect that some sort of deflationary judgment on the body is in play. Rather, he appeals to examples that in his day would have been sources of popular awe and intellectual respect. These include clocks (rare, expensive and much admired as engineering achievements) and complex animal-like automata such as bought by the wealthy elite of seventeenth-century Europe to entertain and impress their most sophisticated guests. (For more on this subject, see Baker and Morris 1996, 92–93.) So when Descartes describes the organic body as a machine, we are supposed to gasp with admiration, not groan with disappointment. In fact we are supposed to be doubly impressed, since Descartes thought that the bodily machine was designed by God, and so is "incomparably better ordered than any machine that can be devised by man, and contains in itself movements more wonderful than those in any such machine" (*Discourse*, Cottingham, Stoothoff, and Murdoch 1985a, p. 139). Our bodies are God's machines and our expectations of them should be calibrated accordingly.

Now that we are properly tuned to Descartes's enthusiasm for "mere" mechanism, we can more reliably plot the limits that he placed on the bodily machine. Here, the standard interpretation of Descartes's position provides an immediate answer: the bodily machine is incapable of conscious experience (see, for example, Williams 1990, 282–83). But is this really Descartes's view? Departing from the traditional picture, Gordon Baker and Katherine Morris (1996) have argued that Descartes held some aspects of consciousness to be mechanizable. This sounds radical, until one discovers that, according to Baker and Morris, the sense in which, for Descartes, certain machines were conscious is the sense in which we can use expressions such as "see" or "feel pain" to designate "(the 'input' half of) fine-grained differential responses to stimuli (from both inside and outside the 'machine') mediated by the internal structure and workings of the machine" (p. 99). Those who favor the traditional interpretation of Descartes might retaliate—with some justification, I think, and in spite of protests by Baker and Morris (see pp. 99–100)—that Descartes would not have considered this sort of differential responsiveness to stimuli to be a form of consciousness at all, at least not in any interesting or useful sense. Indeed, if he had thought of things in this way he would seemingly have been committed to the claim that all sorts of artifacts available in his day, such as the aforementioned entertainment automata, were conscious. It is very unlikely that he would have embraced such a consequence. Nevertheless, in spite of such worries about the Baker and Morris line, I think that some doubt has been cast on the thought that consciousness provides a sufficiently sharp criterion for determining where, in Descartes's view, the limits of mere mechanism lie. It would be nice to find something better.

Time then to explore the passage from the *Discourse* in which Descartes explicitly considers the possibility of machine intelligence. Here it is (Cottingham, Stoothoff, and Murdoch 1985a, p. 140):

> [We] can certainly conceive of a machine so constructed that it utters words, and even utters words which correspond to bodily actions causing a change in its organs (e.g., if you touch it in one spot it asks you what you want of it, if you touch it in another it cries out that you are hurting it, and so on). But it is not conceivable that such a machine should produce different arrangements of words so as to give an appropriately meaningful answer to what is said in its presence, as the dullest of men can do.... [And] ... even though such machines might do some things as well as we do them, or perhaps even better, they would inevitably fail in others, which would reveal that they were acting not through understanding, but only from the disposition of their organs. For whereas reason is a universal instrument which can be used in all kinds of situations, these organs need some particular disposition for each particular action; hence it is for all practical purposes impossible for a machine to have enough different organs to make it act in all the contingencies of life in the way in which our reason makes us act.

Once again Descartes's choice of language may mislead us into thinking that, in his view, any entity which qualifies (in the present context) as a machine must be a look-up table. For example, he tells us that his imaginary robot acts "only from the disposition of [its] organs," organs that "need some particular disposition for each particular action." However, the way in which this robot is supposed to work is surely intended by Descartes to be closely analogous to the way in which the organic bodily machine is supposed to work. (Recall Descartes's enthusiasm for drawing illustrative parallels between the artificial and the biological when describing the workings of the bodily machine.) So we need to guarantee that there is conceptual room for Descartes's imaginary robot to feature the range of processes that, in his account, were found to be possible within the organic bodily machine. In other words, Descartes's imaginary robot needs to be conceived as an integrated collection of special-purpose subsystems, some of which may realize certain simple forms of locally driven intra-lifetime adaptation, learning, and memory. In short, Descartes's robot is a type C machine.

With that clarification in place, we can see the target passage as first plotting the limits of machine intelligence, and then explaining both why these limits exist and how human beings go beyond them. First let's see where the limits lie. Descartes argues that although a machine might be built which is (1) able to produce particular sequences of words as responses to specific stimuli and (2) able to perform individual actions as well as, if not better than, human agents, no mere machine could either (3) continually generate complex linguistic responses that are flexibly sensitive to varying contexts, in the way that all linguistically competent human beings do, or (4) succeed in behaving appropriately in any context, in the way that all behaviorally normal human beings do. Here one might interpret Descartes as proposing two separate human phenomena, generative language use and a massive degree of adaptive behavioral flexibility, both of which are beyond the capacities of any mere machine (for this sort of interpretation, see Williams 1990, 282–83). However, I think that there is another, perhaps more profitable way of understanding the conceptual relations in operation, according to which (1) and (3) ought to be construed as describing the special, linguistic instance of the general case described by (2) and (4). In this interpretation, although it is true that the human capacity for generative language use is one way of marking the difference between mere machines and human beings, the point that no machine (by virtue solely of its own intrinsic capacities) could reproduce the generative and contextually sensitive linguistic capabilities displayed by human beings is actually just a restricted version of the point that no machine (by virtue solely of its intrinsic capacities) could reproduce the unrestricted range of adaptively flexible and contextually sensitive behavior displayed by human beings. This alternative interpretation is plausible, I think, because when Descartes proceeds in the passage to explain why it is that no mere machine is capable of consistently reproducing human-level behavior, he does not mention linguistic behavior at all, but concentrates instead on the nonlinguistic case.

To explain why the limits of machine intelligence lie where they do, Descartes argues as follows: Machines can act "only from the [special-purpose] disposition of their organs." Now, if we concentrate on some individual, contextually embedded human behavior, then it is possible that a machine might be built that incorporated a special-purpose mechanism (or set of special-purpose mechanisms) that would enable the machine to perform that behavior as well as, or perhaps even better than, the human agent. However, it would be impossible to incorporate into any one machine the vast number of special-purpose mechanisms that would be required for that machine to consistently and reliably generate appropriate behavior in all the different situations that make up an ordinary human life. So how do humans do it? What machines lack, and what humans enjoy, is the faculty of understanding or reason, that "universal instrument which can be used in all kinds of situations." In other words, the distinctive and massive adaptive flexibility of human behavior is explained by the fact that humans deploy *general-purpose reasoning processes*.

It is important to highlight two features of Descartes's position here. First, Descartes's global picture is one in which, in human beings, reason and mechanism standardly work together to produce adaptive behavior. To see this, let's return to the case of hunger, introduced previously. As I explained, the first stage in the phenomenon of hunger (as Descartes understands it) involves excitatory mechanical activity in the stomach that, in a way unaided by cognitive processes, initiates bodily movements appropriate to food finding and eating. However, according to Descartes, some of the bodily changes concerned will often lead to mechanical changes in the brain, which in turn cause associated ideas, including the conscious sensation of hunger, to arise in the mind. At this point in the flow of behavioral control, such ideas may prompt a phase of judgment and deliberation by the faculty of reason, following which the automatic movements generated by the original nervous activity may be revised or inhibited.

Second, the pivotal claim in Descartes's argument is that no single machine could incorporate the enormous number of special-purpose mechanisms that would be required for it to reproduce human-like behavior. So what is the status of this claim? Descartes writes (in translation) that "it is *for all practical purposes* impossible for a machine to have enough different organs to make it act in all the contingencies of life in the way in which our reason makes us act" (emphasis added). A lot turns on the expression "for all practical purposes." The French phrase in Descartes's original text is *moralement impossible*—literally, "morally impossible." The idea that something that is morally impossible is something that is impossible for all practical purposes is defended explicitly by Cottingham (1992, p. 249), who cites as textual evidence Descartes's explanation of moral certainty in the *Principles of Philosophy*. There the notion is unpacked as certainty that "measures up to the certainty we have on matters relating to the conduct of life which we never normally doubt, though we know it is possible absolutely speaking that they may be false" (Cottingham, Stoothoff, and Murdoch 1985a, p. 290). I am persuaded by Cottingham's interpretation of the key phrase (despite the existence of alternative readings; see, for example, Baker and Morris 1992, pp. 183–88, especially note 331 on p. 185). And I am equally persuaded by the use that Cottingham makes of that interpretation in his own discussion of the target passage from the *Discourse* (see Cottingham 1992, pp. 249–52). There he leans on his interpretation of *moralement impossible* to argue that Descartes's pivotal claim does not (according to Descartes anyway) have the status of a necessary truth. Rather, it is a scientifically informed empirical bet. Descartes believes that the massive adaptive flexibility of human behavior cannot be generated or explained by the purely mechanistic systems of the body, since, as far as he can judge, it is practically impossible to construct a machine that contains enough different special-purpose mechanisms. However, he is, as far as this argument is concerned, committed to the view that the upper limits of what a mere machine might do must, in the end, be determined by rigorous scientific investigation and not by philosophical speculation. In other words, Descartes accepts that his view is a hostage to ongoing developments in science.

Mechanics and Magic

Suppose one wanted to defend the view that mind may be mechanized, *without exception*. How might one respond to Descartes's argument? Here is a potential line of argument. One might (a) agree that we have reason in Descartes's (general-purpose) sense, but (b) hold that reason (in that sense) can in fact be mechanized, and so (c) hold that the machines that explain human-level intelligence (general-purpose ones) are such as to escape Descartes's tripartite analysis of machine-hood. Let's see how one might develop this case.

Between Descartes and contemporary AI came the birth of the digital computer. What this did, among other things, was effect a widespread transformation in the very notion of a machine. According to Descartes's pre-computational outlook, machines simply were integrated collections of special-purpose mechanisms. To Descartes himself, then, reason, in all its allegedly general-purpose glory, looked staunchly resistant to mechanistic explanation. In the twentieth century, however, mainstream thinking in artificial intelligence was destined to be built in part on a concept that would no doubt have amazed and excited Descartes himself, namely, the concept of a *general-purpose reasoning machine*. The introduction of mechanistic systems that realize general-purpose reasoning algorithms is not something that Descartes himself even considered (how could he have?) but one might argue that the arrival of such systems has shown how general-purpose reason, that absolutely core and, according to Descartes, unmechanizable aspect of the Cartesian mind, might conceivably be realized by a bodily machine. Let's call such a machine a type D machine. Evidence of the importance of type D machines to AI abounds in the literature. It includes massively influential individual models, such as Newell and Simon's (1963) General Problem Solver, a program that used means-end reasoning to construct a plan for systematically reducing the difference between some goal state, as represented in the machine, and the current state of the world, as represented in the machine. And it includes generic approaches to machine intelligence, such as mainstream connectionist theories to be discussed further that think of the engine room of the mind as containing just a small number of general-purpose learning algorithms, such as Hebbian learning and back-propagation.[6]

So is this a good response to Descartes's argument? I don't think so. Why? Because it runs headlong into a long-standing enemy of AI known as the frame problem. In its original form, the frame problem is the problem of characterizing, using formal logic, those aspects of a state that are not changed by an action (see, for example, Shanahan 1997). However, the term has come to be used in a less narrow way, to name a multilayered family of interconnected worries to do with the updating of epistemic states in relevance-sensitive ways (see, for example, the range of discussions in Pylyshyn 1987). A suitably broad definition is proposed by Jerry Fodor, who describes the frame problem as "the problem of putting a 'frame' around the set of beliefs that may need to be revised in the light of specified newly available information" (Fodor 1983, p. 112–13). Here I shall be concerned with the frame problem in its more general form.

To see why the framing requirement described by Fodor constitutes a bona fide problem, as opposed to merely a description of what needs doing, consider the following example (Dennett 1984). Imagine a mobile robot that has the capacity to reason about its world by proving theorems on the basis of internally stored, logic-based representations. (This architecture is just one possibility. Nothing about the general frame problem means that it is restricted to control systems whose representational states and reasoning algorithms are logical in character.) This robot needs power to survive. When it is time to find a power source, the robot proves a theorem such as PLUG-INTO (Plug, Power Source). The intermediate steps in the proof represent subgoals that the robot needs to achieve in order to succeed at its main goal of retrieving a power-source (compare the means-end reasoning algorithm deployed by GPS, as mentioned previously).

Now, consider what might happen when our hypothetical robot is given the task of collecting its power source from a room that also contains a bomb. The robot knows that the power source is resting on a wagon, so it decides (quite reasonably, it seems) to drag that wagon out of the room. Unfortunately, the bomb is on the wagon too. The result is a carnage of nuts, bolts, wires, and circuit boards. It is easy to see that the robot was unsuccessful here because it failed to take account of one crucial side effect of its action—the movement of the bomb. So, enter a new improved robot. This one operates by checking for every side effect of every plan that it constructs. This robot, too, is unsuccessful, simply because it never gets to perform an action. It just sits there and ruminates. What this shows is that it is no good checking for every side effect of every possible action before taking the plunge and doing something. There are just too many side effects to consider, and most of them will be entirely irrelevant to the context of action. For example, taking the power source out of the room changes the number of objects in the room, but in this context, who cares? And notice that the robot needs to consider not only things about its environment that have changed but also things that have not. Some of these will be important some of the time, given a particular context. So the robot needs to know which side effects of its actions and which unchanged facts about its world are relevant, and which are not. Then it can just ignore all the irrelevant facts. Of course, if the context of action changes, then what counts as relevant may change. For instance, in a different context it may be absolutely crucial that the robot takes account of the fact that, as a result of its own actions, the number of objects in the room has changed.

We have just arrived at the epicenter of the frame problem, and it's a place where the idea of mind as machine confronts a number of difficult questions. For example, given a dynamically changing world, how is a purely mechanistic system to take account of those state changes in that world (self-induced or otherwise) that matter, and those unchanged states in that world that matter, while ignoring those that do not? And how is that system to retrieve and (if necessary) to revise, out of all the beliefs that it possesses, just those beliefs that are relevant in some particular context of action? In short, how might a "mere" machine behave in ways that are sensitive to context-dependent relevance?

One first-pass response to these sorts of questions will be to claim that the machine should deploy stored heuristics (rules of thumb) that determine which of its rules and representations are relevant in the present situation. But are relevancy heuristics really a cure for the frame problem? It seems not. The processing mechanisms concerned would still face the problem of accessing just those relevancy heuristics that are relevant in the current context. So how does the system decide which of its stored heuristics are relevant? Another, higher-order set of heuristics would seem to be required. But then exactly the same problem seems to reemerge at that processing level, demanding further heuristics, and so on. It is not merely that some sort of combinatorial explosion or infinite regress beckons here (which it does). A further concern, in the judgment of some notable authorities, is that we seem to have no good idea of how a computational process of relevance-based update might work. As Terence Horgan and John Tienson (1994) point out, the situation cannot be that the system first retrieves an inner structure (an item of information or a heuristic), and then decides whether or not it is relevant, as that would take us back to square one. But then how can the system assign relevance until the structure has been retrieved?

But if the frame problem is such a nightmare, how come AI hasn't simply ground to a halt? According to many front-line critics of the field (including Dreyfus; see chapter 14, this volume), most AI researchers, classical and connectionist, have managed to sidestep the frame problem precisely because they have tended to assume that real-world cognitive problem solving can be treated as a kind of messy and complicated approximation to reasoning or learning in artificially restricted worlds that are relatively static and essentially closed and feature some small number of contexts of action. In such worlds, all the contexts that could possibly arise may be identified and defined, alongside all the factors that could possibly count as relevant within each of them. So the programmer can either take comprehensive and explicit account of the effects of every action or change, or can work on the assumption that nothing changes in a scenario unless it is explicitly said to change by some rule. And if those strategies carried too high an adaptive cost in terms of processing resources, well-targeted relevancy heuristics would appear to have a good chance of heading off the combinatorial explosions and search difficulties that threaten. One might think, however, that the actual world often consists of an indeterminate number of dynamic, open-ended, complex scenarios in which context-driven and context-determining change is common and ongoing and in which vast ranges of cognitive space might, at any time, contain the relevant psychological elements. It is in this world that the frame problem really bites, and in which, it seems, the aforementioned strategies must soon run out of steam.

From what we have seen so far, the frame problem looks to be a serious barrier to the mechanization of mind. Indeed, one possible conclusion that one might draw from the existence and nature of the frame problem is that human intelligence is a matter of magic, not mechanics. However, it is at least arguable that the frame problem is in fact a by-product the conception of mind as a general-purpose (type D) machine, rather than as machine *simpliciter*. What mandates this less extreme conclusion? It's the following line of thought: In the present proposal, what guarantees that "[mechanical] reason is [in principle] a universal instrument which can be used in all kinds of situations" is, at root, that the reasoning mechanism concerned has free and total access to a gigantic body of rules and information. Somewhere in that vast sea of structures lie the cognitive elements that are relevant to the present context. The perhaps insurmountable problem is how to find them in a timely fashion using a process of purely mechanical search. What this suggests is that we might do well to reject the very idea of the bodily machine as a general-purpose reasoning machine, and to investigate what happens to the frame problem if we refuse to accept Descartes's invitation to go beyond special-purpose mechanisms in our understanding of intelligence.

Here is the view from the armchair: a system constructed from a large number of special-purpose mechanisms will simply take the frame problem in its stride. This is because, in any context of action, the special-purpose mechanism that is appropriately activated will, as a direct consequence of its design, have access to no more than a highly restricted subset of the system's stock of rules and representations. Moreover, that subset will include just the rules and representations that are relevant to the adaptive scenario in which the system finds itself. Therefore the kind of unmanageable search space that the frame problem places in the path of a general-purpose mechanism is simply never established. Those are the armchair intuitions. But is there any evidence to back them up? Here is a much-discussed model from the discipline of biorobotics.

Consider the ability of the female cricket to find a mate by tracking a species-specific auditory advertisement produced by the male. According to Barbara Webb's robotic model of the female cricket's behavior, here, roughly, is how the phonotaxis system works (for more details, see Webb 1993 and 1994, and the discussion in Wheeler 2005). The basic anatomical structure of the female cricket's peripheral auditory system is such that the amplitude of her ear-drum vibration will be higher on the side closer to a sound source. Thus, if some received auditory signal is indeed from a conspecific male, all the female needs to do to reach him (all things being equal) is to continue to move in the direction indicated by the ear drum with the higher-amplitude response. How is that the female tracks only the correct stimulus? The answer lies in the activation profiles of two interneurons, one connected to each of the female cricket's ears, that mediate between ear-drum response and motor behavior. The decay rates of these interneurons are tightly coupled with the specific temporal pattern of the male's song, such that signals with the wrong temporal pattern will simply fail to produce the right motor effects.

Why is this robotic cricket relevant to the frame problem? The key idea is suggested by Webb's own explanation of why the proposed mechanism is adaptively powerful (Webb 1993, p. 1092):

Like many other insects, the cricket has a simple and distinctive cue to find a mate, and consequently can have a sensory-motor mechanism that works for this cue and nothing else: there is no need to process sounds in general, provided this specific sound has the right motor effects. Indeed, it may be advantageous to have such specificity built in, because it implicitly provides "recognition" of the correct signal through the failure of the system with any other signal.

A reasonable gloss on this picture would be that the cricket's special-purpose mechanism does not have to start outside of context and find its way in using relevancy heuristics. In the very process of being activated by a specific environmental trigger that mechanism brings a context of activity along with it, implicitly realized in the very operating principles that define its successful functioning. Thus, to repeat the armchair intuition, there is no frame problem here because the kind of unmanageable search space that the frame problem places in the path of a general-purpose mechanism is simply never established.

If one takes the sort of mechanism described by Webb, generalizes the picture so that one has an integrated architecture of such mechanisms, and then looks at the result through historically tinted glasses, then it seems to reflect two of Descartes's key thoughts: that organic bodies are collections of special-purpose subsystems (type C machines), and that such subsystems, individually and in combination, are capable of some pretty fancy adaptive stuff. Moreover, this would seem to be a machine that solves the frame problem, in effect, by not letting it arise. This looks to be a step forward—and it is. Unfortunately, however, it falls short of what we need, because although it solves the frame problem, it doesn't solve Descartes's problem. As we know, Descartes himself argued that there was a limit to what any collection of special-purpose mechanisms could do: no single machine, he thought, could incorporate the enormous number of special-purpose mechanisms that would be required for it to reproduce the massive adaptive flexibility of human behavior. That's why, in the end, Descartes concludes that intelligent human behavior is typically the product of general-purpose reason. Nothing we have discovered so far suggests that Descartes was wrong about that. Here's the dilemma, in a nutshell: If we mechanize general-purpose reason, we get the frame problem; so that's no good. But if we don't mechanize general-purpose reason, we have no candidate mechanistic explanation for the massive adaptive flexibility of human behavior; so that's no good either. The upshot is that if we are to resist Descartes's antimechanistic conclusion, something has to give.

At this juncture let's return to the target passage from the *Discourse*. There is, I think, a tension hidden away in Descartes's claim that (as it appears in the standard English translation) "reason is a universal instrument which can be used in all kinds of situations." Strictly speaking, if reason is a universal instrument then, at least potentially, it ought to be possible for it to be applied unrestrictedly, across the cognitive board. If this is right, then "all kinds of situations" needs to be read as "any kind of situation." However, I don't think we ordinarily use the phrase "all kinds of" in that way. When we say, for example, that the English cricket team, repeatedly slaughtered by Australia during the 2006–7 Ashes tour, is currently having "all kinds of problems," we mean not that the team faces all the problems there are in the world, but rather that they face a wide range of different problems. But now if this piece of ordinary language philosophy is a reliable guide for how we are meant to read Descartes's claim about reason, then that claim is weakened significantly. The suggestion now is only that reason is an instrument that can be used in *a wide range of different situations*.

With this alternative interpretation on the table, one might think that the prospects for an explanation of human reason in terms of the whirrings of a type C machine are improved significantly. The argument would go like this:

Human reason is, in truth, a suite of specialized psychological skills and tricks with domain-specific gaps and shortcomings. That would still be an instrument that can be used in a wide range of different situations. And by

Descartes's own lights, a material system of integrated special-purpose mechanisms (a type C Cartesian machine) ought to be capable of this sort of cognitive profile.

But this is to move too quickly. For even if the claim that reason is a "universal instrument" overstates just how massively flexible human behavior really is, it's undeniably true that human beings are impressively flexible. Indeed, the provisional argument just aired fails to be sufficiently sensitive to the thought that an instrument that really can be used successfully across a wide range of different situations is an instrument that must be capable of fast, fluid, and flexible context switching. Crucially, this sort of capacity for real-time adaptation to new contexts appears to remain staunchly resistant to exhaustive explanation in terms of any collection of purely special-purpose mechanisms. The worry is this: So far, we have no account of the mechanistic principles by which a particular special-purpose mechanism is selected from the vast range of such mechanisms available to the agent and then placed in control of the agent's behavior at a specific time. One can almost hear Descartes's ghost as he claims that we will ultimately need to posit a general-purpose reasoning system whose job it is to survey the options and make the choice. But if that's the "solution," then the door to the frame problem would be reopened, and we would be back to square one, or thereabouts.

Plastic Machines

Our task, then, is to secure adaptive flexibility on a scale sufficient to explain open-ended adaptation to new contexts without going beyond mere mechanism and without a return to Cartesian general-purpose reason. Here is a suggestion—an incomplete one, I freely admit—for how this might be achieved.

Roughly speaking, the term *connectionism* picks out research on a class of intelligent machines in which typically a large number of interconnected units process information in parallel. In as much as the brain, too, is made up of a large number of interconnected units (neurons) that process information in parallel, connectionist networks are "neurally inspired," although usually at a massive level of abstraction. Each unit in a connectionist network has an activation level regulated by the activation levels of the other units to which it is connected, and, standardly, the effect of one unit on another is either positive (if the connection is excitatory) or negative (if the connection is inhibitory). The strengths of these connections are known as the network's weights, and it is common to think of the network's "knowledge" as being stored in its set of weights. In most networks the values of these weights are modifiable, so, given some initial configuration, changes to the weights can be made that improve the performance of the network over time. In other words, within all sorts of limits imposed by the way the input is encoded, the specific structure of the network, and the weight-adjustment algorithm, the network may learn to carry out some desired input-output mapping.

Most work on connectionist networks has tended to concentrate on architectures that in effect limit the range and complexity of possible network dynamics. These features include neat symmetrical connectivity; noise-free processing; update properties that are based either on a global digital pseudo-clock or on methods of stochastic change; units that are uniform in structure and function; activation passes that proceed in an orderly feed-forward fashion; and a model of neurotransmission in which the effect of one neuron's activity on that of a connected neuron will simply be either excitatory or inhibitory, and will be mediated by a simple point-to-point signaling process. Quite recently, however, some researchers have come to favor a class of connectionist machines with richer system dynamics, so-called dynamical neural networks (DNNs).

What we might, for convenience, call Mark I DNNs feature the following sorts of properties (although not every bona fide example of a Mark I DNN exhibits all the properties listed): asynchronous continuous-time processing; real-valued time delays on connections; nonuniform activation functions; deliberately introduced noise; and connectivity that is not only both directionally unrestricted and highly recurrent, but also not subject to symmetry constraints (see, for example, Beer and Gallagher 1992, Husbands et al. 1995). Mark II DNNs add two further twists to the architectural story. In these networks, christened GasNets (Husbands et al. 1998), the standard DNN model is augmented with modulatory neurotransmission (according to which fundamental properties of neurons, such as their activation profiles, are transformed by arriving neurotransmitters), and models of neurotransmitters that diffuse virtually from their source in a cloudlike, rather than a point-to-point, manner, and thus affect entire volumes of processing structures. GasNets thus provide a platform for potentially rich interactions between two interacting and intertwined dynamical mechanisms—virtual cousins of the electrical and chemical processes in real nervous systems. Diffusing "clouds of chemicals" may change the intrinsic properties of the artificial neurons, thereby changing the patterns of "electrical" activity, while "electrical" activity may itself trigger "chemical" activity. Dropping the scare quotes, these biology-inspired machines feature neurotransmitters that not only may transform the transfer functions of the neurons on which they act but also may do so on a grand scale, as a result of the fact that they act by gaseous diffusion through volumes of brain space, rather than by electrical transmission along connecting neural wires.

Systems of this kind have been artificially evolved to control mobile robots for simple homing and discrimination tasks.[7] What does the analysis of such machines tell us? Viewed as static wiring diagrams, many of the successful GasNet controllers appear to be rather simple structures. Typical networks feature a very small number of primitive visual receptors connected to a tiny number of inner and motor neurons by just a few synaptic links. However, this apparent structural simplicity hides the fact that the dynamics of the networks are often highly complex, involving, as predicted, subtle couplings between chemical and electrical processes. For example, it is common to find adaptive use being made of oscillatory dynamical subnetworks, some of whose properties, such as their periods, depend on spatial features of the modulation and diffusion processes, processes that are themselves determined by the changing levels of electrical activity in the neurons within the network (for more details, see Husbands et al. 1998). Preliminary analysis suggests that these complex interwoven dynamics will sometimes produce solutions that are resistant to any modular decomposition. However, there is also evidence of a kind of transient modularity in which, over time, the effects of the gaseous diffusible modulators drive the network through different phases of modular and nonmodular organization (Husbands, personal communication).

What seems clear, then, is that the sorts of machines just described realize a potentially powerful kind of ongoing fluidity, one that involves the functional and even the structural reconfiguration of large networks of components. This is achieved on the basis of bottom-up systemic causation that involves multiple simultaneous interactions and complex dynamic feedback loops, such that the causal contribution of each systemic component partially determines, and is partially determined by, the causal contributions of large numbers of other systemic components and, moreover, those contributions may change radically over time. (This is what Clark [1997] dubs continuous reciprocal causation.) At root, GasNets are mechanisms of significant adaptive plasticity, and it seems plausible that it is precisely this sort of plasticity that, when harnessed and tuned appropriately by selection or learning to operate over different time scales, may be the mechanistic basis of open-ended adaptation to new contexts. It is a moot point whether or not this plasticity moves us entirely beyond the category of type C machines. To the extent that one concentrates on the way Gas-Nets may shift from one kind of modular organization to another (in realizing the kind of transient modularity mentioned previously), the view is compatible with a story in which context switching involves a transition from one arrangement of special-purpose systems to another. Under these circumstances, perhaps it would be appropriate to think of GasNets as type C.5 machines.

Concluding Remarks

In the *Discourse*, Descartes lays down a challenge to the advocate of the mechanization of mind. How can the massive adaptive flexibility of human-level intelligence be explained without an appeal to a nonmechanistic faculty of general-purpose reason? Descartes's scientifically informed empirical bet is that it cannot. Of course, his conclusion is based on an understanding of machine-hood that is linked conceptually to the notion of special-purpose mechanisms. This understanding, and thus his conclusion, has been disputed by the subsequent attempt in AI to mechanize general-purpose reason. However, since this ongoing attempt is ravaged by the frame problem, it does not constitute a satisfactory response to Descartes's challenge. Are plastic machines, as exemplified by GasNets, the answer? So far I know of no empirical work that demonstrates conclusively that the modulatory processes instantiated in GasNets can perform the crucial context-switching function that I have attributed to them. For although there is abundant evidence that such processes can mediate the transition between different phases of behavior within the same task (Smith, Husbands, and O'Shea 2001), that is not the same thing as switching between contexts, which typically involves a reevaluation of what the current task might be. Nevertheless, it is surely a thought worth pursuing that fluid functional and structural reconfiguration, driven in a bottom-up way by low-level neuro-chemical mechanisms, may be at the heart of the more complex capacity. That is my scientifically informed empirical bet, one that needs to be balanced against Descartes's own. At present Descartes's challenge remains essentially unanswered. Never underestimate Descartes. (Have I said that?)

Notes

1. This chapter draws extensively on material from my book *Reconstructing the Cognitive World: The Next Step* (Wheeler 2005), especially chapters 2, 7, and 10. Sometimes

text is incorporated directly, but my reuse of that material here is not simply a rehash of it. The present treatment has some new things to say about Descartes's enduring legacy in the science of mind and contains a somewhat different analysis of the frame problem.

All quotations from, and page numbers for, Descartes's writings are taken from the now-standard English editions of the texts in question. For the texts referred to here, this means the translations contained in Cottingham, Stoothoff, and Murdoch 1985a, 1985b).

2. The first two of these notions are identified in Descartes's work by G. Hatfield (1992, pp. 360–62). The third is not.

3. For Descartes, the essential property of matter is that it takes up space, that is, that it has extension. In effect, mechanics studies changes in manifestations of that property.

4. For the view that useful fictions can be explanatorily powerful, see one common way of understanding Dennett's position on psychological states such as beliefs and desires (Dennett, 1987). Post-Darwin, the overwhelming temptation will be to see natural selection as the source of functional normativity in the case of the bodily machine. In this view, the function of some bodily element will be the contribution that that element has made to survival and reproduction in ancestral populations. Descartes, writing two hundred years before Darwin, didn't have this option in his conceptual tool kit.

5. For a more detailed description of these mechanisms, see Hatfield (1992, p. 346).

6. In the present context, the fact that AI came to mechanize general-purpose reason is plausibly interpreted as a move *against* Descartes, but this is not the only way of looking at things. Aside from its mechanization, nothing about the nature and contribution of reason as a psychological capacity underwent significant transformation in the process of appropriation by AI. Thus, viewed from a broader perspective, one might argue that, by mechanizing general-purpose reason in the way that it did, AI remained within a generically Cartesian framework. For much more on this, see Wheeler (2005, especially chapters 2 and 3).

7. Roughly speaking, design by artificial evolution works as follows: First one sets up a way of encoding potential solutions to some problem as genotypes. Then, starting with a randomly generated population of potential solutions and some evaluation task, one implements a selection cycle such that more successful solutions have a proportionally higher opportunity to contribute genetic material to subsequent generations, that is, to be "parents." Genetic operators analogous to recombination and mutation in natural reproduction are applied to the parental genotypes to produce "children," and, typically, a number of existing members of the population are discarded so that the population size remains constant. Each solution in the resulting new population is then evaluated, and the process starts all over again. Over succes-

sive generations, better solutions are discovered. In GasNet research, the goal is to design a network capable of achieving some task, and artificial evolution is typically allowed to decide fundamental architectural features of that network, such as the number, directionality, and recurrency of the connections, the number of internal units, and the parameters controlling modulation and virtual gas diffusion.

References

Baker, G., and K. J. Morris. 1996. *Descartes' Dualism*. London and New York: Routledge.

Beer, R. D., and J. G. Gallagher. 1992. "Evolving Dynamic Neural Networks for Adaptive Behavior." *Adaptive Behavior* 1: 91–122.

Clark, A. 1997. *Being There: Putting Brain, Body, and World Together Again*. Cambridge, Mass., and London: MIT Press/Bradford Books.

Cottingham, John. 1992. "Cartesian Dualism: Theology, Metaphysics, and Science." In *The Cambridge Companion to Descartes*, edited by John Cottingham. Cambridge: Cambridge University Press.

Cottingham, John, R. Stoothoff, and D. Murdoch, eds. 1985a. *The Philosophical Writings of* Descartes. Volume 1. Cambridge: Cambridge University Press.

———. 1985b. *The Philosophical Writings of* Descartes. Volume 2. Cambridge: Cambridge University Press.

Dennett, Daniel C. 1984. "Cognitive Wheels: The Frame Problem of AI." In *Minds, Machines and Evolution: Philosophical Studies*, edited by C. Hookway. Cambridge: Cambridge University Press.

———. 1987. *The Intentional Stance*. Cambridge, Mass.: MIT Press/Bradford Books.

Fodor, J. A. 1983. *The Modularity of Mind*. Cambridge, Mass.: MIT Press/Bradford Books.

Hatfield, G. 1992. "Descartes' Physiology and Its Relation to His Psychology." In *The Cambridge Companion to Descartes*, edited by John Cottingham. Cambridge: Cambridge University Press.

Horgan, T., and J. Tienson. 1994. "A Nonclassical Framework for Cognitive Science." *Synthese* 101: 305–45.

Husbands, Philip, I. Harvey, and D. Cliff. 1995. "Circle in the Round: State Space Attractors for Evolved Sighted Robots." *Robotics and Autonomous Systems* 15: 83–106.

Husbands, Philip, T. Smith, N. Jakobi, and M. O'Shea. 1998. "Better Living Through Chemistry: Evolving GasNets for Robot Control." *Connection Science* 103–4: 185–210.

Newell, Allen, and Herbert Simon. 1963. "GPS—a Program That Simulates Human Thought." In *Computers and Thought*, edited by E. A. Feigenbaum and J. Feldman. New York: McGraw-Hill.

Pylyshyn, Z., ed. 1987. *The Robot's Dilemma*. Norwood, N.J.: Ablex.

Shanahan, M. 1997. *Solving the Frame Problem: A Mathematical Investigation of the Common Sense Law of Inertia*. Cambridge, Mass.: MIT Press.

Smith, T., Husbands, Philip, and M. O'Shea. 2001. "Neural Networks and Evolvability with Complex Genotype-Phenotype Mapping." In *Advances in Artificial Life: Proceedings of the Sixth European Conference on Artificial Life*, edited by Josef Kelemen and P. Sosik. Berlin and Heidelberg: Springer.

Webb, Barbara. 1993. "Modeling Biological Behaviour or 'Dumb Animals and Stupid Robots.'" In *Pre-Proceedings of the Second European Conference on Artificial Life*.

———. 1994. "Robotic Experiments in Cricket Phonotaxis." In *From Animals to Animats 3: Proceedings of the Third International Conference on Simulation of Adaptive Behavior*, edited by D. Cliff, Philip Husbands, J.-A. Meyer, and S. W. Wilson. Cambridge, Mass.: MIT Press/Bradford Books.

Wheeler, Michael. 2005. *Reconstructing the Cognitive World: The Next Step*. Cambridge, Mass.: MIT Press.

Williams, B. 1990. *Descartes: The Project of Pure Enquiry*. London: Penguin.

14 Why Heideggerian AI Failed and How Fixing It Would Require Making It More Heideggerian

Hubert L. Dreyfus

The Convergence of Computers and Philosophy

When I was teaching at MIT in the early sixties, students from the Artificial Intelligence Laboratory would come to my Heidegger course and say in effect: "You philosophers have been reflecting in your armchairs for over two thousand years and you still don't understand how the mind works. We in the AI Lab have taken over and are succeeding where you philosophers have failed. We are now programming computers to exhibit human intelligence: to solve problems, to understand natural language, to perceive, and to learn."[1] In 1968 Marvin Minsky, head of the AI lab, proclaimed, "Within a generation we will have intelligent computers like HAL in the film *2001*."[2]

As luck would have it, in 1963, I was invited by the RAND Corporation to evaluate the pioneering work of Alan Newell and Herbert Simon in a new field called cognitive simulation (CS). Newell and Simon claimed that both digital computers and the human mind could be understood as physical symbol systems, using strings of bits or streams of neuron pulses as symbols representing the external world. Intelligence, they claimed, merely required making the appropriate inferences from these internal representations. As they put it, "A physical symbol system has the necessary and sufficient means for general intelligent action."[3]

As I studied the RAND papers and memos, I found to my surprise that, far from replacing philosophy, the pioneers in CS had learned a lot, directly and indirectly, from the philosophers. They had taken over Hobbes's claim that reasoning was calculating, Descartes's mental representations, Leibniz's idea of a "universal characteristic"—a set of primitives in which all knowledge could be expressed—Kant's claim that concepts were rules, Frege's formalization of such rules, and Russell's postulation of logical atoms as the building blocks of reality. In short, without realizing it, AI researchers were hard at work turning rationalist philosophy into a research program.

At the same time, I began to suspect that the critical insights formulated in existentialist armchairs, especially Heidegger's and Merleau-Ponty's, were bad news for those working in AI laboratories—that, by combining rationalism, representationalism, conceptualism, formalism, and logical atomism into a research program, AI researchers had condemned their enterprise to reenact a failure.

Symbolic AI as a Degenerating Research Program

Using Heidegger as a guide, I began to look for signs that the whole AI research program was degenerating. I was particularly struck by the fact that, among other troubles, researchers were running up against the problem of representing significance and relevance—a problem that Heidegger saw was implicit in Descartes's understanding of the world as a set of meaningless facts to which the mind assigned what Descartes called values, and John Searle now calls functions.[4]

But, Heidegger warned, values are just more meaningless facts. To say a hammer has the function of being for hammering leaves out the defining relation of hammers to nails and other equipment, to the point of building things, and to the skills required when actually using the hammer—all of which reveal the way of being of the hammer that Heidegger called "readiness-to-hand." Merely assigning formal function predicates to brute facts such as hammers couldn't capture the hammer's way of being nor the meaningful organization of the everyday world in which hammering has its place. "[B]y taking refuge in 'value' characteristics," Heidegger said, "we are...far from even catching a glimpse of being as readiness-to-hand."[5]

Minsky, unaware of Heidegger's critique, was convinced that representing a few million facts about objects, including their functions, would solve what had come to be called the commonsense knowledge problem. It seemed to me, however, that the deep problem wasn't storing millions of facts; it was knowing which facts were relevant in any given situation. One version of this relevance problem was called the frame problem. If the computer is running a representation of the current state of the world and something in the world changes, how does the program determine which of its represented facts can be assumed to have stayed the same, and which would have to be updated?

As Michael Wheeler in his recent book, *Reconstructing the Cognitive World*, puts it:

Given a dynamically changing world, how is a nonmagical system . . . to take account of those state changes in that world . . . that matter, and those unchanged states in that world that matter, while ignoring those that do not? And how is that system to retrieve and (if necessary) to revise, out of all the beliefs that it possesses, just those beliefs that are relevant in some particular context of action?[6]

Minsky suggested that, to avoid the frame problem, AI programmers could use what he called frames—descriptions of typical situations like going to a birthday party—to list and organize those, and only those, facts that were normally relevant. Perhaps influenced by a computer science student who had taken my phenomenology course, Minsky suggested a structure of essential features and default assignments—a structure Edmund Husserl had already proposed and already called a frame.[7]

But a system of frames isn't *in* a situation, so in order to select the possibly relevant facts in the current situation one would need frames for recognizing situations like birthday parties, and for telling them from other situations such as ordering in a restaurant. But how, I wondered, could the computer select from the supposed millions of frames in its memory the relevant frame for selecting the birthday party frame as the relevant frame, so as to see the current relevance of, say, an exchange of gifts rather than money? It seemed to me obvious that any AI program using frames to organize millions of meaningless facts so as to retrieve the currently relevant ones was going to be caught in a regress of frames for recognizing relevant frames for recognizing relevant facts, and that, therefore, the frame problem wasn't just a problem but was a sign that something was seriously wrong with the whole approach.

Unfortunately, what has always distinguished AI research from a science is its refusal to face up to and learn from its failures. In the case of the relevance problem, the AI programmers at MIT in the sixties and early seventies limited their programs to what they called micro-worlds—artificial situations in which the small number of features that were possibly relevant was determined beforehand. Since this approach obviously avoided the real-world frame problem, MIT Ph.D. students were compelled to claim in their theses that their micro-worlds could be made more realistic, and that the techniques they introduced could be generalized to cover commonsense knowledge. There were, however, no successful follow-ups.[8]

The work of Terry Winograd is the best of the work done during the micro-world period. His "blocks-world" program, SHRDLU, responded to commands in ordinary English instructing a virtual robot arm to move blocks displayed on a computer screen. It was the prime example of a micro-world program that really worked—but of course only in its micro-world. So to produce the expected generalization of his techniques, Winograd started working on a new Knowledge Representation Language (KRL). His group, he said, was "concerned with developing a formalism, or 'representation,' with which to describe...knowledge." And he added, "We seek the 'atoms' and 'particles' of which it is built, and the 'forces' that act on it."[9]

But this approach wasn't working. Indeed, Minsky has recently acknowledged in *Wired* magazine that AI has been brain dead since the early seventies, when it encountered the problem of commonsense knowledge.[10] Winograd, however, unlike his colleagues, was scientific enough to try to figure out what had gone wrong. So in the mid-seventies we began having weekly lunches to discuss his problems in a broader philosophical context. Looking back, Winograd says, "My own work in computer science is greatly influenced by conversations with Dreyfus."[11]

After a year of such conversations, and after reading the relevant texts of the existential phenomenologists, Winograd abandoned work on KRL and began including Heidegger in his computer science courses at Stanford. In so doing, he became the first high-profile deserter from what was, indeed, becoming a degenerating research program. John Haugeland now refers to the symbolic AI of that period as good old-fashioned AI—GOFAI for short—and that name has been widely accepted as capturing its current status. Indeed, Michael Wheeler argues that a new paradigm is already taking shape. He maintains that a "Heideggerian cognitive science is...emerging right now, in the laboratories and offices around the world where embodied-embedded thinking is under active investigation and development."[12]

Wheeler's well-informed book could not have been more timely, since there are now at least three versions of supposedly Heideggerian AI that might be thought of as articulating a new paradigm for the field: Rodney Brooks's behaviorist approach at MIT, Phil Agre's pragmatist model, and Walter Freeman's neurodynamic model. All three approaches implicitly accept Heidegger's critique of Cartesian internalist representations, and embrace John Haugeland's slogan that cognition is embedded and embodied.[13]

Heideggerian AI, Stage 1: Eliminating Representations by Building Behavior-Based Robots

Winograd sums up what happened at MIT after he left for Stanford:

For those who have followed the history of artificial intelligence, it is ironic that [the MIT] laboratory should become a cradle of "Heideggerian AI." It was at MIT that

Dreyfus first formulated his critique, and, for twenty years, the intellectual atmosphere in the AI Lab was overtly hostile to recognizing the implications of what he said. Nevertheless, some of the work now being done at that laboratory seems to have been affected by Heidegger and Dreyfus.[14]

Here's how it happened. In March 1986, the MIT AI Lab under its new director, Patrick Winston, reversed Minsky's attitude toward me and allowed, if not encouraged, several graduate students, led by Phil Agre and John Batali, to invite me to give a talk.[15] I called the talk, "Why AI Researchers Should Study 'Being and Time.'" In my talk I repeated what I had written in 1972 in *What Computers Can't Do*: "The meaningful objects...among which we live are not a *model* of the world stored in our mind or brain; *they are the world itself*."[16] And I quoted approvingly a Stanford Research Institute report to the effect that "it turned out to be very difficult to reproduce in an internal representation for a computer the necessary richness of environment that would give rise to interesting behavior by a highly adaptive robot,"[17] and concluded that "this problem is avoided by human beings because their model of the world is the world itself."[18]

The year of my talk, Rodney Brooks, who had moved from Stanford to MIT, published a paper criticizing the GOFAI robots that used representations of the world and problem-solving techniques to plan their movements. He reported that, on the basis of the idea that "the best model of the world is the world itself," he had "developed a different approach in which a mobile robot uses the world itself as its own representation—continually referring to its sensors rather than to an internal world model."[19] Looking back at the frame problem, he writes, "And why could my simulated robot handle it? Because it was using the world as its own model. It never referred to an internal description of the world that would quickly get out of date if anything in the real world moved."[20]

Brooks's approach is an important advance, but Brooks's robots respond only to fixed isolable features of the environment, not to context or changing significance. Moreover, they do not learn. They are like ants, and Brooks aptly calls them "animats." Brooks thinks he does not need to worry about learning, putting it off as a concern for possible future research.[21] But by operating in a fixed world and responding only to the small set of possibly relevant features that their receptors can pick up, Brooks's animats beg the question of changing relevance and so finesse rather than solve the frame problem.

Still, Brooks comes close to an existential insight spelled out by Merleau-Ponty, viz., that intelligence is founded on and presupposes the more basic way of coping we share with animals, when he says:[22]

The "simple" things concerning perception and mobility in a dynamic environment...are a necessary basis for "higher-level" intellect....Therefore, I proposed looking at simpler animals as a bottom-up model for building intelligence. It is soon apparent, when "reasoning" is stripped away as the prime component of a robot's intellect, that the dynamics of the interaction of the robot and its environment are primary determinants of the structure of its intelligence.[23]

Brooks is realistic in describing his ambitions and his successes:

The work can best be described as attempts to emulate insect-level locomotion and navigation....There have been some behavior-based attempts at exploring social interactions, but these too have been modeled after the sorts of social interactions we see in insects.[24]

Surprisingly, the modesty Brooks exhibited in choosing to first construct simple insect-like devices did not deter Brooks and Daniel Dennett from repeating the extravagant optimism characteristic of AI researchers in the sixties. As in the days of GOFAI, on the basis of Brooks's success with insect-like devices, instead of trying to make, say, an artificial spider, Brooks and Dennett decided to leap ahead and build a humanoid robot. As Dennett explained in a 1994 report to the Royal Society of London:

A team at MIT of which I am a part is now embarking on a long-term project to design and build a humanoid robot, Cog, whose cognitive talents will include speech, eye-coordinated manipulation of objects, and a host of self-protective, self-regulatory and self-exploring activities.[25]

Dennett seems to reduce this project to a joke when he adds in all seriousness: "While we are at it, we might as well try to make Cog crave human praise and company and even exhibit a sense of humor."[26]

Of course, the "long-term project" was short-lived. Cog failed to achieve any of its goals and the original robot is already in a museum.[27] But, as far as I know, neither Dennett nor anyone connected with the project has published an account of the failure and asked what mistaken assumptions underlay their absurd optimism. In a personal communication Dennett blamed the failure on a lack of graduate students and claimed that, "progress was being made on all the goals, but slower than had been anticipated."[28]

If progress was actually being made, however, the graduate students wouldn't have left, or others would have continued to work on the project. Clearly some specific assumptions must have been mistaken, but all we find in Dennett's assessment is the implicit assumption that human intelligence is on a continuum with insect intelligence, and that therefore adding a bit of complexity to what has already been done with animats counts as progress toward humanoid intelligence. At the beginning of AI research, Yehoshua Bar-Hillel called this way of thinking the first-step fallacy, and my brother, Stuart Dreyfus, at RAND quipped, "It's like claiming that the first monkey that climbed a tree was making progress towards flight to the moon."

In contrast to Dennett's assessment, Brooks is prepared to entertain the possibility that he is barking up the wrong tree. He soberly comments:

Perhaps there is a way of looking at biological systems that will illuminate an inherent necessity in some aspect of the interactions of their parts that is completely missing from our artificial systems. . . . I am not suggesting that we need go outside the current realms of mathematics, physics, chemistry, or biochemistry. Rather I am suggesting that perhaps at this point we simply do not *get it*, and that there is some fundamental change necessary in our thinking in order that we might build artificial systems that have the levels of intelligence, emotional interactions, long term stability and autonomy, and general robustness that we might expect of biological systems.[29]

We can already see that Heidegger and Merleau-Ponty would say that, in spite of the breakthrough of giving up internal symbolic representations, Brooks, indeed, doesn't get it—that what AI researchers have to face and understand is not only why our everyday coping couldn't be understood in terms of inferences from symbolic representations, as Minsky's intellectualist approach assumed, but also why it can't be understood in terms of responses caused by fixed features of the environment, as in Brooks's empiricist model. AI researchers need to consider the possibility that embodied beings like us take as input energy from the physical universe, and respond in such a way as to open themselves to a world organized in terms of their needs, interests, and bodily capacities without their *minds'* needing to impose meaning on a meaningless given, as Minsky's frames require, nor their *brains'* converting stimulus input into reflex responses, as in Brooks's animats.

Later I'll suggest that Walter Freeman's neurodynamics offers a radically new basis for a Heideggerian approach to human intelligence—an approach compatible with physics and grounded in the neuroscience of perception and action. But first we need to examine another approach to AI contemporaneous with Brooks's that actually calls itself Heideggerian.

Heideggerian AI, Stage 2: Programming the Ready-to-Hand

In my talk at the MIT AI Lab, I introduced Heidegger's nonrepresentational account of the absorption of *Dasein* (human being) in the world.

I also explained that Heidegger distinguished two modes of being: the "readiness-to-hand" of equipment when we are involved in using it, and the "presence-at-hand" of objects when we contemplate them. Out of that explanation, and the lively discussion that followed, grew the second type of Heideggerian AI—the first to acknowledge its lineage.

This new approach took the form of Phil Agre and David Chapman's program, Pengi, which guided a virtual agent playing a computer game called Pengo, in which the player and penguins kick large and deadly blocks of ice at each other.[30] Their approach, which they called "interactionism," was more self-consciously Heideggerian than Brooks's, in that they attempted to capture what Agre called "Heidegger's account of everyday routine activities."[31] In his book, *Computation and Human Experience*, Agre takes up where my talk left off:

I believe that people are intimately involved in the world around them and that the epistemological isolation that Descartes took for granted is untenable. This position has been argued at great length by philosophers such as Heidegger and Merleau-Ponty; I wish to argue it technologically.[32]

Agre's interesting new idea is that the world of Pengo in which the Pengi agent acts is made up, not of present-at-hand objects with properties, but of possibilities for action that trigger appropriate responses from the agent. To program this situated approach, Agre used what he called "deictic representations." He tells us, "This proposal is based on a rough analogy with Heidegger's analysis of everyday intentionality in Division I of *Being and Time*, with objective intentionality corresponding to the present-at-hand and deictic intentionality corresponding to the ready-to-hand."[33] And he explains, "[Deictic representations] designate, not a particular object in the world, but rather a role that an object might play in a certain time-extended pattern of interaction between an agent and its environment."[34]

Looking back on my talk at MIT and rereading Agre's book I now see that, in a way, Agre understood Heidegger's account of readiness-to-hand better than I did at the time. I thought of the ready-to-hand as a special class of *entities*, namely, equipment, whereas the Pengi program treats what the agent responds to purely as *functions*. For Heidegger and Agre the ready-to-hand is not a *what* but a *for-what*.[35] But not just that the hammer is for hammering. As Agre saw, Heidegger wants to get at something more basic than simply a class of objects defined by their use. At his best Heidegger would, I think, deny that a hammer in a drawer has readiness-to-hand as its way of being. Rather, he sees that, for the user, equipment is encountered as a *solicitation to act*, not an *entity* with a function feature. He notes, "When one is wholly devoted to something and 'really' busies oneself with it, one does not do so just alongside the work itself, or alongside the tool, or alongside both of them 'together.'"[36] And he adds, "The peculiarity of what is proximally ready-to-hand is that, in its readiness-to-hand, it must, as it were, withdraw in order to be ready-to-hand quite authentically."[37]

As usual with Heidegger, we must ask: What is the phenomenon he is pointing out? In this case he wants us to see that to observe our hammer or to observe ourselves hammering undermines our skillful coping. We can and do observe our surroundings while we cope, and sometimes, if we are learning, monitoring our performance as we learn improves our performance in the long run, but in the short run such attention interferes with our performance. For example, while biking we can observe passersby, or think about philosophy, but if we start observing how we skillfully stay balanced, we risk falling over.

Heidegger struggles to describe the basic way we are drawn in by the ready-to-hand. The Gestaltists would later talk of "solicitations." In *Phenomenology of Perception* Merleau-Ponty speaks of "motivations" and later, of "the flesh." All these terms point at what is not objectifiable—a situation's way of directly drawing from one a response that is neither caused like a reflex, nor done for a reason.

In his 1925 Marburg lectures, "Logic: The Question of Truth," Heidegger describes our most basic experience of what he later calls "pressing into possibilities" not as dealing with the desk, the door, the lamp, the chair, and so forth, but as directly responding to a "what for":

What is first of all "given"... is the "for writing," the "for going in and out," the "for illuminating," the "for sitting." That is, writing, going-in-and-out, sitting, and the like are what we are a priori involved with. What we know when we "know our way around" and what we learn are these "for-what"s.[38]

It's clear here that, in spite of what some interpreters take Heidegger to be suggesting in *Being and Time*, this basic experience has no as-structure.[39] That is, when absorbed in coping, I can be described objectively as using a certain door as a door, but I'm not *experiencing* the door as a door. Normally there is no "I" and no experiencing of the door at all but simply pressing into the possibility of going out. The important thing to realize is that, when we are pressing into possibilities, there is no experience of an entity doing the soliciting; just the immediate response to a solicitation. (When solicitations don't pan out, what then is disclosed is the world of interconnected equipment, and I can then step back and perceive things as things, and act for reasons.)[40]

But Agre's Heideggerian AI did not try to program this experiential aspect of being drawn in by a solicitation. Rather, with his deictic representations,

Agre objectified both the functions and their situational relevance for the agent. In Pengi, when a virtual ice cube defined by its function is close to the virtual player, a rule dictates a response, namely, kick it. No skill is involved and no learning takes place.

So Agre had something right that I was missing—the transparency of the ready-to-hand—but he nonetheless fell short of programming a Heideggerian account of everyday routine activities. For Heidegger, the ready-to-hand is not a fixed function, encountered in a predefined type of situation that triggers a predetermined response that either succeeds or fails. Rather, as we have begun to see and will soon see further, readiness-to-hand is experienced as a solicitation that calls forth a flexible response to the significance of the current situation—a response that is experienced as either improving one's situation or making it worse.

Moreover, although Agre proposed to program Heidegger's account of everyday routine activities, he doesn't even try to account for how our experience feeds back and changes our sense of the significance of the next situation and what is relevant in it. In putting his virtual agent in a virtual micro-world where all possible relevance is determined beforehand, Agre didn't try to account for how we learn to respond to new relevancies, and so, like Brooks, he finesses rather than solves the frame problem.

Merleau-Ponty's work, on the contrary, offers a nonrepresentational account of the way the body and the world are coupled that suggests a way of avoiding the frame problem. According to Merleau-Ponty, as an agent acquires skills, those skills are "stored," not as representations in the agent's mind but as the solicitations of situations in the world. What the learner acquires through experience is not represented at all but is presented to the learner as more and more finely discriminated situations. If the situation does not clearly solicit a single response or if the response does not produce a satisfactory result, the learner is led to further refine his discriminations, which, in turn, solicit ever more refined responses. For example, what we have learned from our experience of finding our way around in a city is "sedimented" in how that city looks to us. Merleau-Ponty calls this feedback loop between the embodied coper and the perceptual world the intentional arc. He says, "Cognitive life—the life of desire or perceptual life—is subtended by an 'intentional arc' which projects round about us our past, our future, [and] our human setting."[41]

Pseudo-Heideggerian AI: Embedded, Embodied, Extended Mind

As if taking up from where Agre left off with his objectified version of the ready-to-hand, in *Reconstructing the Cognitive World* Wheeler tells us:

Our global project requires a defense of action-oriented representation.... Action-oriented representation may be interpreted as the subagential reflection of online practical problem solving, as conceived by the Heideggerian phenomenologist. Embodied-embedded cognitive science is implicitly a Heideggerian venture.[42]

He further notes, "As part of its promise, this nascent, Heideggerian paradigm would need to indicate that it might plausibly be able either to solve or to dissolve the frame problem."[43] And he suggests, "The good news for the reoriented Heideggerian is that the kind of evidence called for here may already exist, in the work of recent *embodied-embedded cognitive science*."[44] He concludes:

Dreyfus is right that the philosophical impasse between a Cartesian and a Heideggerian metaphysics can be resolved empirically via cognitive science. However, he looks for resolution in the wrong place. For it is not any alleged empirical failure on the part of orthodox cognitive science, but rather the concrete empirical success of a cognitive science with Heideggerian credentials, that, if sustained and deepened, would ultimately vindicate a Heideggerian position in cognitive theory.[45]

I agree that it is time for a positive account of Heideggerian AI and of an underlying Heideggerian neuroscience, but I think Wheeler is the one looking in the wrong place. Merely by supposing that Heidegger is concerned with problem solving and action-oriented representations, Wheeler's project reflects not a step beyond Agre but a regression to aspects of pre-Brooks GOFAI. Heidegger, indeed, claims that that skillful coping is basic, but he is also clear that all coping takes place on the background coping he calls being-in-the-world that doesn't involve any form of representation at all.[46]

Wheeler's cognitivist misreading of Heidegger leads him to overestimate the importance of Andy Clark and David Chalmers' attempt to free us from the Cartesian idea that the mind is essentially inner by pointing out that in thinking we sometimes make use of external artifacts such as pencil, paper, and computers.[47] Unfortunately, this argument for the extended mind preserves the Cartesian assumption that our basic way of relating to the world is by using propositional representations such as beliefs and memories, whether they are in the mind or in notebooks in the world. In effect, while Brooks happily dispenses with representations where coping is concerned, all Chalmers, Clark, and Wheeler give us as a supposedly radical new Heideggerian approach to the human way of being-in-the-world is to note that memories and beliefs are not necessarily inner entities and that, therefore, thinking bridges the distinction between inner and outer representations.

Heidegger's important insight is not that, when we solve problems, we sometimes make use of representational equipment outside our bodies,

but that being-in-the-world is more basic than thinking and solving problems; that it is not representational at all. That is, when we are coping at
our best, we are drawn in by solicitations and respond directly to them, so
that the distinction between us and our equipment—between inner and
outer—vanishes.[48] As Heidegger sums it up:

I *live* in the understanding of writing, illuminating, going-in-and-out, and the like.
More precisely: as Dasein I am—in speaking, going, and understanding—an act of
understanding dealing-with. My being in the world *is* nothing other than this
already-operating-with-understanding in this mode of being.[49]

Heidegger's and Merleau-Ponty's understanding of embedded embodied
coping, then, is not that the mind is sometimes extended into the world
but rather that all such problem solving is derivative, that in our most basic
way of being, that is, as absorbed skillful copers, we are not minds at all but
one with the world. Heidegger sticks to the phenomenon, when he makes
the strange-sounding claim that in its most basic way of being, "Dasein is
its world existingly."[50]

When you stop thinking that mind is what characterizes us most basically but, rather, that most basically we are absorbed copers, the inner/
outer distinction becomes problematic. There's no easily askable question
as to whether the absorbed coping is in me or in the world. According to
Heidegger, intentional content isn't in the mind, nor in some third realm
(as it is for Husserl), nor in the world; it isn't anywhere. It's an embodied
way of being-toward. Thus, for a Heideggerian all forms of cognitivist externalism presuppose a more basic existential externalism, where even to
speak of "externalism" is misleading since such talk presupposes a contrast with the internal. Compared to this genuinely Heideggerian view,
extended-mind externalism is contrived, trivial, and irrelevant.

What Motivates Embedded/Embodied Coping?

But why is Dasein called to cope at all? According to Heidegger, we are constantly solicited to improve our familiarity with the world. Five years before
the publication of *Being and Time* he wrote, "Caring takes the form of a
looking around and seeing, and as this circumspective caring it is at the
same time . . . concerned about developing its circumspection, that is, about
securing and expanding its familiarity with the objects of its dealings."[51]

This pragmatic perspective is developed by Merleau-Ponty, and by
Samuel Todes.[52] These heirs to Heidegger's account of familiarity and coping describe how an organism, animal or human, interacts with what is,

objectively speaking, the meaningless physical universe in such a way as to cope with an environment organized in terms of that organism's *need to find its way around.* All such coping beings are motivated to get a more and more refined and secure sense of the specific objects of their dealings. According to Merleau-Ponty, "My body is geared into the world when my perception presents me with a spectacle as varied and as clearly articulated as possible."[53]

In short, in our skilled activity we are drawn to move so as to achieve a better and better grip on our situation. For this movement toward maximal grip to take place one doesn't need a mental representation of one's goal nor any problem solving, as would a GOFAI robot. Rather, acting is experienced as a steady flow of skillful activity in response to the situation. When one's situation deviates from some optimal body-environment gestalt, one's activity takes one closer to that optimum and thereby relieves the "tension" of the deviation. One does not need to know what the optimum is in order to move toward it. One's body is simply drawn to lower the tension.

That is, if things are going well and I am gaining an optimal grip on the world, I simply respond to the solicitation to move toward an even better grip, and if things are going badly, I experience a pull back toward the norm. If it seems that much of the time we don't experience any such pull, Merleau-Ponty would no doubt respond that the sensitivity to deviation is nonetheless guiding one's coping, just as an airport radio beacon doesn't give a warning signal unless the plane strays off course, and then, let us suppose, the plane gets a signal whose intensity corresponds to how far off course it is and the intensity of the signal diminishes as it approaches getting back on course. The silence that accompanies being on course doesn't mean the beacon isn't continually guiding the plane. Likewise, the absence of felt tension in perception doesn't mean we aren't being directed by a solicitation.

As Merleau-Ponty puts it, "Our body is not an object for an 'I think,' it is a grouping of lived-through meanings that moves towards its equilibrium."[54] Equilibrium is Merleau-Ponty's name for the zero gradient of steady successful coping. Moreover, normally we do not arrive at equilibrium and stop there but are immediately taken over by a new solicitation.

Modeling Situated Coping as a Dynamical System

Describing the phenomenon of everyday coping as being "geared into" the world and "moving toward equilibrium" suggests a dynamic relation between the coper and the environment. Timothy van Gelder calls this dynamic relation between coper and environment coupling, and explains its importance as follows:

The fundamental mode of interaction with the environment is not to represent it, or even to exchange inputs and outputs with it; rather, the relation is better understood via the technical notion of coupling.…

The post-Cartesian agent manages to cope with the world without necessarily representing it. A dynamical approach suggests how this might be possible by showing how the internal operation of a system interacting with an external world can be so subtle and complex as to *defy* description in representational terms—how, in other words, cognition can *transcend* representation.[55]

Van Gelder shares with Brooks the existentialist claim that thinking such as problem solving is grounded in a more basic relation of body and world. As van Gelder puts it:

Cognition can, in sophisticated cases, [such as breakdowns, problem solving, and abstract thought] involve representation and sequential processing; but such phenomena are best understood as emerging from a dynamical substrate, rather than as constituting the basic level of cognitive performance.[56]

This dynamical substrate is precisely the causal basis of the skillful coping first described by Heidegger and worked out in detail by Merleau-Ponty and Todes.

Van Gelder importantly contrasts the rich interactive temporality of real-time on-line coupling of coper and world with the austere step-by-step temporality of thought. Wheeler helpfully explains:

Whilst the computational architectures proposed within computational cognitive science require that inner events happen in the right order, and (in theory) fast enough to get a job done, there are, in general, no constraints on how long each operation within the overall cognitive process takes, or on how long the gaps between the individual operations are. Moreover, the transition events that characterize those inner operations are not related in any systematic way to the real-time dynamics of either neural biochemical processes, non-neural bodily events, or environmental phenomena (dynamics which surely involve rates and rhythms).[57]

Computation is thus paradigmatically austere:

Turing machine computing is digital, deterministic, discrete, effective (in the technical sense that behavior is always the result of an algorithmically specified finite number of operations), and temporally austere (in that time is reduced to mere sequence).[58]

Ironically, Wheeler's highlighting the contrast between rich dynamic temporal coupling and austere computational temporality enables us to see clearly that his appeal to extended minds as a Heideggerian response to Cartesianism leaves out the essential temporal character of embodied embedding. Clark and Chalmers's examples of extended minds manipulating representations such as notes and pictures are clearly cases of temporal austerity—no rates and rhythms are involved.

Wheeler is aware of this possible objection to his backing both the dynamical systems model and the extended-mind approach. He asks, "What about the apparent clash between continuous reciprocal causation and action orientated representations? On the face of it this clash is a worry for our emerging cognitive science."[59] But instead of engaging with the incompatibility of these two opposed models of ground-level intelligence, Wheeler suggests that we must somehow combine them and that "this question is perhaps one of the biggest of the many challenges that lie ahead."[60]

Wheeler, however, hopes he can combine these approaches by appealing to the account of involved problem solving that Heidegger calls dealing with the unready-to-hand. Wheeler's point is that, unlike detached problem solving with its general representations, the unready-to-hand requires situation-specific representations. But, as we have seen, for Heidegger all unready-to-hand coping takes place on the background of an even more basic nonrepresentational holistic coping that allows copers to orient themselves in the world.

Heidegger describes this background as "the background of . . . primary familiarity, which itself is not conscious and intended but is rather present in [an] unprominent way."[61] In *Being and Time* he speaks of "that familiarity in accordance with which Dasein . . . 'knows its way about' [*Kennt sich aus*] in its public environment" (p. 405). This coping is like the ready-to-hand in that it does not involve representations. So Heidegger says explicitly that our background being-in-the-world, which he also calls transcendence, does not involve representational intentionality, but, rather, makes intentionality possible:

Transcendence is a *fundamental determination of the ontological structure of the Dasein.* . . . Intentionality is founded in the Dasein's transcendence and is possible solely for this reason—transcendence cannot conversely be explained in terms of intentionality.[62] To be more exact, background coping is not a traditional kind of intentionality. Whereas the ready-to-hand has conditions of satisfaction, like hammering in the nail, background coping does not have conditions of satisfaction. What would it be to succeed or fail in finding one's way around in the familiar world? The important point for Heidegger, but not for Wheeler, is that *all* coping, including unready-to-hand coping, takes place on the background of this basic nonrepresentational, holistic, absorbed, kind of intentionality, which Heidegger calls being-in-the-world.[63]

This is not a disagreement between Wheeler and me about the relative frequency of dealing with the ready-to-hand and the unready-to-hand in everyday experience. True, Wheeler emphasizes intermittent reflective activities such as learning and practical problem solving, whereas I, like Heidegger, emphasize pervasive activities such as going out the door, walking on the floor, turning the lights on and off, and so forth. The question of the relative frequency of the ready-to-hand and the unready-to-hand modes of being is, Wheeler and I agree, an empirical question.[64]

But the issue concerning the background is not an empirical question. It is an ontological question. And, as we have just seen, Heidegger is clear that the mode of being of the world is not that of a collection of independent modules that define what is relevant in specific situations. It seems to me that Wheeler is on the right track, leaving modular solutions and action oriented representations behind, when he writes (Personal communication):

Where one has CRC [continuous reciprocal causation] one will have a non-modular system. Modularity is necessary for homuncularity and thus, on my account, necessary for representation of any kind. To the extent that the systems underlying intelligence are characterized by CRC, they will be non-representational, and so the notion of action-oriented representation won't help explain them.

Wheeler directly confronts my objection when he adds:

If one could generate the claim that CRC must be the norm at the subagential level from a Heideggerian analysis of the agential level, then the consequence for me would be that, to be Heideggerian, I would have to concede that action-oriented representation will in fact do less explanatory work than I have previously implied.

But Wheeler misses my point when he adds:

However, this takes us back to the points I make above about the prevalence of unreadiness-to-hand. Action-oriented representations will underlie our engagements with the unready-to-hand. In this domain, I suggest, the effects of CRC will be restricted. And, I think, unreadiness-to-hand is the (factual) norm.

We just agreed that this is not an empirical question concerning the frequency of coping with the unready-to-hand but an ontological point about the background of *all* modes of coping. If Wheeler wants to count himself a Heideggerian, he does, indeed, "have to concede that action-oriented representation will in fact do less explanatory work than [he] previously implied."

Wheeler seems to be looking for a neurodynamic model of brain activity such as we will consider in a moment when he writes:

Although there is abundant evidence that (what we are calling) continuous recip-rocal causation can mediate the transition between different phases of behavior within the same task, that is not the same thing as switching between contexts, which typically involves a reevaluation of what the current task might be. Neverthe-less, I am optimistic that essentially the same processes of fluid functional and struc-tural reconfiguration, driven in a bottom-up way by low-level neurochemical dynamics, may be at the heart of the more complex capacity.[65]

Meanwhile, Wheeler's ambivalence concerning which model is more basic, the representational or the dynamic, undermines his Heideggerian approach. For, as Wheeler himself sees, the Heideggerian claim is that action-oriented coping, as long as it is involved (on-line, Wheeler would say) is not representational at all and does not involve any problem solv-ing, and that all representational problem solving takes place off-line and presupposes involved background coping.[66] Showing in detail how the rep-resentational unready-to-hand in all its forms depends upon a background of holistic, nonrepresentational coping is exactly the Heideggerian project and would, indeed, be the most important contribution that Heideggerian AI could make to cognitive science. Indeed, a Heideggerian cognitive science would require working out an ontology, phenomenology, and brain model that deny a basic role to any sorts of representations—even action-oriented ones—and defends a dynamical model like Merleau-Ponty's and van Gelder's that gives a primordial place to equilibrium and in general to rich coupling.

Ultimately, we will have to choose which sort of AI and which sort of neuroscience to back, and so we are led to the questions: Could the brain in its causal support of our active coping instantiate a richly coupled dynamical system, and is there any evidence it actually does so? If so, could this coupling be modeled on a digital computer to give us Heideggerian AI or at least Merleau-Pontian AI? And would that solve the frame problem?

Walter Freeman's Merleau-Pontian Neurodynamics

We have seen that our experience of the everyday world (not the universe) is given as already organized in terms of significance and relevance, and that significance can't be constructed by giving meaning to brute facts, both because we don't normally experience brute facts and because even if we did, no value predicate could do the job of giving them situational significance. Yet all that the organism can receive is mere physical energy.

How can such senseless physical stimulation be experienced directly as significant? All generally accepted neuro-models fail to help, even when they talk of dynamic coupling, since they still accept the basic Cartesian model, namely:

1. The brain receives input from the universe by way of its sense organs (the picture on the retina, the vibrations in the cochlea, the odorant particles in the nasal passages, and so forth).
2. Out of this stimulus information, the brain abstracts features, which it uses to construct a representation of the world.

This is supposedly accomplished either by applying rules such as the frames and scripts of GOFAI—an approach that is generally acknowledged to have failed to solve the frame problem—or by strengthening or weakening weights on connections between simulated neurons in a simulated neural network depending on the success or failure of the net's output as defined by the net designer. Significance is thus added from outside, since the net is not seeking anything. This approach does not even try to capture the animal's way of actively determining the significance of the stimulus on the basis of its past experience and its current arousal.

Both these approaches treat the computer or brain as a passive receiver of bits of meaningless data, which then have to have significance added to them. The big problem for the traditional neuroscience approach is, then, to understand how the brain binds the relevant features to each other. That is, the problem for normal neuroscience is how to pick out and relate features relevant to each other from among all the independent isolated features picked up by each of the independent isolated receptors. For example, is the redness that has just been detected relevant to the square or to the circle shape also detected in the current input? This problem is the neural version of the frame problem in AI: How can the brain keep track of which facts in its representation of the current world are relevant to which other facts? Like the frame problem, as long as the mind/brain is thought of as passively receiving meaningless inputs that need to have significance and relevance added to them, the binding problem has remained unsolved and is almost certainly unsolvable. Somehow the phenomenologist's description of how the active organism has direct access to significance must be built into the neuroscientific model.

Wheeler has argued persuasively for the importance of a positive alternative in overthrowing established research paradigms. Without such a positive account the phenomenological observation that the world is its own best representation, and that the significance we find in our world is constantly enriched by our experience in it, seems to require that the brain be what Dennett derisively calls "wonder tissue."

Fortunately, there is at least one model of how the brain could provide the causal basis for the intentional arc and so avoid the binding problem. Walter Freeman, a founding figure in neurodynamics and one of the first to take seriously the idea of the brain as a nonlinear dynamical system, has worked out an account of how the brain of an active animal can directly pick up and augment significance in its world.[67] On the basis of years of work on olfaction, vision, touch, and hearing in alert and moving rabbits, Freeman has developed a model of rabbit learning based on the coupling of the rabbit's brain and the environment. He maintains that "the brain moves beyond the mere extraction of features.... It combines sensory messages with past experience...to identify both the stimulus and its particular meaning to the individual."[68]

To bring out the structural analogy of Freeman's account to Merleau-Ponty's phenomenological descriptions, I propose to map Freeman's neurodynamic model onto the phenomena Merleau-Ponty has described. Freeman's neurodynamics implies the involvement of the whole brain in perception and action, but for explaining the core of his ideas I'll focus on the dynamics of the olfactory bulb, since his key research was done on that part of the rabbit brain.

Direct Perception of Significance and the Rejection of the Binding Problem

Where all other researchers assume the passive reception of input from the universe, Freeman, like Merleau-Ponty on the phenomenological level, and Gibson on the (ecological) psychology level, develops a third position, between the intellectualist and the empiricist. Merleau-Ponty, Gibson, and Freeman take as basic that the brain is embodied in an animal moving in the environment to satisfy its needs.

Freeman maintains that information about the world is not gained by detecting meaningless features and processing these features step-by-step upward toward a unified representation. The binding problem only arises as an artifact of trying to interpret the output of isolated cells in the receptors of immobilized organisms. Rather, Freeman turns the problem around and asks: Given that the environment is already significant for the animal, how can the animal select a unified significant figure from the noisy background? This turns the binding problem into a selection problem. As we shall see, however, this selection is not among patterns existing in the world but among *patterns in the animal* that have been formed by its prior interaction with the world.

In Freeman's neurodynamic model, the animal's perceptual system is primed by past experience and arousal to seek and be rewarded by relevant experiences. In the case of the rabbit, these could be carrot smells found in the course of seeking and eating a carrot. When the animal succeeds, the connections between those cells in the rabbit's olfactory bulb that were involved are strengthened, according to "the widely accepted Hebbian rule, which holds that synapses between neurons that fire together become stronger, as long as the synchronous firing is accompanied by a reward."[69] The neurons that fire together wire together to form what Hebb called cell assemblies. The cell assemblies that are formed by the rabbit's response to what is significant for it are in effect tuned to select the significant sensory input from the background noise. For example, those cells involved in a previous narrow escape from a fox would be wired together in a cell assembly. Then, in an environment previously experienced as dangerous, the cell assemblies sensitive to the smell of foxes would be primed to respond.

Freeman notes, "For a burst [of neuronal activity] to occur in response to some odorant, the neurons of the assembly and the bulb as a whole must first be 'primed' to respond strongly to that specific input."[70] And he adds, "Our experiments show that the gain [sensitivity to input] in neuronal collections increases in the bulb and olfactory cortex when the animal is hungry, thirsty, sexually aroused or threatened."[71] So if a male animal has just eaten and is ready to mate, the gain is turned down on the cell assemblies responsive to food smells, and turned up on female smells. Thus, from the start the cell assemblies are not just passive receivers of meaningless input from the universe but, on the basis of past experience, are tuned to respond to what is significant to the animal given its arousal state.

Once we see that the cell assemblies are involved in how coping animals respond directly to significant aspects of the environment, we can also see why the binding problem need not arise. The problem is an artifact of trying to interpret the output of isolated cells in the cortex of animals from the perspective of the researcher rather than the perspective of the animal. That is, the researcher, like Merleau-Ponty's intellectualist, interprets the firing of the cells in the sense organ as responding to features of an object-type—features such as orange, round, and tapered that can be specified independent of the object to which they belong. The researcher then has the problem of figuring out how the brain binds these isolated features into a representation of, say, a carrot (and adds the function predicate, good to eat). But, according to Freeman, in an active, hungry animal the output from the isolated detector cells triggers a cell assembly already tuned to detect the relevant input on the basis of past significant experience, which in turn puts the brain into a state that signals to the limbic system "eat this now," without the brain ever having to solve the problem of how the isolated features abstracted by the researchers are brought together into the presentation of an object.

Freeman dramatically describes the brain activity involved:

If the odorant is familiar and the bulb has been primed by arousal, the information spreads like a flash fire through the nerve cell assembly. First, excitatory input to one part of the assembly during a sniff excites the other parts, via the Hebbian synapses. Then those parts re-excite the first, increasing the gain, and so forth, so that the input rapidly ignites an explosion of collective activity throughout the assembly. The activity of the assembly, in turn, guides the entire bulb into a new state by igniting a full-blown burst.[72]

Specifically, after each sniff, the rabbit's olfactory bulb goes into one of several possible states that neural modelers traditionally call energy states. A state tends toward minimum "energy" the way a ball tends to roll toward the bottom of a container, no matter where it starts from within the container. Each possible minimal energy state is called an attractor. The brain states that tend toward a particular attractor, no matter where they start in the basin, are called that attractor's *basin of attraction*. As the brain activation is pulled into an attractor, the brain in effect selects the meaningful stimulus from the background.

Thus the stimuli need not be processed into a representation of the current situation on the basis of which the brain then has to infer what is present in the environment. Rather, in Freeman's account, the rabbit's brain forms a new basin of attraction for each new significant class of inputs. The significance of past experience is preserved in basins of attraction. The set of basins of attraction that an animal has learned form what is called an attractor landscape. According to Freeman, "The state space of the cortex can therefore be said to comprise an attractor landscape with several adjoining basins of attraction, one for each class of learned stimuli."[73] Thus Freeman contends that each new attractor does not represent, say, a carrot, or the smell of carrot, or even what to do with a carrot. Rather, the brain's current state is the result of the sum of the animal's past experiences with carrots. What in the physical input is directly picked up and resonated to when the rabbit sniffs, then, is the affords-eating, and the brain state is directly coupled with (or, in Gibson's terms, resonates to) the affordance offered by the current carrot.[74]

Freeman offers a helpful analogy:

We conceive each cortical dynamical system as having a state space through which
the system travels as a point moving along a path (trajectory) through the state
space. A simple analogy is a spaceship flying over a landscape with valleys resem-
bling the craters on the moon. An expected stimulus contained in the omnipresent
background input selects a crater into which the ship descends. We call the lowest
area in each crater an "attractor" to which the system trajectory goes, and the set of
crater basins of attraction in an attractor landscape. There is a different attractor for
each class of stimuli that the system [is primed] to expect.[75]

Freeman concludes, "The macroscopic bulbar patterns [do] not relate to
the stimulus directly but instead to the *significance* of the stimulus."[76] Indeed, after triggering a specific attractor and modifying it, the stimulus—
the impression made on the receptor cells in the sense organ—has no
further job to perform. Freeman explains:

The new pattern is selected by the stimulus from the internal pre-existing repertoire
[of attractors], not imposed by the stimulus. It is determined by prior experience with
this class of stimulus. The pattern expresses the nature of the class and its significance for the subject rather than the particular event. The identities of the particular neurons in the receptor class that are activated are irrelevant and are not
retained.[77] ... Having played its role in setting the initial conditions, the sense-
dependent activity is washed away.[78]

Thus, as Merleau-Ponty claims and psychological experiments confirm, we
normally have no experience of the data picked up by the sense organs.[79]

Learning and Merleau-Ponty's Intentional Arc

Thus, according to Freeman's model, when hungry, frightened, or in some
other state, the rabbit sniffs around seeking food, runs toward a hiding
place, or does whatever else prior experience has taught it is successful.
The weights on the animal's neural connections are then changed on the
basis of the quality of its resulting experience. That is, they are changed in
a way that reflects the extent to which the result satisfied the animal's current need.

Freeman claims his readout from the rabbit's brain shows that each
learning experience with a previously unknown stimulus, or previously
unimportant stimulus class that is significant in a new way, sets up a new
attractor for that class and *rearranges all the other attractor basins in the
landscape*:

I have observed that brain activity patterns are constantly dissolving, reforming and
changing, particularly in relation to one another. When an animal learns to respond

to a new odor, there is a shift in all other patterns, even if they are not directly involved with the learning. There are no fixed representations, as there are in [GOFAI] computers; there are only significances.[80]

The constantly updated landscape of attractors is presumably correlated with the agent's experience of the changing significance of things in the world, that is, with the intentional arc.

Freeman adds:

I conclude that context dependence is an essential property of the cerebral memory system, in which each new experience must change all of the existing store by some small amount, in order that a new entry be incorporated and fully deployed in the existing body of experience. This property contrasts with memory stores in computers...in which each item is positioned by an address or a branch of a search tree. There, each item has a compartment, and new items don't change the old ones. Our data indicate that in brains the store has no boundaries or compartments....Each new state transition...initiates the construction of a local pattern that impinges on and modifies the whole intentional structure.[81]

Merleau-Ponty likewise concludes that, thanks to the intentional arc, no two experiences of the world are ever exactly alike.[82]

It is important to realize how different this model is from any representationalist account. There is no fixed and independent intentional structure in the brain—not even a latent one. There is nothing that can be found in the olfactory bulb in isolation that represents or even corresponds to anything in the world. There is only the fact that, given the way the nerve cell assemblies have been wired on the basis of past experience, when the animal is in a state of arousal and is in the presence of a significant item such as food or a potential predator or a mate, the bulb will go into a certain attractor state. That activity state in the current interaction of animal and environment corresponds to the whole world of the organism with some aspect salient. The activity is not an isolated brain state but only comes into existence and only is maintained as long as, and in so far as, it is dynamically coupled with the significant situation in the world that selected it, and does not exist apart from it. Whereas, as we have seen, in the cognitivist notion of representations, a representation exists apart from what it represents.

Thus Freeman offers a model of learning that is not an associationist model according to which, as one learns, one adds more and more fixed connections, nor a cognitivist model based on off-line representations of objective facts about the world that enable off-line inferences as to which facts to expect next, and what they mean. Rather, Freeman's model instantiates the causal basis of a genuine intentional arc in which there are no linear casual connections between world and brain nor a fixed library of representations, but where, each time a new significance is encountered, the whole perceptual world of the animal changes so that the significance that is directly displayed in the world of the animal is continually enriched.

The Perception-Action Loop

The brain's movement toward the bottom of a particular basin of attraction underlies the perceiver's perception of the significance for action of a particular experience.[83] For example, if a carrot affords eating the rabbit is directly readied to eat the carrot, or perhaps readied to carry off the carrot, depending on which attractor is currently activated. Freeman tells us, "The same global states that embody the significance provide...the patterns that make choices between available options and that guide the motor systems into sequential movements of intentional behavior."[84] The animal must take account of how things are going and either continue on a promising path or, if the overall action is not going as well as anticipated, the brain must self-organize so the attractor system jumps to another attractor. This either causes the animal to act in such a way as to increase its sense of impending reward, or the brain will shift attractors again, until it lands in one that makes such an improvement. The attractors can change as if they were switching from frame to frame in a movie film, with each further sniff or with each shift of attention. If the rabbit achieves what it is seeking, a report of its success is fed back to reset the sensitivity of the olfactory bulb. And the cycle is repeated.

Freeman's overall picture of skilled perception and action, then, is as follows. The animal, let's say, a rabbit sniffing a carrot, receives stimuli that, thanks to prior Hebbian learning, puts its olfactory bulb into a specific attractor basin—for example, the attractor that has been formed by, and amounts to, the brain's classification of the stimulus as affording eating. Along with other brain systems, the bulb selects a response. The rabbit is solicited to eat this now. It would be too cognitivist to say the bulb *sends a message* to the appropriate part of the brain and too mechanistic to say the bulb *causes* the activity of eating the carrot. The meaning of the input is neither in the stimulus nor in a mechanical response directly triggered by the stimulus. Significance is not stored as a memory representation nor an association. Rather the memory of significance is in the repertoire of attractors as classifications of possible responses—the attractors themselves being the product of past experience.

Once the stimulus has been classified by selecting an attractor that says "Eat this now," the problem for the brain is just how this eating is to be done. On-line coping needs a stimuli-driven feedback policy dictating how to move rapidly over the terrain and approach and eat the carrot. Here, an actor-critic version of Temporal Difference Reinforcement Learning (TDRL) can serve to augment the Freeman model.

According to TDRL, learning the appropriate movements in the current situation requires learning the expected final award as well as the movements. These two functions are learned slowly through repeated experiences. Then the brain can monitor directly whether the expectation of reward is being met as the rabbit approaches the carrot to eat it. If the expected final reward suddenly decreases, owing, for example, to the current inaccessibility of the carrot, the relevant part of the brain prompts the olfactory bulb to switch to a new attractor or perspective on the situation that dictates a different learned action, say, dragging the carrot, with its expected reward.[85] Only after a skill is thus acquired can the current stimuli, plus the past history of responding to related stimuli now wired into cell assemblies, produce the rapid responses required for on-going skillful coping.

Optimal Grip

The animal's movements are presumably experienced by the animal as tending toward getting and maintaining an optimal perceptual take on what is currently significant and, where appropriate, an ongoing optimal bodily grip on it. As Merleau-Ponty says, "Through [my] body I am at grips with the world."[86] Freeman sees his account of the brain dynamics underlying perception and action as structurally isomorphic with Merleau-Ponty's. He explains:

Merleau-Ponty concludes that we are moved to action by disequilibrium between the self and the world. In dynamic terms, the disequilibrium ... puts the brain onto ... a pathway through a chain of preferred states, which are learned basins of attraction. The penultimate result is not an equilibrium in the chemical sense, which is a dead state, but a descent for a time into the basin of an attractor.[87]

Thus, according to Freeman, in governing action the brain normally moves from one basin of attraction to another, descending into each basin for a time without coming permanently to rest in any one basin. The body is thereby led to move toward a maximal grip but the coupled coper, instead of remaining at rest when a maximal grip is achieved, is drawn to move on in response to another affordance that solicits the body to take up the same task from another angle, or to turn to the next task that grows out of the current one.

The selected attractor, together with input from the sense organs, then signals the limbic system to implement a new action with its new expected reward. Then again a signal comes back to the olfactory bulb and elsewhere as to whether the activity is progressing as expected. If so, the current attractor and action will be maintained, but if the result is not as expected, with the formation of the next attractor landscape some other attractor will be selected on the basis of past learning. In Merleau-Ponty's terms, Freeman's model, as we have seen, explains the intentional arc—how our previous coping experiences feed back to determine what action the current situation solicits—while the TDRL model keeps the animal moving toward a sense of minimal tension, that is, a least rate of change in expected reward, and hence toward achieving and maintaining what Merleau-Ponty calls a maximal grip.

Circular Causality

Such systems are self-organizing. Freeman explains:

Macroscopic ensembles exist in many materials, at many scales in space and time, ranging from . . . weather systems such as hurricanes and tornadoes, even to galaxies. In each case, the behavior of the microscopic elements or particles is constrained by the embedding ensemble, and microscopic behavior cannot be understood except with reference to the macroscopic patterns of activity.[88]

Thus, the cortical field controls the neurons that create the field. In Freeman's terms, in this sort of circular causality the overall activity "enslaves" the elements. As he emphasizes:

Having attained through dendritic and axonal growth a certain density of anatomical connections, the neurons cease to act individually and start participating as part of a group, to which each contributes and from which each accepts direction. . . . The activity level is now determined by the population, not by the individuals. This is the first building block of neurodynamics.[89]

Given the way the whole brain can be tuned by past experience to influence individual neuron activity, Freeman can claim, "Measurements of the electrical activity of brains show that dynamical states of neuroactivity emerge like vortices in a weather system, triggered by physical energies impinging onto sensory receptors."[90] Merleau-Ponty seems to anticipate Freeman's neurodynamics when he says:

It is necessary only to accept the fact that the physico-chemical actions of which the organism is in a certain manner composed, instead of unfolding in parallel and independent sequences, are constituted . . . in relatively stable "vortices."[91]

Freeman's Model as a Basis for Heideggerian AI

According to Freeman, the discreteness of global state transitions from one attractor basin to another makes it possible to model the brain's activity on a computer. The model uses numbers to stand for these discrete state transitions. He notes:

> At macroscopic levels each perceptual pattern of neuroactivity is discrete, because it is marked by state transitions when it is formed and ended....I conclude that brains don't use numbers as symbols, but they do use discrete events in time and space, so we can represent them...by numbers in order to model brain states with digital computers.[92]

That is, the states of the model are representations of brain states, not of the features of things in the everyday world. Just as simulated neural nets simulate brain processing but do not contain symbols that represent features of the world, the computer can model the series of discrete state transitions from basin to basin, thereby modeling how, on the basis of past experiences of success or failure, physical inputs are directly perceivable as significant for the organism. But the model is not an intentional being, only a description of such.

Freeman has actually programmed his model of the brain as a dynamic physical system, and so claims to have shown what the brain is doing to provide the material substrate for Heidegger's and Merleau-Ponty's phenomenological accounts of everyday perception and action. This may well be the new paradigm for the cognitive sciences that Wheeler proposes to present in his book but which he fails to find. It would show how the emerging embodied-embedded approach could be step toward a genuinely existential AI. Although, as we shall see, it would still be a very long way from programming human intelligence. Meanwhile, the job of phenomenologists is to get clear concerning the phenomena that must to be explained. That would include an account of how human beings, unlike the so-called Heideggerian computer models we have discussed, neither just ignore the frame problem nor solve it, but show why it doesn't occur.

How Heideggerian AI Would Dissolve Rather Than Avoid or Solve the Frame Problem

As we have seen, Wheeler rightly thinks that the simplest test of the viability of any proposed AI program is whether it can solve the frame problem. We've also seen that the two current supposedly Heideggerian approaches to AI avoid rather than solve the frame problem. Brooks's empiricist-behaviorist approach, in which the environment directly causes responses, avoids it by leaving out significance and learning altogether, while Agre's action-oriented approach, which includes only a small fixed set of possibly relevant responses, fails to deal with the problem of *changing* relevance.

Wheeler's own proposal, however, by introducing flexible action-oriented representations, like any representational approach has to face the frame problem head on. To see why, we need only slightly revise his statement of the frame problem (quoted earlier), substituting "representation" for "belief":

Given a dynamically changing world, how is a nonmagical system...to retrieve and (if necessary) to revise, out of all the *representations* that it possesses, just those *representations* that are relevant in some particular context of action?[93]

Wheeler's frame problem, then, is to explain how his allegedly Heideggerian system can determine in some systematic way which of the action-oriented representations it contains or can generate are relevant in a current situation, and keep track of how this relevance changes with changes in the situation.

Given his emphasis on problem solving and representations, it is not surprising that the concluding chapter of Wheeler's book, where he returns to the frame problem to test his proposed Heideggerian AI, offers no solution or dissolution of the problem. Instead, he asks us to "give some credence to [his] informed intuitions,"[94] which I take to be on the scent of Freeman's account of rabbit olfaction, that nonrepresentational causal coupling must play a crucial role. But I take issue with his conclusion that

...*in extreme cases* the neural contribution will be *nonrepresentational* in character. In other cases, *representations* will be active partners alongside certain additional factors, but those representations will be action oriented in character, and so will realize the same content-sparse, action-specific, egocentric, context-dependent profile that Heideggerian phenomenology reveals to be distinctive of online *representational* states at the agential level.[95]

But for Heidegger, *all* representational accounts are part of the problem. Wheeler's account, so far as I understand it, gives no explanation of how on-line dynamic coupling is supposed to dissolve the on-line frame problem. Nor does it help to wheel in, as Wheeler does, action-oriented representations and the extended mind. Any attempt to solve the frame problem by giving any role to any sort of representational states, even on-line ones, has so far proved to be a dead end. It looks like nonrepresentational neural activity can't be understood to be the "extreme case." Rather, such activity must be, as Heidegger, Merleau-Ponty, and Freeman contend, our basic way of responding directly to relevance in the everyday world, so that the frame problem does not arise.

Heidegger and Merleau-Ponty argue that, and Freeman demonstrates how, thanks to our embodied coping and the intentional arc it makes possible, we directly respond to relevance and our skill in sensing and responding to relevant changes in the world is constantly improved. In coping in a particular context, say, a classroom, we learn to ignore most of what is in the room, but if it gets too warm, the windows solicit us to open them. We ignore the chalk dust in the corners and the chalk marks on the desks but we attend to the chalk marks on the blackboard. We take for granted that what we write on the board doesn't affect the windows, even if we write, "open windows," and what we do with the windows doesn't affect what's on the board. And as we constantly refine this background know-how, the things in the room and its layout become more and more familiar, take on more and more significance, and each thing draws us to act when an action is relevant. Thus we become better able to cope with change. Given our experience in the world, whenever there is a change in the current context we respond to it only if in the past it has turned out to be significant, and even when we sense a significant change we treat everything else as unchanged except what our familiarity with the world suggests might also have changed and so needs to be checked out. Thus, for embedded-embodied beings a local version of the frame problem does not arise.

But the frame problem reasserts itself when we consider changing contexts. How do we sense when a situation on the horizon has become relevant to our current task? When Merleau-Ponty describes the phenomenon, he speaks of one's attention being drawn by an affordance on the margin of one's current experience: "To see an object is either to have it on the fringe of the visual field and be able to concentrate on it, or else respond to this *summons* by actually concentrating on it."[96] Thus, for example, as one faces the front of a house, one's body is already being *summoned* (not just *prepared*) to go around the house to get a better look at its back.[97]

Merleau-Ponty's treatment of what Husserl calls the inner horizon of the perceptual object—its insides and back—applies equally to our experience of a situation's outer horizon of other potential situations. As I cope with a specific task in a specific situation, other situations that have in the past been relevant are right now present on the horizon of my experience as potentially (not merely possibly) relevant to my current situation.

If Freeman is right, our sense of familiar-but-not-currently-fully-present aspects of what is currently ready-to-hand, as well as our sense of other potentially relevant familiar situations on the horizon of the current situation, might well be correlated with the fact that brain activity is not simply in one attractor basin at a time but is influenced by other attractor basins in the same landscape, as well as by other attractor landscapes that, under what have previously been experienced as relevant conditions, are ready to draw current brain activity into themselves. According to Freeman, what makes us open to the horizonal influence of other attractors is that the whole system of attractor landscapes collapses and is rebuilt with each new rabbit sniff, or in our case, presumably with each shift in our attention. And after each collapse, a new landscape may be formed on the basis of new significant stimuli—a landscape in which, thanks to past experiences, a different attractor is active.[98] This presumably underlies our experience of being summoned.

And, once one correlates Freeman's neurodynamic account with Merleau-Ponty's description of the way the intentional arc feeds back our past experience into the way the world appears to us, so that the world solicits from us ever-more-appropriate responses to its significance, we can see that we can be directly summoned to respond appropriately not only to what is relevant in our current situation, but we may be summoned by other familiar situations on the horizon of the present one. Then the fact that we can deal with changing relevance by anticipating what will change and what will stay the same no longer seems unsolvable.

But there is a generalization of the problem of relevance, and thus of the frame problem, that still seems intractable. In *What Computers Can't Do* I gave an example of the possible relevance of everything to everything. In placing a racing bet we can usually restrict ourselves to such relevant facts as the horse's age, jockey, and past performance, but there are always other factors such as whether the horse is allergic to goldenrod or whether the jockey has just had a fight with the owner, which in some cases can be decisive. Human handicappers are capable of noticing such anomalies when they come across them.[99] But since anything in experience could be relevant to anything else, for representational or computation AI such an ability seems incomprehensible. Jerry Fodor follows up on my pessimistic example:

"The problem," [Dreyfus] tells us, "is to get the structure of an entire belief system to bear on individual occasions of belief fixation." We have, to put it bluntly, no computational formalisms that show us how to do this, and we have no idea how such formalisms might be developed.... If someone—a Dreyfus, for example—were to ask us why we should even suppose that the digital computer is a plausible mechanism for the simulation of global cognitive processes, the answering silence would be deafening.[100]

But if we give up the cognitivist assumption that we have to relate isolated meaningless facts and events to each other and see that all facts and events are experienced on the background of a familiar world, we can see the outline of a solution. The handicapper has a sense of which situations are significant. He has learned to ignore many anomalies, such as an eclipse or an invasion of grasshoppers that have so far not turned out to be important, but, given his familiarity with human sports requiring freedom from distraction, he may well be sensitive to these anomalies. Of course, given his lack of experience with the new anomaly, it will not show its relevance on its face and summon an immediate appropriate response. Rather, the handicapper will have to step back and *figure out* whether the anomaly is relevant and, if so, how. Unfamiliar breakdowns require us to go off-line and think.

In his deliberations, the handicapper will draw on his background familiarity with how things in the world behave. Allergies and arguments normally interfere with one's doing one's best. Of course, given his lack of experience with this particular situation, any conclusion he reaches will be risky, but he can sense that a possibly relevant situation has entered the horizon of his current task and his familiarity with similar situations will give him some guidance in deciding what to do. While such a conclusion will not be the formal computational solution required by cognitivism, it is correlated with Freeman's claim that on the basis of past experience, attractors and whole landscapes can directly influence each other.[101] This suggests that the handicapper need not be at a loss; that this extreme version of the frame problem, like all the simpler versions, is an artifact of the atomistic cognitivist/computational approach to the mind/brain's relation to the world.

Conclusion

It would be satisfying if we could now conclude that, with the help of Merleau-Ponty and Walter Freeman, we can fix what is wrong with current allegedly Heideggerian AI by making it more Heideggerian. There is, however, a big remaining problem. Merleau-Ponty's and Freeman's accounts of how we directly pick up significance and improve our sensitivity to relevance depends on our responding to what is significant *for us*, given our needs, body size, ways of moving, and so forth, not to mention our personal and cultural self-interpretation. If we can't make our brain model responsive to the significance in the environment as it shows up specifically for human beings, the project of developing an embedded and embodied Heideggerian AI can't get off the ground.

Thus, to program Heideggerian AI, we would not only need a model of the brain functioning underlying coupled coping such as Freeman's; we would also need—and here's the rub—a model of our particular way of being embedded and embodied such that what we experience is significant for us in the particular way that it is. That is, we would have to include in our program a model of a body very much like ours with our needs, desires, pleasures, pains, ways of moving, cultural background, etc.

So, according to the view I have been presenting, even if the Heideggerian–Merleau-Pontian approach to AI suggested by Freeman is ontologically sound in a way that GOFAI and subsequent supposedly Heideggerian models proposed by Brooks, Agre, and Wheeler are not, a neurodynamic computer model would still have to be given a detailed description of a body and motivations like ours if things were to count as significant for it so that it could learn to act intelligently in *our* world.[102] We have seen that Heidegger, Merleau-Ponty, and Freeman offer us hints of the elaborate and subtle body and brain structures we would have to model and how to model some of them, but this only makes the task of a Heideggerian AI seem all the more difficult and casts doubt on whether we will ever be able to accomplish it.[103]

We can, however, make some progress toward animal AI. Freeman has actually used his brain model to model intelligent devices.[104] Specifically, he and his coworkers have modeled the activity of the brain of the salamander sufficiently to simulate the salamander's foraging and self-preservation capacities. The model seeks out the sensory stimuli that make available the information it needs to reach its goals. Presumably such a simulated salamander could learn to run a maze and so have a primitive intentional arc and avoid a primitive frame problem. Thus, one can envisage a kind of animal artificial intelligence inspired by Heidegger and Merleau-Ponty, but that is no reason to believe, and there are many reasons to doubt, that such a device would be a first step on a continuum toward making a machine capable of simulating human coping with what is significant.

Notes

1. This isn't just my impression. Philip Agre, a Ph.D. candidate at the AI Lab at that time, later wrote:

I have heard expressed many versions of the propositions ... that philosophy is a matter of mere thinking whereas technology is a matter of real doing, and that philosophy consequently can be understood only as deficient.

See Philip E. Agre, *Computation and Human Experience* (Cambridge: Cambridge University Press, 1997), p. 239.

2. Marvin Minsky, quoted in a 1968 MGM press release for Stanley Kubrick's *2001: A Space Odyssey.*

3. Allen Newell and Herbert A. Simon, "Computer Science as Empirical Inquiry: Symbols and Search," in *Mind Design*, edited by John Haugeland (Cambridge, Mass.: MIT Press, 1988).

4. John R. Searle, *The Construction of Social Reality* (New York: Free Press, 1995).

5. Martin Heidegger, *Being and Time*, translated by J. Macquarrie and E. Robinson (New York: Harper & Row, 1962), pp. 132, 133.

6. Michael Wheeler, *Reconstructing the Cognitive World: The Next Step* (Cambridge, Mass.: MIT Press/Bradford Press, 2007), p. 179.

7. Edmund Husserl, *Experience and Judgment* (Evanston: Northwestern University Press, 1973), p. 38. To do the same job, Roger Schank proposed what he called scripts such as a restaurant script. "A script," he wrote,

is a structure that describes appropriate sequences of events in a particular context. A script is made up of slots and requirements about what can fill those slots. The structure is an interconnected whole, and what is in one slot affects what can be in another. A script is a predetermined, stereotyped sequence of actions that defines a well-known situation.

See Roger C. Schank and R. P. Abelson, *Scripts, Plans, Goals and Understanding: An Inquiry into Human Knowledge Structures* (Hillsdale, N.J.: Lawrence Erlbaum, 1977), p. 41. Quoted in John Preston and Mark Bishop, eds., *Views into the Chinese Room: New Essays on Searle and Artificial Intelligence* (Oxford: Clarendon Press, 2002).

8. After I published *What Computers Can't Do: A Critique of Artificial Reason* (Cambridge, Mass.: MIT Press, 1972) and pointed out this difficulty among many others, my MIT computer colleagues, rather than facing my criticism, tried to keep me from getting tenure on the grounds that my affiliation with MIT would give undeserved credibility to my "fallacies," and so would prevent the AI Lab from continuing to receive research grants from the Defense Department.

The AI researchers were right to worry. I was considering hiring an actor to impersonate an officer from DARPA (Defense Advanced Research Projects Agency) having lunch with me at the MIT Faculty Club. (A plan cut short when Jerry Wiesner, the President of MIT, after consulting with Harvard and Russian computer scientists, and reading my book himself, personally granted me tenure.) I did, however, later get called to Washington by DARPA to give my views, and the AI Lab did lose DARPA support during what has come to be called the AI Winter.

9. Terry Winograd, "Artificial Intelligence and Language Comprehension," in *Artificial Intelligence and Language Comprehension* (Washington, D.C.: National Institute of Education, 1976), p. 9.

10. *Wired*, August 2003, p. 8.

11. Mark Wrathall, ed., *Heidegger, Coping, and Cognitive Science: Essays in Honor of Hubert L. Dreyfus*, volume 2 (Cambridge, Mass.: MIT Press, 2000), p. iii.

12. Wheeler, *Reconstructing the Cognitive World*, p. 285.

13. John Haugeland, "Mind Embodied and Embedded," *Having Thought: Essays in the Metaphysics of Mind* (Cambridge, Mass.: Harvard University Press, 1998), p. 218.

14. Terry Winograd, "Heidegger and the Design of Computer Systems," talk delivered at Applied Heidegger Conference, Berkeley, California, September 1989. Cited in Dreyfus, *What Computers Still Can't Do*, p. xxxi.

15. Not everyone was pleased. One of the graduate students responsible for the invitation reported to me, "After it was announced that you were giving the talk, Marvin Minsky came into my office and shouted at me for ten minutes or so for inviting you."

16. Dreyfus, *What Computers Still Can't Do*, pp. 265–66.

17. Ibid., p. 300.

18. Ibid.

19. Rodney A. Brooks, "Intelligence Without Representation," in *Mind Design*, edited by John Haugeland (Cambridge, Mass.: MIT Press, 1988), p. 416 (Brooks's paper was originally published in 1986). Haugeland explains Brooks's breakthrough using as an example Brooks's robot, Herbert:

> Brooks uses what he calls "subsumption architecture," according to which systems are decomposed not in the familiar way by local functions or faculties, but rather by global *activities* or *tasks*.... Thus, Herbert has one subsystem for detecting and avoiding obstacles in its path, another for wandering around, a third for finding distant soda cans and homing in on them, a fourth for noticing nearby soda cans and putting its hand around them, a fifth for detecting something between its fingers and closing them, and so on ... fourteen in all. What's striking is that these are all complete input/output systems, more or less independent of each other.

See Haugeland, *Having Thought: Essays in the Metaphysics of Mind* (Cambridge, Mass.: Harvard University Press, 1998), p. 218.

20. Brooks, in "Intelligence Without Representation," gives me credit for "being right about many issues such as the way in which people operate in the world is intimately coupled to the existence of their body" (p. 42) but he denies the direct influence of Heidegger (p. 415):

> In some circles, much credence is given to Heidegger as one who understood the dynamics of existence. Our approach has certain similarities to work inspired by this German philosopher

(for instance, Agre and Chapman 1987) but our work was not so inspired. It is based purely on engineering considerations.

See Brooks, "Intelligence without Representation," p. 415, and Brooks, *Flesh and Machines: How Robots Will Change Us* (New York: Vintage Books, 2002), p. 168.

21. "Can higher-level functions such as learning occur in these fixed topology networks of simple finite state machines?" he asks. But he offers no response (Brooks, "Intelligence without Representation," p. 420).

22. See Maurice Merleau-Ponty, *The Structure of Behavior*, translated by A. L. Fisher, 2nd ed. (Boston: Beacon Press, 1966).

23. Brooks, "Intelligence Without Representation," p. 418.

24. Rodney A. Brooks, "From Earwigs to Humans," *Robotics and Autonomous Systems* 20: 291(2007).

25. Daniel Dennett, "The Practical Requirements for Making a Conscious Robot," *Philosophical Transactions of the Royal Society of London* (series A) 349: 133–46(1994).

26. Ibid., p. 133.

27. Although, as of summer 2007, you couldn't tell it from the Cog web page (www.ai.mit.edu/projects/humanoid-robotics-group/cog/).

28. Private communication, October 26, 2005 (emphasis added).

29. Brooks, "From Earwigs to Humans," p. 301. (The missing idea may well be Walter Freeman's. See below.)

30. Philip E. Agre, "The Dynamic Structure of Everyday Life," MIT AI Technical Report no. 1085, October 1988, chapter 1, section A.1.a, p. 9.

31. Agre, *Computation and Human Experience*, p. 243. His ambitious goal was to "develop an alternative to the representational theory of intentionality, beginning with the phenomenological intuition that everyday routine activities are founded in habitual, embodied ways of interacting with people, places, and things in the world."

32. Ibid., p. xi.

33. Ibid., p. 332.

34. Ibid., p. 251. As Beth Preston sums it up in her paper "Heidegger and Artificial Intelligence," *Philosophy and Phenomenological Research* 53, no. 1(March): 43–69(1993):

What results is a system that represents the world not as a set of objects with properties, but as current functions (what Heidegger called in-order-tos). Thus, to take a Heideggerian example, I experience a hammer I am using not as an object with properties but as an in-order-to-drive-in-this-nail.

35. Heidegger himself is not always clear about the status of the ready-to-hand. When he is stressing the holism of equipmental relations, he thinks of the ready-to-hand as equipment, and of equipment as things like lamps, tables, doors, and rooms that have a place in a whole nexus of other equipment. Furthermore, he holds that breakdowns reveal that these interdefined pieces of equipment are made of present-at-hand stuff that was there all along (*Being and Time*, p. 97.) At one point Heidegger even goes so far as to include the ready-to-hand under the categories that characterize the present-at-hand (p. 70):

We call *"categories"* characteristics of being for entities whose character is not that of Dasein.... Any entity is either a *"who"* (existence) or a what (present-at-hand in the broadest sense).

36. Ibid., p. 405.

37. Ibid., p. 99.

38. Martin Heidegger, *Logic: The Question of Truth*, translation, by Thomas Sheehan, of Martin Heidegger, *Gesamtausgabe* (Bloomington: Indiana University Press, 20TK), volume 21, p. 144.

39. Heidegger goes on immediately to contrast the total absorption of coping he has just described with the as-structure of thematic observation:

Every act of having *things* in front of oneself and *perceiving them* is held within [the] disclosure of those *things*, a disclosure that things get from a *primary meaningfulness* in terms of the what-for. Every act of *having something in front of oneself and perceiving it* is, in and for itself, a "having" *something as something.*

To put it in terms of *Being and Time*, the as-structure of equipment goes all the way down in *the world,* but not in the way the world shows up in our absorbed coping. It is poor phenomenology to read the self and the as-structure into our experience when we are coping at our best.

40. There is a third possible attitude. Heidegger calls it responding to signs. Then I am sensitive to possibly relevant aspects of my environment and take them into account as I cope. We normally do this when driving in traffic, and the master potter, for example, is alert to the way the pot she is making may be deviating from the normal.

41. Maurice Merleau-Ponty, *Phenomenology of Perception*, translated by C. Smith (London: Routledge & Kegan Paul, 1962), p. 136.

42. Wheeler, *Reconstructing the Cognitive World*, pp. 222–23.

43. Ibid., 187.

44. Ibid., 188.

45. Ibid., 188–89.

46. Merleau-Ponty says the same (*Phenomenology of Perception*, 139.):

To move one's body is to aim at things through it; it is to allow oneself to respond to their call, which is made upon it independently of any representation.

47. See A. Clark and D. Chalmers, "The Extended Mind," *Analysis* 58, no. 1: 7–19, 1998.

48. As Heidegger puts it, "The self must forget itself if, lost in the world of equipment, it is to be able 'actually' to go to work and manipulate something" (*Being and Time*, p. 405).

49. Heidegger, *Logic*, p. 146. It's important to realize that when he uses the term "understanding," Heidegger explains (with a little help from the translator) that he means a kind of know-how:

In German we say that someone can *verstehen* something—literally, stand in front of or ahead of it, that is, stand at its head, administer, manage, preside over it. This is equivalent to saying that he *versteht sich darauf*, understands in the sense of being skilled or expert at it, has the know-how of it.

See Martin Heidegger, *The Basic Problems of Phenomenology*, translated by A. Hofstadter (Bloomington: Indiana University Press, 1982), p. 276.

50. Heidegger, *Being and Time*, 416. To make sense of this slogan, it's important to be clear that Heidegger distinguishes the human world from the physical universe.

51. Martin Heidegger, *Phenomenological Interpretations in Connection with Aristotle*, in *Supplements: From the Earliest Essays to Being and Time and Beyond*, edited by John van Buren (State University of New York Press, 2002), p. 115 (emphasis added). This way of putting the source of significance covers both animals and people. By the time he published *Being and Time*, however, Heidegger was interested exclusively in the special kind of significance found in the world opened up by human beings who are defined by the stand they take on their own being. We might call this meaning. In this chapter I'm putting the question of uniquely human meaning aside to concentrate on the sort of significance we share with animals.

52. See Samuel Todes, *Body and World* (Cambridge, Mass.: MIT Press, 2001). Todes goes beyond Merleau-Ponty in showing how our world-disclosing perceptual experience is structured by the structure of our bodies. Merleau-Ponty never tells us what our bodies are actually like and how their structure affects our experience. Todes points out that our body has a front-back and up-down orientation. It moves forward more easily than backward, and can successfully cope only with what is in front of it. He then describes how, in order to explore our surrounding world and orient ourselves in it, we have to balance ourselves within a vertical field that we do not produce, be effectively directed in a circumstantial field (facing one aspect of that field rather than another), and appropriately set to respond to the specific thing we are encountering within that field. For Todes, then, perceptual receptivity is an embodied, normative, skilled accomplishment, in response to our need to orient

ourselves in the world. Clearly, this kind of holistic background coping is not done for a reason.

53. Merleau-Ponty, *Phenomenology of Perception*, 250 (translation modified).

54. Ibid., 153 (emphasis added).

55. Timothy van Gelder, "Dynamics and Cognition," in *Mind Design II*, edited by John Haugeland (Cambridge, Mass.: MIT Press/Bradford Books, 1997), pp. 439, 448.

56. Ibid.

57. Michael Wheeler, "Change in the Rules: Computers, Dynamical Systems, and Searle," in *Views into the Chinese Room: New Essays on Searle and Artificial Intelligence*, edited by John Preston and Mark Bishop (Oxford: Clarendon Press, 2002), p. 345.

58. Ibid., pp. 344, 345.

59. Wheeler, *Reconstructing the Cognitive World*, p. 280.

60. Ibid.

61. Martin Heidegger, *History of the Concept of Time*, translated by T. Kisiel (Bloomington: Indiana University Press, 1985), p. 189.

62. Heidegger, *Basic Problems of Phenomenology*, p. 162.

63. Moreover, the background solicitations are constantly enriched, not by adding new bits of information, as Wheeler suggests, but by allowing finer and finer discriminations that show up in the world by way of the intentional arc.

64. We agree, too, that both these modes of encountering the things in the world are more frequent and more basic than an appeal to general-purpose reasoning and goal-oriented planning.

65. Wheeler, *Reconstructing the Cognitive World*, p. 279.

66. Ibid., p. 134.

67. Wheeler (*Reconstructing the Cognitive World*, p. 91–93) explains:

For the purposes of a dynamical systems approach to Cognitive Science, a dynamical system may be defined as any system in which there is *state-dependent change*, where systemic change is state dependent just in case the future behavior of the system depends causally on the current state of the system. . . .

Nonlinear dynamical systems exhibit a property known as *sensitive dependence on initial conditions*, according to which the trajectories that flow from two adjacent initial-condition-points diverge rapidly. This means that a small change in the initial state of the system becomes, after a relatively short time, a large difference in the evolving state of the system. This is one of the distinguishing marks of the phenomenon of chaos. . . .

[Consider] the case of two theoretically separable dynamical systems that are bound together, in a mathematically describable way, such that some of the parameters of each system either are, or are functions of, some of the state variables of the other. At any particular time, the state of each of these systems will, in a sense, fix the dynamics of the other system. Such systems will evolve through time in a relation of complex and intimate mutual influence, and are said to be *coupled.*

68. Walter J. Freeman, "The Physiology of Perception," *Scientific American,* February 1991, p. 78.

69. Ibid., p. 81.

70. Ibid., p. 82.

71. Ibid.

72. Ibid., p. 83.

73. Walter Freeman, *How Brains Make Up Their Minds* (New York: Columbia University Press, 2000), p. 62. (Quotations from Freeman's books have been reviewed by him and sometimes modified to correspond to his latest vocabulary and way of thinking about the phenomenon.)

74. Thus Freeman's model might well describe the brain activity presupposed by Gibson's talk of "resonating" to affordances.

75. Walter J. Freeman, "Nonlinear Dynamics of Intentionality," *Journal of Mind and Behavior* 18: 291–304(1997). The attractors are abstractions relative to what level of abstraction is significant given what the animal is seeking.

76. Walter Freeman, *Societies of Brains: A Study in the Neuroscience of Love and Hate* (Hillsdale, N.J.: Lawrence Erlbaum, 1995), p. 59 (emphasis added).

77. Ibid., p. 66 (emphasis added).

78. Ibid., p. 67.

79. Sean Kelly, "Content and Constancy: Phenomenology, Psychology, and the Content of Perception," *Philosophy and Phenomenological Research,* forthcoming.

80. Freeman, *How Brains Make Up Their Minds,* p. 22.

81. Freeman, *Societies of Brains,* p. 99 (emphasis added).

82. Merleau-Ponty, *Phenomenology of Perception,* p. 216.

83. See Sean Kelly, "The Logic of Motor Intentionality," Unpublished paper. Also, Corbin Collins describes the phenomenology of this motor intentionality and spells out the logical form of what he calls instrumental predicates. See, "Body Intentionality," *Inquiry,* December 1988.

84. Freeman, *How Brains Make Up Their Minds,* p. 114.

85. See Stuart Dreyfus, "Totally Model-Free Learned Skillful Coping," *Bulletin of Science, Technology and Society* 24, no. 3(June): 182–87(2004). This article does not,

however, discuss the role of a controlling attractor or the use of expected reward to jump to a new attractor.

86. Merleau-Ponty, *Phenomenology of Perception*, 303.

87. Freeman, *How Brains Make Up Their Minds*, 121.

88. Ibid., p. 52.

89. Ibid., p. 53.

90. Freeman, *Societies of Brains*, 111.

91. Merleau-Ponty, *Structure of Behavior*, p. 153.

92. Freeman, *Societies of Brains*, p. 105.

93. Wheeler, *Reconstructing the Cognitive World*, p. 179.

94. Ibid., p. 279.

95. Ibid., p. 276 (emphasis added).

96. Merleau-Ponty, *Phenomenology of Perception*, p. 67 (emphasis added).

97. Sean D. Kelly, "Seeing Things in Merleau-Ponty," in *The Cambridge Companion to Merleau-Ponty* (Cambridge: Cambridge University Press, 2004).

98. We do not experience these rapid changes of attractor landscapes anymore than we experience the flicker in changes of movie frames. Not everything going on in the brain is reflected in the phenomena.

99. Dreyfus, *What Computers Still Can't Do*, p. 258.

100. Jerry A. Fodor, *The Modularity of Mind* (Cambridge, Mass.: MIT Press/Bradford Books, 1983), pp. 128–29.

101. Freeman writes: "From my analysis of EEG patterns, I speculate that consciousness reflects operations by which the entire knowledge store in an intentional structure is brought instantly into play each moment of the waking life of an animal, putting into immediate service all that an animal has learned in order to solve its problems, without the need for look-up tables and random access memory systems" (*Societies of Brains*, p. 136).

102. Dennett sees the "daunting" problem, but he is undaunted. He optimistically sketches out the task:

Cog ... must have *goal-registrations* and *preference-functions* that map in rough isomorphism to human desires. This is so for many reasons, of course. Cog won't work at all unless it has its act together in a daunting number of different regards. It must somehow delight in learning, abhor error, strive for novelty, recognize progress. It must be vigilant in some regards, curious in others, and deeply unwilling to engage in self-destructive activity.

See Dennett, "Consciousness in Human and Robot Minds," in *Cognition, Computation and Consciousness*, edited by Masao Ito, Yasushi Mayashita, and Edmund T. Rolls (Oxford: Oxford University Press, 1997).

103. Freeman runs up against his own version of this problem and faces it frankly: "It can be shown that the more the system is 'open' to the external world (the more links there are), the better its neuronal correlation can be realized. However, in the setting up of these correlations also enter quantities which are intrinsic to the system, they are *internal* parameters and may represent (parameterize) subjective attitudes. Our model, however, is not able to provide a dynamics for these variations." See Walter J. Freeman and G. Vitiello, "Nonlinear Brain Dynamics as Macroscopic Manifestation of Underlying Many-Body Field Dynamics," *Physics of Life Reviews* 3, no. 2: 21(2006).

104. Freeman writes in a personal communication: "Regarding intentional robots that you discuss in your last paragraph, my colleagues Robert Kozma and Peter Erdí have already implemented my brain model for intentional behavior at the level of the salamander in a Sony AIBO (artificial dog) that learns to run a simple maze and also in a prototype Martian Rover at the JPL in Pasadena." See Robert Kozma, Walter J. Freeman, and Peter Erdí, "The KIV Model—Nonlinear Spatio-Temporal Dynamics of the Primordial Vertebrate Forebrain," *Neurocomputing* 52: 819–26(2003), available at http://repositories.cdlib.org/postprints/1049; Robert Kozma and Walter Freeman, "Basic Principles of the KIV Model and Its Application to the Navigation Problem," *Journal of Integrative Neuroscience* 2: 125–45(2003); and Robert Kozma, "Dynamical Approach to Behavior-Based Robot Control and Autonomy," *Biological Cybernetics* 92(6): 367–79(2005).

Figure 15.1
John Maynard Smith. Image courtesy of University of Sussex.

15 An Interview with John Maynard Smith

John Maynard Smith (1920–2004), FRS, was born in London and educated at Eton and Cambridge, where he studied aeronautical engineering. After the Second World War, during which he worked on military aircraft design, he changed career direction and studied fruit fly genetics under J. B. S. Haldane at University College, London. In 1965 he became the founding dean of biological sciences at the University of Sussex, where he stayed for the rest of his career. He was one of the great evolutionary biologists, making many important contributions, including the application of game theory to understanding evolutionary strategies, and a clear definition of the major transitions in the history of life. He won numerous awards and honors, including the highly prestigious Crafoord Prize in 1999 and the Kyoto Prize in 2001.

This is an edited transcript of an interview conducted on May 21, 2003, in John Maynard Smith's office in the John Maynard Smith Building at the University of Sussex, which houses the life sciences. The discussion centered on John's interactions with people involved in cybernetics and early AI.

John Maynard Smith: Shall I tell you about my meeting with Turing?
Philip Husbands: Please.
JMS: It was when I was a graduate student of Haldane at University College, London; very soon after I started, so we're talking about 1952, and I was counting fruit flies. But one of the other things I had been doing, inevitable given my past in aeronautical engineering, was to think about animal flight. And I wrote various papers on that; I was thinking particularly about stability and control of animal flight. I was influenced by John Pringle's work, of course; he did this very, very beautiful empirical work showing that the halteres of the fruitfly, or indeed of any fly—all flies have halteres—are involved in control of the horizontal plane and the yaw.[1] Anyway, Haldane came into the lab, where I was sitting counting flies,

with this rather nice-looking dark small chap, and said, "Maynard Smith!"
—no, "Smith!," he never got round to calling me Maynard Smith—"This is
Dr.——," and I didn't catch the name. "He would be interested in what
you have been doing recently on flight." And I remember thinking, "Oh
God, not another of these biologists who doesn't know a force from an
amoeba; I'm going to have to go very very slowly."

So I started explaining to him some stuff I'd been doing on instability. I
was at the time interested in the fact that primitive flying animals had long
tails—you know, dinosaurs such as *Archaeopteryx*. This had always been
explained away as just an evolutionary hangover: they had long tails
when they were on the ground and hadn't had time to get rid of them.
And that is part of the truth, obviously, but it occurred to me that the
more interesting truth was that they actually needed them for stability. I
proposed that it was only after their nervous system evolved a bit to control
flight more, that they were able to fly with short tails. In fact, my first published paper was called "*The Importance of the Nervous System in the Evolution
of Animal Flight*," and it discusses this problem with a lot of criticism of
previous claims, and basically I think I still believe it.[2] Anyway, I started
explaining this to this poor buffoon with some diagrams. He listened
patiently without saying anything, then he held out his hand for my pen
and changed the direction of one of my arrows. And when I looked at it
he was obviously right; I'd made a mistake in a force diagram. I thought,
"Oh shit," because I'd really been talking to him like a two-year-old, and
so I said, "Look, I'm so sorry but I didn't actually catch your name." And
he said "Well my name is Turing." "Oh shit!" I thought again!

Anyway, I have this other interest in Turing. As you know, he wrote this
very remarkable paper on the chemical basis of morphogenesis describing a
reaction-diffusion-based model.[3]

PH: Yes, that paper must have just come out [in 1952], or was about to be
published.

JMS: Yes, it'd come out just before. So we talked about that for quite a
while, several hours. We talked about what kinds of observations might be
made in the field in connection with the theory, and I've been interested in
reaction-diffusion systems ever since. When I came down here [Sussex Uni-
versity] it was one of the topics that I hoped we'd investigate—the relationship between chemical gradients and development and so on. And so that
is why I invited Brian Goodwin to come here. It was really his interest in
morphogenesis and development that led to the invitation. Anyway, I
didn't get to know Turing, as I only met him on this one occasion.

PH: So there wasn't really a scientific interaction, but some influence?

JMS: Yes, through his morphogenesis work he had a lasting influence on me and what I thought was important in biology. But there was no real scientific interaction. Various young people like myself were influenced by his ideas and followed them up later, and the ideas are still very much in currency. But at the time we were just postgraduates. I don't remember the idea being discussed at the time at mainstream biology meetings. Embryology was a *very* empirical science, a very nonmathematical branch of biology, back then. I don't think it had much impact at the time because of that.

Now Pringle I interacted with a bit more because he'd done this work on flight. But he was already a rather senior figure. He wasn't a lot older than me but he'd started younger and was already an established figure and I was just a graduate student. We talked about the control of flight and stability. It's interesting to learn he was a member of the Ratio Club. I hadn't realized.

PH: Yes, he was one of the founding members and it was he who suggested Turing, whom he knew quite well, should become a member, as they needed some mathematicians to keep the biologists in order.

JMS: Yes, I can well understand that! Of course most people who were doing theoretical biology at that time had worked on radar during the war and had worked with engineers and mathematicians and so appreciated what they could contribute.

PH: Indeed. Pringle and many of the other biologists involved in the Ratio Club seemed to have had an inclination towards theoretical work before the war, but it was greatly strengthened during the war due to deeper exposure to and involvement with engineering. Pringle, for instance, worked on airborne radar development; in fact he was in charge of it for a while. This kind of wartime work seemed to profoundly influence the subsequent careers of quite a few biologists. But you were very close to the whole thing, of course. Does that seem right to you?

JMS: Oh I'm sure of it. Absolutely sure. Of course I came into biology from engineering, but on the other hand not from electrical engineering or control theory.

PH: You had studied aeronautical and mechanical engineering?

JMS: I was basically a mechanical engineer. Though, curiously enough, during the war it had occurred to me that, at least in principle, if an automatic pilot was sensitive enough and quick enough it would be able to control an unstable aircraft, whereas a pilot couldn't. Things would happen too quickly for a human; they would be dead before they'd learnt. You see, there were certain real advantages, aerodynamically, in having an unstable aircraft. It wasn't just that such an aircraft could maneuver quicker, but also landing speeds could be increased and things of that kind. But it also became clear to me very quickly that at that time electrical control was simply not fast enough. But the idea of automatic pilots and control of instability were in my mind and so when I started thinking about insect flight, after the war, it came back to the fore.

PH: Obviously, you became aware of cybernetics, but how underground or mainstream was it, as far you can remember, in the late 1940s, early 1950s.

JMS: Well, you know, I don't think I was explicitly aware of cybernetics until later. That early, I think only a small number of scientists were involved. So, not mainstream.

In fact I remember being rather annoyed when I read about cybernetics a few years later. One of the problem we had in aircraft design was to predict, before the structure of the aircraft was built, what its natural modes of vibration would be. How was it possible to find out? Now—I'm rather proud of this, actually—it occurred to me that you could build an electrical analogue of any mechanical systems if you knew what the masses and stiffnesses and so on were. So we could build an electrical analogue of the structure of the aircraft that oscillated and get its fundamental modes from that. And we did! Anyway, it was rather useful at the early design stage. Of course I wasn't the only person it occurred to, mind you, or the first, but at the time we hadn't come across the idea.

So this was actually used. It was rather exciting, because what you then did was to build the aeroplane and discover what its actual modes of vibration were and if they agreed. By the way, we did the actual measurements using a variable-speed electric motor that drove a wheel and you could shake the thing at any frequency you liked. You bolted this to the frame and you gradually speeded it up until you got the whole structure singing, very dramatic. Anyway, this was the kind of way that people with a little bit of mathematics in aircraft were thinking. Then some years later, after the war, I read [W. Ross] Ashby's *Introduction to Cybernetics*[4]—an interesting book—and in it he describes electrical analogue computers. And I thought, "Christ, I've been going along all this time without knowing what I'd done." And it was rather annoying. [Chuckles.]

Anyway, that aircraft-modeling work is typical of the kind of thinking and problem solving that was in the air. There weren't, on the other hand, too many of us with the necessary technical skills, and of course we brought some of what we learned into our work after the war.

PH: Yes, that's very interesting, and presumably this need to be imaginative, as well as the mixing of biologists and engineers and mathematicians, played an important part in developing theoretical biology, or at least pushing more theoretical thinking into biology?

JMS: Yes, I think that's right. Some biologists became more theoretical following their war work, but there was an effect on mathematicians, too. That happened because people with mathematics, of which Turing is an example, had been drafted into all kinds of technical work during the war, applying their mathematics. In Turing's case, of course, this was mainly decoding work, and for others it was radar research and development. So they became used to thinking about practical problems and at the end of the war they had this interest in applying their mathematics to the real world, and biology was one of the obvious places to do it.

But now when it comes to theoretical biology I'm quite intrigued. Earlier I'd say it was mainly population dynamics. The first burst of theoretical biology was from those two guys [Vito] Volterra and [Alfred J.] Lotka, making models of population growth. That would have been in the thirties or a little before. And the second burst was [Ronald] Fisher, [Sewall] Wright, and Haldane's work on population genetics, and that was actually very important in biology at that time. That was also in the 1930s, and prior to that there was really a complete dichotomy between the Darwinists and the Mendelians. The Mendelians thought that evolution happened when a mutation occurred and the Darwinists were doubtful about Mendel and thought it was all a matter of selection. And the extraordinary thing is, certainly looking at it now, that these views were regarded as incompatible. And indeed Haldane, Wright, and Fisher showed that actually they were completely compatible. That's a very important example of how theory and mathematical thinking can really advance biology. Now of course all this happened when I was a schoolboy, before the war; but the amount of theoretical work increased after the war. Of course there were parallel developments in neurobiology, but I think they were probably very largely independent.

PH: Yes, I think that's right. Interestingly, many of those involved in the rich interaction between cybernetics and neurobiology had worked on radar in the war.

JMS: Yes, right. Now I never had any contact with the radar people during the war, which is probably why I didn't get very involved in these things later. Eventually I saw myself as an evolutionary biologist working in population genetics.

The people whom I had contact with, and who had influence, and I think were involved with some of the people in the Ratio Club, were C. H. Waddington and Joseph Needham.

PH: Waddington was certainly involved a little in the British cybernetics scene. For instance, he gave a talk at the Ratio Club on development as a cybernetic process.

JMS: Waddington, who was an interesting man, didn't actually use his mathematics at all. I knew him fairly well, from those curious meetings he used to run on theoretical biology.[5] He encouraged young mathematically inclined biologists and by bringing us together he helped by making us feel less like loners. He was interested in relating development to evolution, and he liked ideas. But the point of mathematics is to use it. What good is it unless you are doing something that couldn't be done without it? I don't think that ever filtered through to him. Needham I hardly knew. I met him as an undergraduate and he was an awfully formidable figure. But again it's not quite clear to me what he actually did. I think there are these great gray eminences who people don't understand so they think their work must be very important indeed. He was rather like that.

Now what about Donald Michie? We worked in the same lab for years and years when he was a geneticist, and then he became very involved in artificial intelligence, as you know. Of course he was a close colleague of Turing during the war. He told me many entertaining tales of those times. Turing's gold, for instance. According to Donald, in nineteen thirty-whatever, when it looked as if the Germans were going to invade, Turing decided that what he was going to turn all his money into gold. And so he did and he buried it in the corner of a field somewhere. Donald didn't know about this at the time, but he became involved much later, in the war, when it was fairly clear the Germans were not going to invade, and Turing decided he was going to dig it up. Donald has this dramatic story about how they built a home-made metal detector and spent their weekends tearing around the Home Counties looking for this bloody gold. As far as I know they never found it. He wouldn't tell me what area it was in! But it gives you an image of Turing. Great story.

When we were all at University College London, Michie and his then wife, Anne McLaren, who is a very distinguished, but not theoretical, biologist, were working on perpendicular fertilization. One of the happiest evenings of my life was spent with these two in a pub after they had first managed to take an egg out of a mouse, fertilize it, pop it back into the same mouse, and get a baby mouse! Now to do something for the first time is bloody hard. So they were basically experimental embryologists. That work couldn't have been done without Anne, who is someone for whom I have an immense admiration.

I don't know how Donald got into Bletchley. He's an extremely bright guy, no question about it, but not formally mathematical, not back then. I think he was a classics scholar at that stage. But right from the early days he was interested in artificial intelligence. Donald and I played one of the very first games between two chess computers. You see we both had an interest in inventing rules to govern games and processes. During the war we had each produced a set of rules, an algorithm, to play chess. If you carefully carried out the calculations, which you could do by hand, the rules specified what your next move should be. And mine was called SOMA, for Smith's One-Move Analyzer; it didn't look at all deep. His was called Machiavelli, for reasons I'm not quite clear about. Machi, because it's like Michie, and his collaborator was someone whose name ended in "velli," or something like that, and the obvious reference to Machiavelli, I suppose! Anyway, we spent a long weekend playing these two sets of rules against each other with my older son as referee, because neither of us trusted the other one![6] You know, because if the obvious move was pawn to king, it had to be the rules that made the moves, not the humans. It was even published.[7]

PH: How much do you think science had changed from those heady postwar days, when there seemed to be a tremendous energy and an enthusiasm for innovation?

JMS: Well, obviously the particular part of science I work in has been dramatically transformed by technical advances, many involving computers, so that it is easier as well as cheaper and cheaper to obtain data. So we get submerged in data these days.

PH: What about the way science operates? Maybe this is purely illusionary, but it seems to me there might have been a bit more freethinking around at that period. Do you think people were less hemmed in by discipline boundaries or very specific kinds of methodologies?

JMS: I'm not sure. There are plenty of young people today who seem to me to be very capable and imaginative and able to tackle these sorts of problems and are not too constrained, and one way or another manage to get the job done and the message out, with some fairly way-out topics and speculative research. However, there is much more money, and that brings red tape. But money, and the need to get it, takes over more and more today, and that is a great pity. Our relationship to the funding has changed a great deal; we hardly had to think about money at all then. We weren't constantly brooding about how to keep research funding up. So in that sense we were freer to get on with the science.

Acknowledgments

We are very grateful to the Maynard Smith family for giving permission to publish this interview, and to Tony Maynard Smith for valuable comments on an earlier draft.

Notes

1. G. Fraenkel and J. W. S. Pringle, "Halteres of Flies as Gyroscopic Organs of Equilibrium," *Nature* 141: 919–21(1938); J. W. S. Pringle, "The Gyroscopic Mechanism of the Halteres of Diptera," *Philosophical Transactions of the Royal Society of London* (series B) 233: 347–84(1948).

2. J. Maynard Smith, "The Importance of the Nervous System in the Evolution of Animal Flight," *Evolution* 6: 127–29(1952).

3. Alan M. Turing, "The Chemical Basis of Morphogenesis," *Philosophical Transactions of the Royal Society of London* (series B) 237: 37–72(1952).

4. W. Ross Ashby, *An Introduction to Cybernetics* (London: Chapman & Hall, 1956).

5. C. H. Waddington, ed., *Towards a Theoretical Biology*, volume 1: *Prolegomena* (Edinburgh: Edinburgh University Press, 1968). Several volumes in this series were published.

6. Said referee, Tony Maynard Smith, recalls that his umpiring may have been less than perfect, as hand calculation was too slow and boring for a teenager to put up with!

7. An article on the match between these chess machines appeared in the popular science magazine *New Scientist*: J. Maynard Smith and D. Michie, "Machines That Play Games," *New Scientist*, 9 November 1961, pp. 367–69. The article records the result of the match thus: "Move 29: Draw agreed. NOTE: Combatants exhausted; in any case, neither machine is programmed to play a sensible end game."

Figure 16.1
John Holland. Image courtesy of John Holland.

16 An Interview with John Holland

John Holland was born in 1929 in Indiana. After studying physics at MIT, he worked for IBM, where he was involved in some of the first research on adaptive artificial neural networks. He went on to the University of Michigan for graduate studies in mathematics and communication sciences and has remained there ever since; he is professor of psychology and professor of electrical engineering and computer science there. Among many important contributions to a number of different fields, mostly related to complex adaptive systems, he developed genetic algorithms and learning classifier systems, foundation stones of the field of evolutionary computing. He is the recipient of a MacArthur Fellowship, a fellow of the World Economic Forum, and a member of the Board of Trustees and Science Board of the Santa Fe Institute.

This is an edited transcript of an interview conducted on May 17, 2006.

Philip Husbands: Could you start by saying something about your family background?

John Holland: My father's family came from Amsterdam, way back, so the Holland has some relation to my origins. My mother's family originally came from Alsace in France. My father owned several businesses that all had to do with soybean processing and my mother often worked as his accountant. She was quite adventurous; in her forties she learned to fly.

PH: Were there any particular influences from early school days or from your family that led you to a career in science?

JH: Not particularly, although my parents always encouraged me and supported my interest, from the first chemistry set they bought me, and in those days these were much more explosive than they are now, through to high school and beyond. But I grew up in a very small town, with a population of less than nine thousand, so there wasn't much in the way of direct encouragement in science.

PH: You went on to study at MIT; can you say a bit about your time there? Were there particular people you came across who influenced the intellectual direction you took?

JH: There was one person who was very important. I was in physics. At MIT in those days, I think it's still true, you had to do a dissertation for your bachelor's degree. I decided that I wanted to do something that was really quite new: work with the first real-time computer, Whirlwind. Work on Whirlwind was largely classified, but I knew someone who was involved: Zdenek Kopal. He had taught me a course on what was then called numerical analysis; now it would be called algorithms. He was an astronomer, working in the Electrical Engineering Department, so I went and knocked on his door and he agreed to be the director of my dissertation. He helped me get double the usual number of hours and I wrote a dissertation—using Whirlwind, getting the necessary security clearances and everything—on solving Laplace's equation using Southwell's Relaxation Method. He later took the first chair of astronomy at Manchester University and had a very distinguished career.

PH: What year was this?

JH: This would be 1949.

PH: So very early, as far as modern digital computing is concerned.

JH: Indeed. Whirlwind was only recently operational and was, as far as I know, the first computer to run in real time with a video display. It was being used for such things as air-traffic control, or at least that's what we were told, but it was obvious that it was also being used relative to missile detection and all that kind of stuff.

PH: During your undergraduate days did you come across [Norbert] Wiener, [Warren] McCulloch or any of the other cybernetics people?

JH: Oh yes, and that had an influence, but a rather distant one. Wiener we saw all the time. He was often called Peanuts because he'd walk down the hall flipping peanuts into the air and catching then in his mouth. So there was some influence there, and I also took a course on Bush's Differential Analyzer, which of course got me more interested in computers as well.

PH: What happened next? Did you go straight to graduate school?

JH: No. Because of the Whirlwind, I was offered a very interesting position at IBM in what was then their main research lab at Poughkeepsie, New York. The job was in the planning group for their first commercial computer, the 701. They called it the Defense Calculator. I was one of a group of about eight. For such a young guy it was quite an eye opener. We did the logical planning for the organization of the 701. This was in the very early days of commercial computing—the 701 laboratory models had cathode-ray tubes for storage and used punched cards for input. The engineers were building the prototype during the day, and we were testing it using our programs at night. There was a rush because Remington-Rand was also racing to produce a commercial programmed computer. Arthur Samuels and I worked coincidentally at night—I was doing neural nets and he was working on his checkers player.[1] A major part of our logical planning was to make sure that the machine was readily programmable in machine language (remember, this was before FORTRAN). Rochester convinced the lab director that these unusual programs (the checkers player and the neural net) gave the machine a good workout. The neural net research came about after J. C. R. Licklider, from ARPA [Advanced Research Projects Agency], who knew Hebb's theory of adaptation in the nervous system very well, came through and lectured on it at IBM.[2] Nathaniel Rochester, my boss, and I became quite interested in this and did two separate models, which were later published in a single paper.[3] We went back and forth to Montreal at least six or seven times to see [Donald] Hebb at McGill University while we were developing the model.

PH: Did you interact much with Samuels while you were at IBM?

JH: Yes, I did. We met with him regularly at lunch and once every other week we met at his house to play, in rotation, poker, Kriegspiel, and Go.

That time at IBM was obviously an influence on me. I worked for them for eighteen months and then decided I really did want to go to graduate school. IBM was good enough to offer me a consulting contract to help pay my way for four years of graduate school. I would go to school in the winter and go to IBM in the summer. So I came to the University of Michigan, which had one of the best math departments in the country; they had a couple of members of the Bourbaki group—the influential movement who were trying to rigorously found all mathematics on set theory—and things of that sort. Also, and not totally incidentally, they had a lot of co-eds.

So anyhow I did math and I had actually started writing a dissertation in mathematics—on cylindrical algebras, algebras that extended Boolean algebras to predicate logic with quantifiers—when I met Art Burks. He is certainly one of the big influences in my life. He and others were starting a new program called Communication Sciences, which went all the way from language and information theory through to the architecture of computers. Both MIT and Michigan had Communication Sciences programs, and in both cases they later became Computer Science departments. Art convinced me that this was of great interest to me, and indeed it was, and so I stopped writing my math dissertation and took another year of courses in areas such as psychology and language, and then did my dissertation within the Communication Sciences program.

PH: What was the topic of that thesis?

JH: It was called "Cycles in Logical Nets." [Arthur] Burks and others had set up a kind of abstract logical network, related to McCulloch and Pitts networks, and I wanted to see if I could characterize the kinds of changes you got if you allowed the network to contain cycles, feedback in other words. The thesis was finished in 1959.

PH: Who were the people you interacted with during that time, apart from Burks?

JH: There were quite a few. Someone who was in the same cohort as me was Bill Wang, who later went to Berkeley and became a world-renowned linguist. Actually quite recently, within the last four or five years, Bill and I have got back together again to build agent-based models of language acquisition; so that was a kind of long-range boomerang. George Lakoff, the linguist who's done work on metaphor at the logical level, among other things, was here at Michigan. Gunnar Hok, a man who is not so well known but wrote an important book on information theory, was also someone I interacted with. Anatol Rapoport, well known in game theory and several other areas, was also here at that time. So there was a good spread of people with a real knowledge of many aspects of what we would now probably call complexity.

PH: Yes, and it sounds as if there was also quite a strong flavor of what would become cognitive science.

JH: Oh yes, definitely. Anatol Rapoport, especially, was developing ideas in that direction.

PH: During this period did the group at Michigan interact much with other groups in the U.S., for instance at MIT?

JH: Yes. We had summer courses in what was called automata theory; after the first year I directed them. Herb Simon, Al Newell, Marvin Minsky, John McCarthy, they all came and lectured on them, so there was quite a bit of interaction. John McCarthy was also at IBM during the same summer periods as me, so we got to know each other pretty well. In fact, he taught me how to play Go. That would be about 1954 or 1955. John was editing the *Automata Studies* book with Claude Shannon at that time.[4]

PH: Do you remember what the spirit was like at the time? What were the expectations of people working in your area?

JH: There were already differences in expectations. But two things that I remember are that there was a fair amount of camaraderie and excitement, and also a bit of challenge back and forth between us. Who knows how much this is colored by memory, but I think of it as a heady time. I enjoyed it.

PH: These were heady times, as you said, but do you remember if people's expectations were naive, at least in hindsight, or if the difficulties of the problems were appreciated from the start?

JH: Let me make some observations. By this time, the mid-1950s, there was already a strong belief that you could program intelligence into a computer. There was already an interesting nascent division, which later became much more prevalent, between what came to be known as the Neats and the Scruffies. The Scruffies were on the East Coast, strangely enough, since we tend to think of people from there as pretty neat, and they were going to hack it all in. The Neats were on the West Coast and they wanted to do it by logic—the logic of common sense and all that—and make it provably correct. The Scruffies didn't believe the problem was tractable using logic alone and were happy to put together partly ad hoc systems. To some extent this was a split in approaches between John McCarthy, in the West, and Marvin Minsky in the East. Interestingly enough, as this unfolded there was very little interest in learning. In my honest opinion, this held up AI in quite a few ways. It would have been much better if Frank Rosenblatt's Perceptron work, or in particular Samuels's checkers-playing system, or some of the other early machine learning work, had had more of an impact.[5] In particular, I think there would have been less of this notion that you can just put it all in as expertise.

PH: The alternative to that, adaptive systems, seem to have been the focus of your attention right from the start of your career. Is that right?

JH: Yes, certainly. A major influence on me in that respect was Fisher's book *On the Genetical Theory of Natural Selection*.[6] That was the first time I really realized that you could do mathematics in the area of biological adaptation.

PH: Was that the starting point for genetic algorithms?

JH: Yes. I came across the book when I was browsing in the open stacks of the math library. That must have been somewhere around 1955 or 1956. Computer programming was already second nature to me by that time, so once I saw his mathematical work it was pretty clear immediately that it was programmable.

PH: Were you initially thinking in terms of computational modeling of the biology or in terms of more abstract adaptive systems?

JH: Well, probably because of exposure to Rapoport and others, I began to think of selection in relation to solving problems as well as the straight biological side of it. In fact, by the time I was doing the final writing up of my thesis I had already gone heavily in the direction of thinking about genetics and adaptive systems. So the thesis became pretty boring to me and I wanted to move on to the new stuff.

PH: Once you'd finished your thesis, how did you get to start work on what became genetic algorithms? Were you given a postdoc position or something?

JH: Well, this is where I had a great piece of luck and Art Burks was just superb. The stuff I wanted to do was not terribly popular—the typical comment you'd get was "Why would want to use evolution to try and solve problems; it's so slow"—but Art always stood up for me and said, "This is interesting work. Let him get on with it." So I got a job where I was teaching a couple of courses—logic for the philosophy department, and so on—and doing my research. Within a year they made me an assistant professor and in those days you got promoted pretty rapidly, so things went on very quickly and I settled in at Michigan.

PH: Almost hidden away in some of the cybernetics writing of the 1940s and 1950s there are several, usually fairly vague, mentions of the use of artificial evolution. For instance, Turing in his 1950 "Mind" paper.[7] So the idea was floating around to some extent. Were you aware of any of these? Were they a kind of background influence?

JH: Oh yes. One thing that I came across in retrospect and under analysis from others, at IBM actually, was [Richard] Friedberg's work on evolving programs.[8] This was a really important piece of work, but it was flawed. One of the people in his own group, Dunham I think, later wrote a paper with him showing that this evolutionary process was slower than random search.[9] Still, the idea was there; Friedberg was a smart guy. That was of great interest to me because you could see why it didn't work.

PH: Did you come across this after you'd started work on developing your evolutionary approach?

JH: Yes, I'd already read Fisher and gotten interested. A bit later Fogel, Owens and Walsh wrote their book on using evolutionary techniques to define finite state machines for simple predictive behaviors.[10] Again, you could easily see why it could go wrong, but it was influential in helping to show the way. So there was something in the wind at the time.

PH: It took a long time for genetic algorithms, which developed into the field of evolutionary computing, to become mainstream.[11] When it did, were you surprised? What were your feelings when, in about 1990, it suddenly became enormous?

JH: Yes, it did seem almost explosive at that time. It was surprising. By that time I'd had a lot of graduate students who had finished their degrees with me, so there was a local sphere of influence and we knew there were kinds of problems that could be solved with evolutionary methods that couldn't be solved easily in other ways. But I think that the tipping point, as we'd call it nowadays, was when it became more and more obvious that the kinds of expert systems that were being built in standard AI were very brittle. Our work offered a way around that. So suddenly, partly because people were looking elsewhere for alternatives, and because some of my students had became reasonably well known by then, the whole thing just took off.

PH: That must have been gratifying.

JH: Yes, but there were pluses and minuses to it. It was nice to see after all that time, but on the other hand, you begin to get too many phone calls!

PH: Rewinding back to the 1950s, your name is mentioned on the proposal for funding for the Dartmouth Conference as someone who would be invited. But did you actually go?

JH: No, I did not. At that time I had heavy commitments at Michigan. I can't remember why I didn't go, because I planned to. But I did not and that was my great loss, because that was a very important meeting.

PH: Yes, and of course it was very influential in advocating what became known as symbolic AI. That seems to go against the grain of the kind of work you have always been involved in. Were you aware at the time that the tide was turning in that direction?

JH: I would say within the year I was aware of it. Not immediately, but fairly quickly, because, even though I wasn't there, there was a lot of back and forth. Notions like adaptation simply got shoved off to one side so any conversation I had along those lines was sort of bypassed. My work and, for instance, Oliver Selfridge's work on Pandemonium, although often cited, no longer had much to do with the ongoing structure of the area.[12]

PH: Why do you think that happened? Why was the work on adaptive systems and learning sidelined?

JH: John McCarthy and Marvin Minsky are both very articulate and they both strongly believed in their approach. That was part of it. Herb Simon and Al Newell had worked on their Logic Machine, so they were oriented in that direction and they were influential.[13] McCulloch and Pitts's network model, even though it was connected to neural networks, was itself highly logical—Pitts was a brilliant logician. This was the time when symbolic logic had spread from philosophy to many other fields and there was great interest in it. But even so, it's still not absolutely clear to me why the other approaches fell away. Perhaps there was no forceful advocate.

PH: Putting yourself back into the shoes of the graduate student of the 1950s, are you surprised how far things have come, or haven't come?

JH: Well, let's see. If I look back and think of expectation from that time, I am surprised. I really believed that by now we would be much better at things like pattern recognition or language translation, although I didn't think we'd get there the logic way. Partly because I had worked with Art Samuels, and had great respect for him, I really believed that taking his approach to playing games, developing it, and spreading it into things that were game-like would make tremendous advances within a decade or two. But what we have today is Deep Blue,[14] which doesn't use pattern recognition at all, and we still don't have a decent Go playing program.

PH: Why do you think that is? Because the difficulty of the problems was underestimated?

JH: I think that's part of it. In my opinion, those problems can't be solved without something that looks roughly like the human ability to recognize patterns and learn from them.

PH: We've already discussed the sudden popularity of genetic algorithms, but a lot of other related topics came to the fore in the late 1980s and early 1990s. The rise of artificial life, complex-systems theory, nouvelle AI, and the resurgence of neural networks all happened at about that time, and there was the founding of the Santa Fe Institute in the mid-1980s. You were involved in most of those things. At least in AI, the switch from the mainstream to topics that had been regarded as fringe for a long time seemed quite sudden. Was there a shift in scientific politics at this time, or some successful lobbying? Or something else?

JH: I think the Santa Fe Institute is a good way to look at this tipping point; I think its founding says a lot about what was happening. Let me make a comparison. Just before World War II, there was this really exceptional school of logic in Poland, the Lwów-Warsaw School of Logic, and many of the best logicians in the world came out of there. The Santa Fe Institute seemed similar to me in that it depended a lot on a very few people. George Cowan, a nuclear chemist from Los Alamos, had the idea to set up the institute. He thought there were a group of very important problems, that weren't being solved, which required an interdisciplinary approach—what we would now call complexity. He recruited Murray Gell-Mann and together they brought in three other Nobel laureates and they decided they should start an institute that wasn't directly connected to Los Alamos so there would be no classification and security problems. It was originally called the Rio Grande Institute. About a year and a half later they decided it should be located in Santa Fe and renamed it the Santa Fe Institute. That really did start something. As I often say to graduate students, research at the Santa Fe Institute was how I imagined research would be when I was a young assistant professor just starting out, but it wasn't until the institute was set up that I really got engaged in the way that I had dreamed of as a young guy. It still is an extremely exciting place. The first major impact we had was when we got a group of people together to discuss how we might change economics. The group included people like John Reed, who was the CEO of Citicorp; Ken Arrow, Nobel laureate in economics; Phil Anderson, Nobel laureate in theoretical physics, who had a real interest in economics; a bunch of computer scientists; and some others. We got together for a week and produced some interesting ideas about viewing the economy as a complex system.

PH: Did you have much of an interest in economics before that? You've done quite a bit of work in that area since.

JH: I got interested in economics as an undergraduate at MIT where I took the first course that Paul Samuelson offered in the subject. Samuelson was a great teacher—his textbook became a huge classic—as well as a great economist (he went on to get a Nobel Prize). But I hadn't really done any work in the area until the Santa Fe meeting. Ken Arrow was a great influence on me—the Arrow-Debreu model is the basis of so much of modern economics, but Ken was ready to change it. He said, "Look, there's this wrong and this wrong." Interacting with him was really good. Anyway, I think the energy and intellectual excitement of the Santa Fe Institute, which involved some highly regarded and influential people, played an important part in shifting opinion and helped to catalyze changes in outlook in other areas. That was a very exciting period.

PH: In my opinion, the spirit of your work has always seemed close to that of some parts of cybernetics, perhaps not surprisingly, given when you started. It often reminds me of the work of people like Ross Ashby. Is that a fair link?

JH: Yes. I certainly read his books avidly and there was a group of people, including Ashby, Bertalanffy, and the General Systems theorists, Rapoport, Rashevsky, who was here for a while—not to mention von Neumann and Art Burks—who created a whole line of thought that was influential for me. Art Burks actually edited von Neumann's papers on cellular automata, so we were seeing that stuff before it was published.

PH: Those are some of the main names we would associate with the beginnings of systems theory and complexity.

JH: That's right. Someone else I should mention is Stan Ulam, the great mathematician who invented the Monte Carlo method, among many other things. At the time the Santa Fe Institute was founded he was still alive, but he died soon thereafter and his wife donated his library to the institute. For a while all his books were collected together on a few shelves so you could go in and pick them out. He had a habit of making notes in the margins. This was about the first time I'd been able to almost see into someone's mind, following the way it works. Ulam was just exceptional.

PH: Let's concentrate on the present for the final part of this interview. What do you think are the most import problems in evolutionary computing today?

JH: Well I think a really deep and important problem is what has come to be called evo-devo: evolutionary development. I think a lot of the framework we have is relevant to that problem in biology. At the moment most of the discussion on evo-devo is sort of like evolutionary biology pre-Fisher—a broad framework, some useful fact, but nothing like Fisher's mathematical framework.

PH: That's very interesting. So you think there is a bigger role for evolutionary computing in theoretical biology?

JH: I think it's quite possible, especially when combined with agent-based modeling of complex adaptive systems. A major effort at the Santa Fe Institute, and one I am involved in, is developing those kinds of studies of complex adaptive systems involving multiple agents that learn. Evo-devo has got to be heavily related to that. You can really think of developmental processes, where the cells in the body modify themselves and so on, as a complex adaptive system where agents are interacting—some agents stop others from reproducing and things like that. It seems to be a natural framework for development.

PH: Extrapolating a bit, do you think that if we are going to use evolutionary methods to develop machine intelligence, development will have to be taken seriously? That it will be an important part of the story?

JH: Yes I do, very much so. A nice basic project in that direction might be to try and develop a seed machine—a self-replicating machine out of which more complex systems could develop—or at least the theory for one. NASA have already put a lot of money into this kind of thing, and it won't be easy or happen quickly, but I think it should be doable and would be a good goal to set up in looking at evo-devo.

PH: Related to this, more generally what do you think the relationship between computer science and biology should be? Should they get closer or be wary of that?

JH: I'm a very strong advocate of cross-disciplinary research. My own idiosyncratic view is that the reason many scientists burn out early is that they
dig very deep in one area and then they've gone as far as it's humanly possible at that time and then can't easily cross over into other areas. I think
at the heart of most creative science are well-thought-out metaphors, and
cross-disciplinary work is a rich source of metaphor. Although you've got
to be careful; metaphors can be overhyped.

PH: A slightly different angle, particularly in relation to AI, is the notion
that the only way we are ever going to make significant progress is to learn
from biological systems.

JH: In a way I do agree with that. If we go back to when Ashby and Grey
Walter, Wiener, Selfridge, and all the others were looking at these problems, they used biological metaphor in a rich and careful way. I do not
think that simply making a long list of what people know and then putting
it into a computer is going to get us anywhere near to real intelligence. So
then you have to ask yourself, what are the alternatives? Artificial neural
nets are one possibility, and another is to try to work with a mix of cognitive science and agent-based modeling, which could be very fruitful for
AI and computer science in general. This allows you to work at a more abstract level than trying to reverse-engineer biology, as some people, I think
wrongly, advocate. I become very cautious when I hear people claiming
they are going to use evolution and they're going to download human
brains into computers within twenty years. That seems to me to be at least
as far-fetched as some of the early claims in AI. There are many rungs to
that ladder and each of them looks pretty shaky!

PH: Without imposing any timescales, how do you see the prospects for
AI? Where is it going?

JH: As I mentioned before, it seems to me that very central to this is what
we loosely call pattern recognition, and also building analogies and metaphors. My views on this owe a big debt to Hebb. I think that we can, and
must, get a better grasp on these rather broad, vague things. My personal
view on how to go about this is through agent-based modeling. We
have some of the pieces but we need to understand how to take things further. I think Melanie Mitchell's work with Doug Hofstadter on Copycat
points the way to a much different approach to notions such as analogy.[15]
Central is the need to get much better at recognizing patterns and structures that repeat at various levels. One thing that we haven't done much
with so far is tiered models, where the models have various layers. I think
all of these things fall roughly under the large rubric of complex adaptive
systems

PH: Looking back at all the work you have been involved in, is there one piece that stands out?

JH: I guess I really feel good about the mixture of rule-based systems and genetic algorithms that I called classifier systems.[16] That injection of flexibility into rule-based systems was something that really appealed to me at the time. Many of the people working with production systems, and rule-based systems in general, knew that they were brittle. The notion that you could take rules but make them less brittle, able to adapt to changes, was very pleasing. In a way, classifier systems were the genesis of the agent-based modeling work at Santa Fe. When we started on the economic modeling work, economists like Brian Arthur and Tom Sargent started using classifier systems. I tried to collect many of these ideas, in a form available to the general, science-interested reader in the inaugural set of Ulam Lectures, published as *Hidden Order*.[17]

Notes

1. Arthur Samuels, "Some Studies in Machine Learning Using the Game of Checkers," *IBM Journal of Research and Development* 3, no. 2: 210–29(1959). This is a landmark paper in machine learning and adaptive approaches to game-playing systems.

2. Donald O. Hebb, *The Organization of Behavior* (New York: Wiley, 1949). This highly influential book introduced, among many other things, Hebb's description of a fundamental adaptive process postulated to occur in the nervous system: connections between neurons increase in efficacy in proportion to the degree of correlation between pre- and postsynaptic activity.

3. N. Rochester, J. Holland, L. Haibt, and W. Duda, "Tests on a Cell Assembly Theory of the Action of the Brain, Using a Large Scale Digital Computer," *IRE Transactions of Information Theory* IT-2: 80–93(1956). This was one of the very first papers on an artificial neural network, simulated on a computer, that incorporated a form of Hebbian learning.

4. Claude E. Shannon and John McCarthy, *Automata Studies* (Princeton: Princeton University Press, 1956).

5. F. Rosenblatt, "The Perceptron: A Probabilistic Model for Information Storage and Organization in the Brain," *Psychological Review* 65, no. 6: 386–408(1958).

6. R. A. Fisher, *On the Genetical Theory of Natural Selection* (Oxford: Clarendon Press, 1930).

7. Alan M. Turing, "Computing Machinery and Intelligence," *Mind* 49: 433–60(1950).

8. R. M. Friedberg, "A Learning Machine," part 1, *IBM Journal of Research and Development* 2, no. 1: 2–13(1958).

9. Ibid., part 2, *IBM Journal of Research and Development* 3, no. 3: 282–87(1959).

10. L. Fogel, A. Owens, and M. Walsh, *Artificial Intelligence Through Simulated Evolution* (New York: Wiley, 1966).

11. John H. Holland, *Adaptation in Natural and Artificial Systems: An Introductory Analysis with Applications to Biology, Control, and Artificial Intelligence* (Ann Arbor: University of Michigan Press, 1975; 2nd ed., Cambridge, Mass.: MIT Press, 1992). This seminal work on genetic algorithms was the culmination of more than a decade of research.

12. Oliver G. Selfridge, "Pandemonium: A Paradigm for Learning," in *The Mechanisation of Thought Processes*, edited by D. Blake and Albert Uttley, volume 10, *National Physical Laboratory Symposia* (London: Her Majesty's Stationery Office, 1959).

13. Allen Newell and Herbert A. Simon, "The Logic Theory Machine," *IRE Transactions on Information Theory* IT-2: 61–79(1956).

14. M. Campbell, A. Hoane, and F. Hsu, "Deep Blue," *Artificial Intelligence* 134, no. 1–2: 57–83(2002).

15. M. Mitchell, *Analogy-Making as Perception: A Computer Model* (Cambridge, Mass.: MIT Press, 1993); D. R. Hofstadter and M. Mitchell, "The Copycat Project: A Model of Mental Fluidity and Analogy-Making," in *Fluid Concepts and Creative Analogies*, edited by D. Hofstadter and the Fluid Analogies Research Group (New York: Basic Books, 1995), chapter 5.

16. John H. Holland and J. S. Reitman, "Cognitive Systems Based on Adaptive Algorithms," in *Pattern-Directed Inference Systems*, edited by D. A. Waterman and F. Hayes-Roth (New York: Academic Press, 1978); John Holland, K. J. Holyoak, R. E. Nisbett, and P. R. Thagard, *Induction: Processes of Inference, Learning, and Discovery* (Cambridge, Mass.: MIT Press, 1986).

17. John Holland, *Hidden Order: How Adaptation Builds Complexity* (Redwood City, Calif.: Addison-Wesley, 1995).

Figure 17.1
Oliver Selfridge. Image courtesy of Oliver Selfridge.

17 An Interview with Oliver Selfridge

Oliver Selfridge was born in 1926 in London. He studied mathematics at MIT under Norbert Wiener and went on to write important early papers on pattern recognition and machine learning. His 1958 paper on the Pandemonium system is regarded as one of the classics of machine intelligence. He has worked at MIT's Lincoln Laboratory, the BBN laboratory at Cambridge, and the GTE laboratory, where he was a chief scientist. He has served on various advisory panels to the White House and numerous national committees. As well as his scientific publications, he has written several books for children.

This is an edited transcript of an interview conducted on May 8, 2006.

Philip Husbands: Could you start by saying a little about your early years? Were there any particular influences from home or school that put you on the road to a career in science and engineering?

Oliver Selfridge: Well, an important part of my education was my father. Without knowing any mathematics himself, he was wildly enthusiastic about my interest in it, which started at quite an early age—seven or eight. As was usual in England back then, I went away to school when I was ten. At the age of thirteen, I entered Malvern College, one of the (so-called) public schools. I remember we spent the year of 1940 in Blenheim Palace because the Royal Radar Establishment (RRE) had taken over the school. While at Malvern I covered calculus to the standard you'd reach after the first two years of a degree at MIT. One of the great things about education back then, and I am not sure that it's true anymore, is that if you were good in one subject they'd move you ahead in that subject. You didn't have to worry about being good in both mathematics and French (which I was very bad at). So I'm very grateful to the English school system, although I didn't know it then; and I hated going away to school, of course, as I think everybody did. After Malvern I came to this country [the United States] and started at MIT after a year and a half at Middlesex School in Concord, Massachusetts.

PH: What year was this? You were quite young when you started at MIT, weren't you?

OS: This was 1942. I was just sixteen and the youngest in my class by more than a year. Last year we had a sixtieth reunion—the class of '45.

PH: What brought you to MIT? Did you go to the States because of the war in England?

OS: The Selfridges originally came from this country. My grandfather was born in Ripon, Wisconsin. He worked for a big store in Chicago called Marshall Field's and became executive vice president at an early age because he was, I guess, smart as hell. He went on to own another store, which he sold, and then he moved to London, where he opened Selfridge's, a department store on Oxford Street. He borrowed a million pounds in 1906 or 1907, which was a lot of cash back then, and the store opened in 1909.

PH: And is still going strong.

OS: Still going strong, although there are no Selfridges in it! We lived in Kensington and then out in Norwood Green; there were four of us siblings. But then we came to this country because my father and grandfather were kicked off the board of directors of Selfridge's at the end of the 1930s or thereabouts. My father came back to the States, because he had always been an American citizen; my grandfather had switched and become a British citizen in 1934. My father ended up working for a firm here called Sears Roebuck. Anyway, I went to MIT more or less by accident because I was very interested in mathematics and science. So I entered MIT at just sixteen and graduated at nineteen, having specialized in mathematics. I went through the V12 program, which meant I joined the [U.S.] Navy as a junior when I turned seventeen, or something like that, and they kept me at MIT, paying all the bills, which was wonderful, and then I went and got a commission in the Navy just after Japan surrendered. After the Navy I went back to MIT, to graduate school. I was working with Norbert Wiener, and my friends Walter Pitts, Warren McCulloch, and Jerry Lettvin were also there. By the way, I recommend the recent book *Dark Hero* as a good source of information on Norbert Wiener.[1] Anyway, by this time Walter had written the very important paper with Warren McCulloch, who was already a very well known neurophysiologist, showing how a neural net could do computations.[2] That came out in 1943, when Walter was only nineteen or twenty. I was very lucky to have met these people and then of course at graduate school I was introduced to a lot of others.

After that I joined Lincoln Lab, which was also a part of MIT, where we built the first spread-spectrum system, under Bill Davenport. Let me explain what that is. Communications theory, channel capacity, and ideas like that, had just started; Shannon had written about them in 1948. The notion was that you needed a certain bandwidth to carry a certain amount of information. A spread-spectrum system uses a much bigger bandwidth for that amount of information, and that helps to protect the signal, making it difficult to track or jam. We built the first system, which was classified, and the next ones weren't built for another twenty years. They are becoming more and more widely used now.

At that point, that would be 1953, I met Marvin Minsky, who had just got through Princeton and was a junior fellow at Harvard. We were both very interested in what became known as artificial intelligence. He worked for me at Lincoln Lab for a couple of summers before he became a professor at MIT. Marvin and I ran the first meeting on artificial intelligence a year before the Dartmouth conference at the Western Joint Computer Conference.[3] At about this time, 1954 I think it was, I met a psychologist from Carnegie Mellon University at the Rand Corporation in Santa Monica: Allen Newell. After talking for a couple of hours we had dinner that evening and he really appreciated what we were trying to do and he turned on fully to AI and started working on symbolic AI, which was different from what we'd been doing. Of course Allen, who died, alas, some time ago, became very well known, a very powerful guy. He was incredibly bright. Allen was terrific. He gave one of the papers at our 1955 meeting.

PH: I'd like to come back to Dartmouth and early AI later, but can we rewind slightly at this point to talk a bit about the origins of your celebrated Pandemonium system?

OS: I first presented that at the Teddington conference.[4] Do you know where the word comes from?

PH: I believe you took it from Milton's *Paradise Lost*.

OS: That's right. From the Greek for all the demons. It's mentioned in the first couple of pages of *Paradise Lost*, which was written in 1667, I think. I wasn't alive then, it just sounds as if I were.

PH: The Teddington Mechanisation of Thought Processes Symposium was in 1958, but when did you start working on the system? Was it much before that?

OS: Well, we had been thinking about the general techniques of cognition for a while. The first AI paper I'd written was on pattern recognition, elementary pattern recognition and how to do it, and we spent a lot of time talking about it and getting people interested, and actually I work incredibly slowly. The cognition aspect was first sort of tackled by McCulloch and Pitts in their papers in 1943 and 1947.[5] So I talked with Walter a lot about certain things in cognition and the first paper on my work on pattern recognition systems was at the 1955 Western Joint Computer Conference.[6] So Pandemonium incorporated many of the ideas I'd been developing. It's an idea that is very powerful and people like it, but nobody uses it.

PH: It's a really impressive piece of work; the paper pulled together a lot of very important ideas in a coherent way—parallel distributed processing, adaptive multilayered networks, feature detectors, and so on. I'm curious about the influences, the currents that came together in that paper. For instance, you knew Jerry Lettvin very well and during that same period he was working with Humberto Maturana, McCulloch, and Pitts on the research that produced the landmark paper "What the Frog's Eye Tells The Frog's Brain," which gave a detailed functional account of part of the frog's visual system and demonstrated the existence of various kinds of visual feature detectors suggestive of "bug detectors."[7] It seems to me there are quite a lot of connections between that work and Pandemonium. Is that right?

OS: Oh absolutely. In fact if you look at their paper there is an acknowledgement to me, and I acknowledge Jerry in the Pandemonium paper. They were influenced by my pattern-recognition work and the ideas behind it, which were to do with cognition. We regularly discussed the work. The question is about cognition—what does the frog do when he sees. Many people still think that the retina merely detects pixels and ships them off to the brain, which of course is just not true. The frog's-eye paper was published in the *Proceedings of the IRE*, now the IEEE, because the *Journal of Neurophysiology* wouldn't accept it; they said it didn't have real data in it, like numbers. Well of course it didn't—it was much better than that. I remember we laughed about it. Jerry built the first microelectrode needles for reading from single axons in the frog's optic nerve. It was an absolutely brilliant piece of work in terms of both the ideas and the experimental manipulations.

Of course Jerry and I were roommates while I was in graduate school, along with Walter Pitts. It was always exciting.

PH: That was quite a combination.

OS: Well, I had a good time indeed. Walter and I often went places together. One summer, I think it was '48, we climbed the Tetons in Wyoming just before spending the rest of the summer with Norbert in Mexico city.

PH: The frog's-eye paper is often quoted as containing the first full statement about low-level feature detectors in a vision system—moving-edge detectors, convexity detectors, and so on—building on Barlow's earlier work giving evidence for "fly detectors."[8] This notion became very important in vision science. Did you play a part in that, since you were using the idea of feature detectors in your pattern-recognition systems?

OS: Well in some sense I probably did, but a lot of other people came to it independently. My first paper on pattern recognition included the question of how you recognize a square. It described how the features of a square include a corner and a line and asks how do you detect a line against a noisy background, and so on. So, yes, I was the first one to put it in specific enough terms that it could be computerized, as far as I know, but I think a lot of others came up with it independently... of course this was fifty-three, fifty-four years ago, so not quite B.C., but getting that way.

PH: So maybe the idea was floating around to some extent, but it seems you made a very important contribution and obviously influenced the Lettvin-Maturana work.

OS: Thank you. Well, Jerry and I have always been on very good terms and I knew Maturana quite well, but he went off and had an independent life of some notoriety. Walter Pitts, of course, fell apart. That was tragic, really tragic. I'll tell you the story very briefly. In 1952 Norbert Wiener accused us—Warren McCulloch, Walter, and me—of corrupting his daughter, Barbara Wiener, who was a year younger than I, based on what Norbert's wife, Margaret, told him. She didn't like us because she thought we were too free and so on. The accusation was absolutely false. Norbert then turned against us and wouldn't speak to us or acknowledge our existence for the rest of his life, which was a great tragedy. Now Walter fell to pieces because of that, because he was dependent on Wiener. Walter had the highest IQ of anyone I've ever met, but he was fragile. When Walter was about eighteen or nineteen he bumped into Norbert Wiener and greatly impressed him with his mathematical ability—he corrected something Norbert showed him—and so he started working with Wiener and they became very close. Anyway, you can read more about their relationship in *Dark Hero*. Then after Norbert wouldn't speak to us, I remember being at a party somewhere in Cambridge with Walter and he said, "I wonder why people smoke. I'd better try." Two weeks later he was two packs a day. So he sort of fell apart and he played with drugs of all kinds and fifteen or so years later he died, essentially of overdoses. He was a total genius but he didn't know how to handle himself at all in a social way. It was just terrible.

PH: Pitts is reported to have destroyed most of his work from that time, so many of his ideas never saw the light of day. Is that true or did some of his work live on through his influence on people like you, who worked with him?

OS: Well, it's pretty much true. But he did a lot of other interesting things. For instance, there was at MIT a professor called Giorgio De Santillana, a historian and philosopher of science, whom Walter spent a lot of time working with later when he had his personal problems. His inspiration for Jerry was quite real. The full list of authors on the frog's-eye paper is Lettvin, Maturana, McCulloch, and Pitts. The work was done in '56 and '57 and he still had a real input at that time. Incidentally, something that pissed everyone off, including me, was that [David] Hubel and [Torsten] Wiesel took the genius of the ideas and the genius of the microelectrodes and the experimental setup, and they got a Nobel Prize. In their Nobel Prize speeches they did not give any credit to Jerry. That was rotten manners, putting it very mildly.

PH: During that period, in the 1940s and 1950s, you interacted a lot with at least two people who have had very important influences in neuroscience: Jerry Lettvin and Warren McCulloch. Was this more by accident than design or did you deliberately work in an interdisciplinary way?

OS: Sort of both. The number of people interested in these things in the mid-1950s wasn't very large, and so we tended to know each other and talk to each other. Norbert and Warren and others had initiated interdisciplinary ways of thinking and that was still around. AI had only just started at this point and new people, such as John McCarthy, were coming in. Claude Shannon was still interested, although he soon stopped. Von Neumann was interested, although he'd written all his papers by this time. He became a devout Roman Catholic in 1955, when he was suffering from cancer. Warren McCulloch kept going, although his papers got less specific and I think less useful, too general. By the late fifties he was drinking a quart of scotch a day, and you can't do that and keep your mind working as well; at least he couldn't. Maybe it's a good way to go.

PH: You seem to be making a clear distinction between AI and cybernetics. Is that how you see it?

OS: Yes, very much. Cybernetics obviously preceded AI; in fact, Jerry Lettvin and I are probably the only two people left alive who are specifically mentioned in Wiener's *Cybernetics*.[9] The notions of cybernetics are in AI but the focus is different. Cybernetics turned out to be much more an engineering business than AI. There is a great deal of engineering in AI and all the major thrusts that we now have are based on mathematics, but that is not what AI is about. A lot of AI you will see expressed in mathematical terms, but many of those aspects pretty much ignore what to me is the key power of AI, which is learning. Learning is central to intelligence.

PH: In a nutshell, how would you define AI?

OS: I think it's about trying to get computers, or pieces of software, to exhibit the intellectual powers of a person. That's a vast range of things, so to me the deep key is learning. It comes out in three very different aspects: the actual actions you take, the cognition, and the memories of experience. There is a special action part of experience, which is planning. In an intellectual sense planning is done only by people. A key thing that we are working on now is the essence of control as part of action. How do we learn how to control things? I think the essence of control is purpose—you want to do something. It's not just that you like beauty or you like good art or something, it's that you have a whole structure of purposes. If you're right-handed and you hurt your right hand so you have to use your left, you can still pick up a cup of coffee without thinking about it. The purpose is to get coffee to your mouth. This means you have subpurposes of finding where the cup is, moving your arm and so on and so forth: it's purposes all the way down and also all the way up. But those purposes change all the time and the essence of control is trying something and improving it. As Marvin Minsky said, "The best is the enemy of the good." Because "the best" implies a static universe, but it ain't static. The problem with a lot of the mathematical treatments is that generally they are looking for formalistic presentations of processes that can then be optimized. But we don't optimize, we improve. To me that should be part of the essence of AI.

But AI, like any other science, is a very complicated thing. In physics, Newtonian mechanics is a perfectly adequate way of expressing many processes, such as shooting a gun or something like that. But it turns out that in a deep sense Newtonian mechanics is just wrong. But looked at another way it isn't wrong, exactly; for certain purposes it gets improved. The same thing is true with AI. For certain purposes the simple memories we have about what we did, and why we wanted to do it, are adequate. But often, next time we do something we are trying to do it differently, or we modify it. It's very hard to think of something that we don't do better the second time. Likewise it's very hard to think of a computer task that the computer *does* do better the second time. So I think that in AI we should work on developing software which will notice what it does, remember the experiences and what it wants during the experiences, and be able to improve. Not just the actions, but the cognition and the planning too.

PH: You put learning and adaptation at the heart of AI, so looking back over the past fifty years do you think the trajectory of the field has been reasonably sensible or do you think there have been some disastrous directions?

OS: No, not disastrous. We've done a lot of powerful things, but they're missing a great deal of what I'm interested in now: purpose. You raise your children by encouraging and motivating them, but how do you encourage a computer program? To use a high-tech Americanism, the program doesn't give a shit. Well, your children all did and still do. So that is what I'm working on now and what I think is important. Marvin's *Society of Mind* discussed some of these issues, in very different terms, twenty odd years ago.[10] But I'm trying to be more specific and we'll see if I live long enough to get these ideas in any kind of shape.

Learning and adaptation have certainly been constant themes throughout my work. Adaptation I regard as a special case of learning, an affirmative case. For instance, the motor cortex makes a muscle move without affecting it directly—there is a loop out from the spinal cord to, say, a finger muscle with the signal coming back to the spine, so that we have a control circuit. The motor cortex modifies the gain of that circuit so it's adaptation all the way down, so to speak. We don't necessarily need to go as far as that; indeed I think copying all the details of neurophysiology is a silly error, but understanding what happens and why is the thing. Most people in AI don't do that. When I give a talk many people agree with me but then they go back and do the old things. Most computer programs are full of errors with no way to correct them. Well I want a piece of software that can limit its errors by learning, and thereby try to correct them.

PH: During the cybernetic period and in the early days of AI there was a lot of interest in adaptive and learning systems, but that seems to have greatly diminished by the late sixties and the pattern continued throughout the seventies. Why was that?

OS: It was regarded as too hard. When Edward Feigenbaum and Joshua Lederberg developed expert systems in the late sixties there was almost no learning involved. The learning was confined to the people. I have a very high regard for what they did and don't object to it, but my feeling was and is that learning is the key, and a lot of the deep questions were ignored. But work like that did bring a lot of people into AI, and I want more of those people to turn to basic research questions again. As I've said, I think purpose and motivation are the deepest requirement that we need in AI now: you want the software to care. People might say, "Well my system has the goal of winning as many games as possible, isn't that caring?"

Well, yes, sort of, but it's only the beginning; why do we stop there? We still can't really usefully praise or reward a system.

PH: Looking ahead and speculating, do you think the sorts of architectures and methods used in AI today will have to be abandoned or radically changed to make significant progress?

OS: Well, I think we will get to the point where AI has some sort of reward structure that enables it to learn in a more sophisticated way and then we won't so much program our systems as educate them. That will work and it will work spectacularly well. Communication will be very important as pieces of software will also teach each other. Getting motivation and caring and being able to adapt on multiple levels will be big breakthroughs, but it will require more than that—there won't be just one thing. But we need to get started. It will also have to make money for someone, because funding for pure basic research is very hard to come by today, certainly in this country.

PH: Do you think AI will need to get closer to biology to make these advances, or maybe move further away?

OS: Well, I don't think we need to move further away. There is a big effort now in neurobiology, and computational methods are playing a part in that. A lot of the effort is looking at single neurons in detail. I'm not sure that will help us get AI. There are too many steps from understanding a single neuron to having intelligence. That isn't to say that we can't learn some very important lessons and take very useful ideas from understanding more about how the brain works—just as happened, for instance, with Jerry Lettvin's work—but I think it has to be at a higher level than single neurons. Of course, the picture keeps changing in neuroscience anyway. The recent discovery of the important functional role of glial cells is an example; in essence they really have to start thinking all over again and come up with a new explanation.

PH: So you think detailed modeling is too ambitious, but taking inspiration at a more abstract level is useful?

OS: Yes. Detailed modeling is too ambitious and won't work. But more abstract inspiration is very important. Absolutely. Two important biologically inspired areas are of course neural nets and John Holland's genetic algorithms. There is a lot of stuff going on in both areas and a lot of it is very successful at solving problems, but there are great limitations and simplifications in these areas as they stand today. One fault is the emphasis on a single evaluation function. You need multiple purposes at different levels and multiple ways of evaluating these at different levels. It's time to try and tackle issues like that.

PH: Related to this, how would you say your interests are divided between developing artificial intelligence and understanding natural intelligence?

OS: Oh equally. I'm interested in both, that's always been the case.

PH: This year is the fiftieth anniversary of the Dartmouth conference and there is a lot of talk again about its being the birthplace of AI and all that. From what you've already said here and elsewhere, that's obviously an oversimplification, as the basic ideas were already around or being developed. You had your West Coast meeting in 1955, the name AI had already being used by some of you and so on. So do you think anything much actually came out of Dartmouth itself or was it more a part of an ongoing process?

OS: Both. Dartmouth generated a spectacular amount of interest because it got a lot of publicity. People were persuaded to look at new problems, and Allen Newell convinced a lot of people that symbolic processing and reasoning was important. So it was a very effective step; it got national interest, much more than Marvin and I had got for our earlier meeting, and it spread the message around. There were a lot of interesting and powerful people there: John McCarthy was a founding trigger of the meeting; there was Nat Rochester from IBM, and many others.

PH: Presumably the publicity and interest were helpful in generating funding.

OS: Well, funding didn't follow particularly speedily, but yes, Dartmouth did help in that respect, it opened various people's minds to the possibilities.

PH: Finally, is there any particular piece of work of the many that you have been involved in that stands out for you.

OS: Well not exactly, but I suppose the Pandemonium work is special to me because it helped me to finally nail a lot of issues.

Notes

1. Jim Siegelman and Flo Conway, *Dark Hero of the Information Age: In Search of Norbert Wiener—Father of Cybernetics* (New York: Basic Books, 2004).

2. Warren McCulloch and Walter Pitts, "A Logical Calculus of the Ideas Immanent in Nervous Activity," *Bulletin of Mathematical Biophysics* 5: 115–33(1943).

3. The Western Joint Computer Conference, Los Angeles, March 1–3, 1955.

4. Oliver G. Selfridge, "Pandemonium: A Paradigm for Learning, in *The Mechanisation of Thought Processes*, edited by D. Blake and Albert Uttley, volume 10, pp. 511–29 (proceedings of the symposium held at the National Physical Laboratory, Ted-

dington, in 1958), *National Physical Laboratory Symposia* (London: Her Majesty's Stationery Office, London, 1959).

5. See McCulloch and Pitts, "Logical Calculus," and Walter Pitts and Warren McCulloch, "How We Know Universals: The Perception of Auditory and Visual Forms," *Bulletin of Mathematical Biophysics* 9: 127–47(1947).

6. Oliver G. Selfridge, "Pattern Recognition in Modern Computers," in *Proceedings of the Western Joint Computer Conference* (New York: ACM, 1955).

7. Jerry Lettvin, H. R. Maturana, Warren S. McCulloch, and Walter Pitts, "What the Frog's Eye Tells the Frog's Brain," *Proceedings of the IRE* 47: 1940–59(1959).

8. Horace B. Barlow, "Summation and Inhibition in the Frog's Retina," *Journal of Physiology* 119: 69–88(1953).

9. Norbert Wiener, *Cybernetics, or Control and Communication in the Animal and the Machine* (Cambridge, Mass.: MIT Press, 1948).

10. Marvin Minsky, *The Society of Mind* (New York: Simon & Schuster, 1986).

Figure 18.1
Horace Barlow. Image courtesy of Cambridge University.

Horace Barlow, FRS, was born in 1921 in Chesham Bois, Buckinghamshire, England. After school at Winchester College he studied natural sciences at Cambridge University and then completed medical training at Harvard Medical School and University College Hospital, London. He returned to Cambridge to study for a Ph.D. in neurophysiology and has been a highly influential researcher in the brain sciences ever since. After holding various positions at Cambridge University he became professor of physiological optics and physiology at the University of California, Berkeley. He later returned to Cambridge, where he was Royal Society Research Professor of Physiology, and where he is a fellow of Trinity College. He has made numerous important contributions to neuroscience and psychophysics, both experimental and theoretical, mainly in relation to understanding the visual system of humans and animals. His many awards include the Australia Prize and the Royal Medal of the Royal Society.

This is an edited transcript of an interview conducted on July 20, 2006.

Philip Husbands: Would you start by saying a little about your family background, in particular any influences that may have led you towards a career in science.

Horace Barlow: I come from a scientific family; my mother was Nora Darwin, Charles Darwin's granddaughter, and she was very scientifically inclined herself. In fact, she worked with William Bateson on genetical problems in the early days of genetics at Cambridge and has one or two papers to her name in that field, although she never got a degree or anything. She was not only a good botanist, so to speak, but had a very scientific way of looking at things and kept asking herself and us children questions about why things were the way they were. So she undoubtedly had an influence in directing me towards science. She was instrumental in reviving Charles Darwin's reputation in the middle of the twentieth century, publishing an unexpurgated version of his autobiography and editing several collections of letters and notes.[1]

Two of my elder brothers became doctors and they also had strong scientific interests. My father, Alan Barlow, was a senior civil servant and had read classics at Oxford. He was very keen on words and origins and that kind of thing but wasn't scientifically inclined. But his father, Thomas Barlow, was a very successful doctor in Victorian times, in fact he was physician to Queen Victoria's household and had a disease named after him. He was one of the people who was very keen on medicine becoming more scientific and had numerous medical publications, so there's some science on that side too.

PH: You went to school at Winchester College. Were there any particular influences there?

HB: Yes. The teaching of science there was very good as you can tell from the fact that amongst my contemporaries were Freeman Dyson, the famous theoretical physicist; Christopher Longuet-Higgins, who made outstanding contributions to theoretical chemistry and cognitive science; James Lighthill, who was an important applied mathematician; and many others who became distinguished scientists. One person who certainly had an influence on me was the biology teacher, whose name was Lucas. Because at that stage I wanted to go on and study medicine, biology was important, but it did mean, because of the way the timetable was structured, that I was restricted to doing what was called four-hour mathematics rather than seven-hour mathematics, which I very much regret, actually. There were some very good mathematics teachers; I particularly remember Hugh Alexander, who was British chess champion and who went on to work with Turing at Bletchley Park during the war.

PH: After Winchester you went to Cambridge to study natural sciences. Can you say a bit about your undergraduate days there?

HB: Well, one of the big influences on me there was someone who later became a fellow member of the Ratio Club: the neurophysiologist William Rushton. He was quantitatively inclined and an inspiring teacher. He was my and Pat Merton's (another Ratio Club member) director of studies at Trinity College. A fascinating character. He was a very good musician, playing the bassoon and viola. But he was also extremely knowledgeable about music and took a highly intellectual approach to it. He was also a marvelous person to talk to because he would always encourage any pupil who came up with a bright idea; I can see him now turning towards you and getting you to say more and help you to relate your ideas to him. At that time his work was on the electrical properties of nerves. He did some important work in that area and in a sense he was a precursor of Hodgkin and Huxley and I think they did both acknowledge him. So in a way he was a bridge between the old Adrian and the new Hodgkin and Huxley.[2] His work in this area was highly regarded but not very widely known. But later he went into vision and become one of the world's top ranking visual physiologists.

Another person I had supervisions from, and who made a big impression, was Wilhelm Feldberg, who had worked on cholinergic transmission with Henry Dale in the early days. The other thing that had a big influence on me when I was an undergraduate was the clubs. I was a member of the Natural Science Club, which consisted of about twenty people, roughly half undergraduate and half graduate students, with maybe one or two people of postdoc status. We met about four times a term and gave talks to each other on various subjects. That had a big effect on me and was a great means of teaching and learning without any staff being involved!

PH: During this time at Cambridge did you still have a clear career path in mind? Did you still intend to go into medicine?

HB: Yes. When I was at school I was rather inclined towards physics, but being in the same school, and on occasions in the same class, as Freeman Dyson and James Lighthill, I realized there was a disparity in our mathematical abilities, so I thought perhaps biology would be more appropriate for me! I did the natural sciences tripos in anatomy, physiology, pharmacology, biochemistry, etc., which was the normal thing for medical students at Cambridge, and then went on to do my clinical work.[3] I was lucky enough to get a Rockefeller studentship to go and do that at Harvard. This was in the middle of the war and the Rockefeller Foundation realized that medical education in Britain was disrupted, and furthermore they couldn't get the postdoctoral researchers they usually supported to do work in the States because they were all engaged in the war effort, so they spent the money on medical studentships instead. There were about twenty or thirty of us. Before I started at Harvard I worked for the summer of 1943 at the Medical Research Council's lab in London at Mount Vernon. I was working on problems of diving in relation to the war. In fact I stayed there for a year; I delayed the start of my clinical studies in America to continue this work. That was my first proper laboratory science job. The lab was run by G. L. Brown and at first we were concerned with oxygen poisoning related to breathing oxygen under pressure, and then later on we worked on some problems with the essentially scuba-diving gear used for some operations. They used a self-contained system rather than the flow-through type, so that far fewer bubbles were produced and the divers were less easily detected. But that kind of system has its own dangers, and the equipment they were using was inadequate in some ways and we helped to sort that out.

PH: Did you come across Kenneth Craik at all during that period?

HB: Yes, I did actually meet Craik when I was working at Mount Vernon. I was working for the Royal Navy but he was doing the equivalent work for the Air Force and they had mutual inspection visits, so our paths crossed. I remember putting him on a bicycle ergometer to measure the oxygen consumption while using one of the self-contained diving sets we were working on. So it was only a rather brief meeting but of course I was very much aware of his work. His book, *The Nature of Explanation*, had appeared by then and his work in vision was very interesting because he had a very different approach from what was prevalent in psychology at the time.[4]

PH: Once you got to Harvard did you meet anyone who was a particular influence?

HB: Well there were a lot of very interesting people at Harvard Medical School at that time; this would have been 1944. One of them was Carroll Williams, who was doing some very interesting molecular biology, as we'd now call it, on silkworms. He seemed a good deal older than the rest of us, but at the beginning of the war he had decided to take up medicine. Anyway, he was a fascinating chap who became a distinguished scientist and was later professor of biology at Harvard. I did a research project with two fellow medical students, Henry Kohn and Geoff Walsh, on vision. We investigated the effect of magnetic fields on the eye. This resulted in my second scientific paper; I'd already published one with William Rushton from my undergraduate days, but it was my first work in vision.[5] The three of us also published some work on dark adaptation and light effects on the electric threshold of the eye.[6]

PH: By that time was it clear you wanted to continue as a research neurophysiologist?

HB: Yes. What I planned to do, and actually did do, was to complete a full medical qualification on my return to the UK and then try my hand at a research position. In those days you could get a full medical qualification without having to do any "house jobs"—internships, as they are called in North America—so when I got back I did a few more months' additional clinical work at University College Hospital, London, and was fully qualified. I then wanted to try research before I had to embark on many years of internships, which was the way ahead in the medical profession. In 1947 I managed to get a Medical Research Council research studentship at Cambridge under E. D. Adrian, who later became Lord Adrian but was universally known simply as Adrian both before and after his elevation to the peerage.

Pinning Adrian down was never easy, so finding him to explore the possibility of a studentship took some doing. I knew he was in Cambridge, and often in the Physiological Laboratory, but whenever I called he was not in his office. After several visits his secretary rather reluctantly admitted that he was probably downstairs in his lab, but when I asked if I could find him there her jaw dropped and she said "Well, er..." I got the message, but went down to look for him all the same. The entrance was guarded by his assistant, Leslie, who said, "He's in there with an animal and does not want any visitors." This time I took the hint, but as I was leaving I met one of my former lecturers [Tunnicliffe] and explained my problem. He told me I was not alone in finding it difficult to catch Adrian, but, he said, "He usually goes to Trinity on his bicycle around lunchtime, and if you stand in front of him he won't run you down." So I lurked around the lab entrance for a few lunchtimes, and the tactic worked: as I stood triumphantly over the front wheel of his bike he said, "Come to my office at two o'clock."

There he asked if I had any ideas I wanted to work on. My proposals, which were really hangovers from my undergraduate physiology days, included one on looking at the oscillations you sometimes get in nerve fibers. I thought that would be interesting to work on, but Adrian brushed that aside rather quickly, along with my other ideas, but then said I might like to look at the paper by Marshall and Talbot on small eye movements to see if there was anything in their idea,[7] and, by the way, he thought he could get me a research studentship from the Medical Research Council. The total duration of the interview was certainly no more than five minutes; Adrian believed in getting a lot done in his time outside the lab as well as inside it.

When I reported for work a few days later, Adrian seemed surprised to see me, and even more surprised when I asked him what I should do, but said something like "We've discussed that—Marshall and Talbot, don't you know?"

PH: You were already interested in vision before you started your Ph.D.; can you pinpoint when that interest started?

HB: Well, while I was an undergraduate one of the talks I gave to the Natural Sciences Club was on color vision. I read up on the subject for the talk and found it interesting and could understand what it was about. So that was an important point where I got interested, and then talking about it with William Rushton developed that further. Another piece of work that particularly interested me as an undergraduate was Selig Hecht and

Maurice Pirenne's research on the absolute threshold of vision.[8] Maurice
Pirenne was at Cambridge working with William Rushton at the time, so
the three of us discussed this topic at length. I think I gave a talk about
the statistical evidence for the visual system's sensitivity to single quanta
of light that Hecht had obtained. I was interested in finding out more
about the statistical aspect of this and William pointed me at R. A. Fisher's
books, which I read very keenly and learnt a great deal from. Of course the
absolute threshold of vision was a topic I was to return to a little later in my
career.[9]

But I didn't have to make a decision on my research area until I'd done
my clinical stuff and come back to Cambridge three years later. During
that time I'd worked on visual problems with Geoff Walsh and Henry
Kohn, which furthered my interest. The reason I was interested in vision
was because what we knew about the quantitative aspects of the integrative
action of neurons and so on was derived from Sherrington's work on the
spinal cord, and most of that was done with electrical stimuli delivered to
nerves, which of course doesn't produce patterns of excitation that are at
all like anything which occurs naturally. With vision you are in the position to control quantitatively the properties of the stimulus. You can
change its color, size, shape, duration, and so on. And you can try and
match it to natural stimuli. So I was interested in how the neurons in the
retina would deal with these quantitative aspects of the stimuli. This was
something you could do with vision and to some extent with hearing,
too, although just exactly what happens in the cochlea was not clear then
and is still not quite clear now! When I started on my Ph.D., Rushton was
moving into vision and we talked together a lot, which was very helpful.

PH: So after Adrian took you on, did you initially work on the eye-
movement problem, as he suggested?

HB: Yes. The Marshall and Talbot paper suggested that the small oscillatory movements of the eyes were actually important in generating visual
responses, playing a role in hyperacuity. So I spent six months or so working on eye movement and came to the conclusion that their suggestion
was not a very good one and there was no good evidence that small eye
movements played a role in hyperacuity, rather they might impair it
through motion blur. But I developed a method for measuring small eye
movements and was able to show that there is great variation in the fine
oscillations from subject to subject but that they didn't have any effect on
the ability to resolve fine gratings, didn't seem to have any effect on acuity.
But what struck me was that in the patterns of eye movements recorded
there were fixational pauses where the movement of the eyes was remark-

able small, the fixation was extremely stable—almost the opposite of what Marshall and Talbot suggested. So I dropped the eye movement research and switched to working on the frog's retina.

PH: Of course the frog retina work, where you gave the first account of lateral inhibition in the vertebrate retina, and suggested the idea of cells acting as specialized "fly detectors," was the first piece of your research to become very well known and it is recognized as being very important in the history of neuroscience.[10] It seems that even that early your work was strongly theoretically driven, there were theoretical notions behind the kinds of empirical work you were doing. Would you agree with that?

HB: Yes that's right. There were two theoretical inspirations behind that work. One was from the ethologists Konrad Lorenz and Niko Tinbergen, who suggested that at least the simpler reactions of birds, amphibians, reptiles, and so on could be understood in terms of quite primitive discriminatory mechanisms occurring at early stages in the sensory pathways.[11] So it occurred to me that the kind of sensitivity that the ganglion cells in the frog retina had might well be suitable for making frogs react to small moving objects. This was probably the basis for them snapping at flies and things like that—hence the idea of specialized fly detectors that I introduced in my 1953 paper.[11]

PH: So you were looking for evidence for that from the start?

HB: Yes. But the problem was that beyond pointing out that the best stimulus for some of these retinal ganglion cells is a small moving object, I couldn't see any way of following that up further. The kind of things one might think of would be to ask whether this was any different in toads, for example, which look for slower-moving objects such as worms and larvae rather than fast-moving objects like flies. But it was going to be very hard work to build up a comparative case like that, so I rather shied off it.

But the other theoretical area where my interests were developing, and which influenced the frog retina work, was the signal-to-noise problem. Tommy Gold was always an interesting person to talk to about that at Ratio Club meetings and other times when we met. I was interested in making quantitative measurements of, for example, the area threshold curves— measuring the sensitivity of the retinal ganglion cells as a function of the size of the stimulating spot [of light]. That was what led to the discovery of lateral inhibition in the frog retina, because I found that the sensitivity decreases as the spot gets bigger and spreads onto the inhibitory surround. That interest in the quantitative aspects was very much inspired by William Rushton, who was always keenly interested in that aspect of things.

PH: What are your memories of Adrian as a supervisor?

HB: Well he would poke his head around the corner now and again to see how I was getting on and would occasionally point me to a useful reference or give me some advice. He could be quite a distant character. I wouldn't say he was exactly encouraging in his supervision! I remember his advice when I wanted to switch to working on the frog's retina. I was convinced there was something funny about Hartline's results on the size of the receptive fields for the retinal ganglion cells, because they were very large, which would have given really rather poor visual performance, and if that was all the frog had it was very difficult to account for their actual performance.[12] Well, Adrian wasn't having any of that and he said, "Oh, I wouldn't do that—Hartline is a very clever chap, you know. It would be a mistake to try and prove him wrong." Well, of course he was quite right on one level, but I was right, too—there was more to be discovered. Anyway, I persisted and got his permission to go to London to buy the equipment I needed. At that time one could buy war surplus electronic equipment at absurd prices—it was sold by weight, and one could buy a photon-multiplier complete with all circuitry for a few shillings.

PH: Did you ever discuss with him later the fact that it turned out to be a very good change in direction?

HB: We never went back over the question of whether it was a wise move or not, but he certainly agreed that the results were very interesting, particularly the evidence for inhibition. But he was not at all theoretically based; his attitude was that we had the means of recording from nerve fibers and we should just see what happens. Of course he was absolutely brilliant at teasing out the first simple facts but then he never enquired further along any of the theoretical lines that were opening up.

I remember when I had first got the apparatus for the frog retina experiments assembled and in sometimes-working condition, Adrian made one of his unannounced visits to my lab, on this occasion with a visitor smoking a large cigar and speaking completely incomprehensible English.

A few minutes before I had dropped an electrode on to the floor; I had just remounted it and was lowering it onto a frog retina, without much hope of success, when they came in, so I turned on the light and started explaining what I was trying to do, and how. At this point the visitor was standing under the room light, and took a deep puff from his cigar. As he exhaled the smoke, its shadow fell across the preparation and it gave a long and vigorous "off" discharge. Ragnar Granit,[13] for that was who it turned out to be, was astonished; so was I, and his English became at least partly intelligible as we discussed the technicalities of what makes a good electrode and so forth.

Adrian spent a lot of time in his laboratory, where he definitely did not like visitors. I only recall making one very brief visit, when Adrian was actually doing an experiment. Now, whenever he was in the Physiology Lab [the university department] Adrian was always moving, never at rest, always reacting. His body movements were like saccadic eye movements, jerking incessantly from one object of attention to another. Ordinarily these movements, while much more frequent than most people's, were quite well spaced out, and synchronized with other events occurring around him, so that he surprised one with an unexpected shift of attention only, say, once every thirty seconds or so. But in his own laboratory they seemed to occur every second, and each of one's own movements elicited a response. When I went there he was doing an experiment on a monkey that was infected with amoebic dysentery—the reason, he explained, why he was able to get hold of it. If I turned towards the table to ask a question he seemed to jump to intervene between me and the infected monkey, and my attention was so riveted by his heightened state of reactivity that I could take in nothing about his laboratory or the experiment he was conducting.

William Rushton also had a rather alarming experience on one of his rare visits to Adrian's lab. It was near the beginning of his postgraduate research under Adrian—about the mid-1920s. Most students in his position were, to put it mildly, awe-struck by the great man. So it was with some trepidation that Rushton ventured in one afternoon to borrow a galvanometer. There was no one in the lab, so he set about searching. He eventually located one amidst all the clutter and went to pick it up. As his hand grasped the instrument, Adrian's voice suddenly boomed out of nowhere, "Put that down, Rushton!" He was perched in a small dark cupboard at the back of the lab where he liked to shut himself in to think. He could see the whole lab through a crack in the door.

PH: During the early part of your Ph.D., before the Ratio Club started, did you have any interactions with people at Cambridge who were interested in cybernetics and machine intelligence?

HB: That mainly started with the Ratio Club, but before that I did interact with some psychologists who were developing interests in that direction. W. E. Hick, famous for Hick's law and later a member of the Ratio Club, was one of them. Another character in psychology, C. G. Grindley, a physics-based psychologist, was a very interesting person. Unfortunately he was an alcoholic. I saw quite a lot of him because I'd often go for an after-work drink with Geoffrey Harris, who worked in the room next door to me. At six o'clock we'd go to the Bun Shop, a bar which was very close to the lab. Grindley was usually already there and I talked quite a lot to him about problems in psychology.

PH: Did you have an interest in the more psychological side of the brain sciences before that?

HB: I did. In fact in my final year as an undergraduate I had considered specializing in psychology rather than physiology—the way the natural sciences tripos is arranged at Cambridge involves studying many topics for part 1 and then specializing for part 2. The professor of psychology at the time, Frederick Bartlett, ran a course of seminars—fire side chats, they were—in the long vacation term. We met once a week and discussed various problems in psychology. I was never very happy with the material we covered. The concepts and thinking seemed to me to be very strongly verbally based whereas I think in a much more model-based and quantitative way. At any rate, at the end of that course I stayed behind and told Bartlett that I had to choose between psychology and physiology and asked for his advice. I explained to him some of the problems I had with psychology—that it seemed to me that in order to make progress in understanding the brain you had to get behind the words, you couldn't possibly explain it all in words. He agreed with that and said that the scientific advance that had done more for psychology than anything from within psychology over the past few decades had been Adrian's work in physiology, and no doubt there was going to be a lot more physiology-based work that would have a big influence in psychology. And that was what tipped the balance for me in favor of physiology.

PH: Did you interact with Hodgkin and Huxley during the period when you were doing your Ph.D.?

HB: Oh yes. I remember many teatime conversations with them. I remember Alan Hodgkin explaining to me about the noise limit when recording through an electrode and how the resistance isn't actually in the electrode itself but in the sphere of saline surrounding the tip. I remember after I'd written up my work on eye movement, Andrew Huxley read it through and pointed out various things about the statistical treatment that could be improved. I had a lot of useful conversations with them.

PH: It was during your Ph.D. studies that you became involved in the Ratio Club.[14] How did that happen?

HB: It was through Pat Merton. Pat worked with John Bates, who organized the club, at the National Hospital in Queen's Square. As I mentioned earlier, Pat and I had known each other since undergraduate days and he suggested me to Bates.

PH: How important was the club in the development of your ideas?

HB: Oh, very. It gave me an opportunity to hear and talk to people who were leading experts in this area. Probably most important to my work were Tommy Gold and Philip Woodward. Tommy was a wonderful person. He started life as an engineer and then switched to physics. He had a very distinguished career and was extraordinarily versatile. He is well known as one of the founders of the steady-state theory of the universe, for overseeing the construction and operation of the Arecibo dish, the world's largest radio telescope, and for many contributions to astrophysics, but he did much more than that and at the time of the Ratio Club he was working on hearing in the Zoology Department at Cambridge. He argued that there was a positive feedback mechanism involved in hearing. It was many years before he was proved right. He was very useful to talk to about signal-to-noise problems and statistical matters. He wasn't a particularly statistical sort of person himself but he knew it all, as an engineer, essentially, and he was very keenly interested in applying engineering ideas in biology. Anyway, he was always tremendously good value and always has an original point of view. Never listened to anyone else!

Philip Woodward was a marvelous person to interact with. He had a very deep understanding of information theory and could communicate it very clearly; he gave extremely good talks and his book on information theory applied to radar was very helpful.[15] I learned a lot from him.

There were two other members who were particularly influential, as far as I was concerned. One was Donald Mackay, who was a wonderful speaker; his talks were always brilliant expositions of ideas which often subsequently proved to be important. The other was Albert Uttley, who was at TRE [Telecommunications Research Establishment], Malvern, and then NPL [National Physical Laboratory]. He had some very interesting ideas. Pat Merton was very keen on him because he had developed one of those servo feedback devices for controlling gun turrets and so on during the war. But there was always something difficult to understand about his ideas and he wasn't a very clear expositor of them! This meant he could be given short shrift by some of the more precise members of the club. But I think that some of the ideas he had were very good. He had an idea about unitary representation which I think is the same basic concept as sparse representation, sparse coding, but was ahead of it.[16] I think it was an important idea, but he didn't really get it across to us successfully.

The meetings were always very enjoyable and stimulating and I learned a great deal.

PH: Are there any particular meetings that stick in your mind?

HB: I remember the very first meeting, where Warren McCulloch spoke. I think it's fair to say that he deeply failed to impress us. For many of us this was our first exposure to him. As we saw more of him that view tended to change, as we got to appreciate his style. Donald Mackay and others went on to form close friendships with him; Donald went to visit him often and they collaborated on various pieces of research.

My memories are probably more of people and ideas rather than specific meetings. I remember Alan Turing talking about how patterns could be generated from reaction-diffusion systems and how this might play a part in morphogenesis. One particular phrase of his really stuck in my mind as somehow summing him up. He was talking at one meeting about the brain and about looking at pictures of groups of hundreds of neurons and how they seem to be partially randomly determined; he said, "I don't know how to put this but they are not very accurately determined; they are more like a tree than a horse." That was very much the way he thought. Very expressive but not very precisely formulated ideas. Of course he was more than capable of formulating them precisely when it came to the crunch, but in getting the initial idea across he didn't try to.

PH: Do you remember if there was any debate in the Ratio Club about whether brains should be viewed as digital or analogue or mixed digital-analogue devices?

HB: Yes, there was a lot of discussion of that. I think the general consensus was that if it was digital it wasn't digital in the way that computers are. I think there was a general agreement that the fact that conduction down nerve fibers was by impulses rather than by graded potentials was because digital coding is more error-resistant. Having an all or nothing impulse is in fact the same as one aspect of using digital, as opposed to analogue, systems—the all-or-nothing response means that you can eliminate one kind of noise. But that is more or less where the similarity ends, basically because of the very great asymmetry between the presence and absence of an impulse in a nervous system compared with digital coding as used in engineering, where there is symmetry between the 1 and 0—they both have the same information capacity and in many cases they are used that way. So I think we understood that the way impulses were used in nervous systems was very different from in digital electronic systems. William Rushton wrote a paper on some of these issues at about that time.[17] This was before the idea of sparse coding and its implications, although, as I mentioned earlier, I think Albert Uttley was actually onto that idea even though he couldn't get it across to us.

PH: In the late forties, when the Ratio Club started, what was the typical view of cybernetics within neurophysiology, or neuroscience, as it was becoming?

HB: Well, I think for some of us information theory seemed to be a great new thing—here was something else to follow in the brain other than just impulses and electric currents and chemical metabolism. Here was a definable quantity that was obviously important in the kinds of things the brain did. So there was a great deal of enthusiasm for that cybernetic approach among a group of us who made up a fairly small section of the neurophysiological community. But a lot of people regarded it as airy-fairy theoretical nonsense. Neurophysiology was very untheoretical at that time; most of the important advances were made by people who took a very empirical approach, like Adrian, for example. William Rushton wrote a kind of scientific autobiography in his later years in which he says essentially that throughout his early years he was much too strongly theoretical and was trying to browbeat nature into behaving as he wanted it to, rather than eliciting how it actually was.[18]

PH: What is your view on that question, and has it changed since those early days?

HB: Well, I have two views which are to some extent in conflict. One is that the purely empirical approach still has a very important role in neuroscience. A lot of advances will still occur because of the development of new techniques that enable you to have access to something else in the brain that had hitherto been hidden from view. The technique will be used to find out what goes on and people will be guided in what they do by the discoveries they make—just as has happened since Adrian and before. William Rushton described it as thinking with your fingers. I think this is just a fact of life in neuroscience because we understand so little theoretically about how the brain works. It is not like a well-developed science, where theory explains ninety-five percent of what you are confronted with, so you have to use that theory; in contrast, theory in neuroscience explains five percent or less so you have to make use of other approaches and tools. The other view is that neuroscience is so badly fragmented that it is really not one community but half a dozen different ones who hardly understand what each other are saying. So there must be some kind of unification through a shared approach to trying to find a common coherent understanding of what the brain is doing. At this stage this might not be very theoretically elaborate, but it is crucial.

PH: I believe that sometime in the mid-fifties Oliver Selfridge and Marvin Minsky, and maybe others, were trying to organize an international conference on AI—it would have been the first such event—and they were interested in holding it at Cambridge University. I understand that you were involved in trying to make that happen. Is that right?

HB: Yes indeed. Oliver Selfridge and I went to see Maurice Wilkes, head of the Computer Laboratory, to try and get his support, as we would need a senior person in the university involved. He took an extremely negative view of it. He dismissed us with a comment like "Oh, an international conference—that's just a way of getting unpublishable papers published without being refereed." Of course such considerations couldn't have been further from our minds, but that was that, because he was the obvious person in the university whose support we needed. Of course the other anti-AI person at Cambridge was James Lighthill, who some years later wrote a rather damning report on the area for the UK science research council.

PH: One concept whose development in neuroscience you have been involved in is that of feature detectors. The idea of object detectors originated in your 1953 paper, where you postulate fly detector neurons in the frog retina.[19] The idea of feature detectors, where features refer to more primitive constituent properties of objects—edge detectors, convexity detectors, and so on—built on this, coming later. The idea is certainly present in Lettvin et al.'s 1959 paper.[20] Were you thinking in terms of feature detectors before that? What's your take on where the idea came from?

HB: I think it originated more in computer science, in early work in machine intelligence. Early work on pattern recognition, particularly on systems for automatically recognizing handwritten or printed letters and text, used the idea of features. Oliver Selfridge was working on it in the States and Dick Grimsdale and Tom Kilburn in Britain. Oliver Selfridge influenced Jerry Lettvin on this, I think. The computer work is certainly where I first became aware of the idea and then thought it was very likely that feature detectors were used in biological vision. I was certainly influenced by the fact that this early work in pattern recognition showed that the problem was much harder than had been thought; the nature of the difficulties was very illuminating.

PH: Let's talk a bit about information theory in neuroscience. You wrote some influential papers on the idea of redundancy reduction in the nervous system.[21] I think your first paper on that was at the Mechanization of Thought Processes symposium in 1958.[22] Could you say a bit about how the ideas developed? I suppose you had been thinking about it for some time before that.

HB: Yes I had. Actually the first time I talked about that was at one of a series of meetings on "Problems in Animal Behavior" organized by Thorpe and Zangwill. My talk was in 1955, although it wasn't published until 1961, when a book based on the meetings appeared.[23]

I also talked about it at a great meeting on "Sensory Communication" in 1959 at MIT, held at Endicott House.[24] I remember it being a very interesting meeting, over several days, and also one of the first international meetings I went to, so I particularly enjoyed it. They had a very good swimming pool and a very good bar! It was also notable as the first time I got to speak to certain people for any length of time. For instance, this was where I first met Jerry Lettvin—one of the amazing personalities from that era—and got to visit his lab. I also renewed my acquaintance with Warren McCulloch and got to know him better. Later the proceedings from the meeting were translated into Russian for a Soviet edition. My contribution was the only one that was expunged; for some reason it was thought to be too subversive!

PH: Had you discussed it at the Ratio Club?

HB: I don't recall giving a talk on it at the Ratio Club but I do remember trying to discuss it with Donald Mackay. He was extremely good at expressing his own ideas but he wasn't always terribly eager to learn about other people's. I got nowhere at all with him, except for him to say something like he'd already thought about it years ago and that kind of thing. But earlier talks and discussion at the club would have influenced the development of the idea.

I was very enthusiastic about how information was now something we could measure, but when you are actually confronted with doing an experiment on a physiological preparation, the prevalent techniques were all based on classical statistical measures rather than Shannon information, as was most of signal detection theory. So there was a problem in using it practically. I think this is part of the reason the idea rather fizzled out in neuroscience, to be reintroduced again in the 1980s by people like Simon Laughlin. Another reason may have been that important empirical advances were coming from people like Hubel and Wiesel who, like Adrian, were antitheoretical. Of course now information theory and other statistical ideas are quite strong in some areas of neuroscience.

PH: Sometime later you moved toward the idea of redundancy exploitation.[25] Can you say a bit about how you changed your mind?

HB: Well, initially I thought the idea of redundancy reduction was a perfectly plausible supposition because there were so many cells in the brain and, for instance, in the cortex it appeared most are very rarely active. It was only really when people started recording from awake behaving monkeys, and particularly when they started recording from MT [middle temporal cortex], which has much higher maintained discharge rates than elsewhere, it became pretty difficult to hang on to the notion that the mean firing rate in the brain is so low that the information capacity dictated by that supported the idea of redundancy reduction. I probably hung on to the idea, which is a kind of extreme version of sparse coding, for too long.

PH: Of course your famous 1972 neuron doctrine paper relates to these issues.[26] In that paper you propose the influential idea of sparse, or economical, coding in which "the sensory system is organized to achieve as complete a representation of the sensory stimulus as possible with the minimum number of active neurons." In all you laid down five speculative dogmas. How do you think that paper has stood the test of time?

HB: Oh, reasonably well. There were some new ideas that needed to be discussed and thought about and I don't think I was too wide of the mark with most of the ideas.

PH: One of the things you pointed out was the complexity of single neurons and the potential complexity of the processing they are capable of. Since then considerably more complexity has been revealed with the discovery of mechanisms such as volume signaling, and now intracellular processes are starting to be probed.

HB: Indeed, and I think that there is probably a great future in that direction—intra-neural processing may well turn out to be very important. It was work on *E. coli* that really opened my eyes to that possibility. Intracellular mechanisms successfully run their lives with all the important decisions being made by biochemical networks inside a single cell about the size of a bouton in the cortex.[27] If all that can go on in one bouton, one has to wonder if we're missing something about what a pyramidal cell can do. Maybe over the next decade or so we shall find out a bit more.

PH: Some people have remarked that the neuroscience establishment never really showed researchers like you and Jerry Lettvin the kind of appreciation you deserved, partly because they thought you were too theoretical.

HB: Well, they would be dead right up to a point. But it shouldn't be either/or. In this context I'm reminded of something Rutherford was supposed to have said in the 1930s when Jews were under threat in Germany and scientists like Einstein were looking to get out. In many ways Cambridge was an obvious place for Einstein to go, but it is claimed Rutherford said something like "Einstein's theories are all very well, but I think we can manage without him." So it wasn't just in neurophysiology that there was this prevailing antitheoretical attitude.

PH: I'd like to finish with a few rather general questions. First, looking back at the development of neuroscience over the sixty or so years you have been involved in it, has it turned out very different from what you might have imagined going back to the start of your career?

HB: I think it is a pity that more attention is not paid to trying to find simple preparations that exemplify particular cognitive or brain tasks. I think we have much more chance of ironing out the basic principles by studying these simpler systems. For instance, there seems to be some progress in understanding the cerebellum partly because people have found electric fish and things like that where it is possible to do observations and experiments which actually reveal what the cerebellum is doing. This ties in with what I was saying earlier about the need for a coherent theoretical framework.

On more theoretical developments, I'm a bit critical of what has happened in some areas of computer modeling. I don't think many, if any, of the neural network models are good models in the sense that the Hodgkin-Huxley model was—that dealt with quantities that could be defined and measured in a single cell. The neural network models tend to be considerably removed from anything you could measure at the cell level. I think they have got to be pulled down to a more biophysical basis. I'm more interested in the Bayesian approaches because I think that there they are getting much closer to realistic models of what certain quantities (here probabilities rather than simple physical values) might actually represent. The emphasis on probabilistic inference in certain strands of modern modeling is very good; I think that has a future.

I remember that when I was starting out in my research career there was quite a bit of optimism about how quickly some form of machine intelligence would be developed. Those members of the Ratio Club more involved in that area were very hopeful. But a computer wouldn't beat a grand master at chess until the 1990s. None of us would have predicted that back then; we all thought it would be much sooner. I remember being more sceptical than many at how much progress would be made, but obviously not sceptical enough. But of course there have been tremendous advances in processing power and miniaturization of electronics and so on, much of which most of us wouldn't have foreseen, which has meant that the use of computer-based technology has had a big impact on neuroscience. Initially this was more for data collection and analysis. But now, and even more in the future, the important thing is that you can test whether a theoretical idea, or a model mechanism, can actually perform in the way the real brain performs. That's a fantastic advance.

PH: Are you surprised at how much progress has or hasn't been made in neuroscience during your career?

HB: It's come a long way in one sense. When I was a graduate student, in neurophysiological circles the idea of being able to understand what was going on in the cortex was dismissed as being utterly impossible. It was just too complex. Of course we don't believe that now; we think we'll find out all about it next week. That's equally far from the truth, but the outlook is much more hopeful. I was one of the few people back then who thought we would understand these things physiologically, and I think we have made a lot of progress, but there is still a hell of a way to go.

PH: Finally, you've made a lot of important contributions, but is there any particular piece of your research that stands out for you?

HB: I'm always rather disappointed by the general response to the attempts that I've made to measure the actual statistical efficiency of both psychophysical performance and neural performance,[28] because it does seem to me that when you can say that the brain is using whatever percentage it may be of the statistical information that is available in the input, this has an importance for understanding the brain comparable with being able to say that a muscle uses whatever percentage it is of available chemical energy in generating mechanical movement. I think this is a big step forward in getting to grips with one aspect of what the brain actually does. Imagine how we would regard intelligence tests if they were of this nature, if they were actual measures of mental efficiency at performing some task, which they obviously are not; they're ad hoc plastered-up God knows what.

Notes

1. See Nora Barlow, ed., *The Autobiography of Charles Darwin, 1809-1882: With Original Omissions Restored* (London: Harcourt Brace & World/Collins, 1958; reprint, New York: Norton, 1958), and Emma Nora Barlow, "Darwin's Ornithological Notes," *Bulletin of the British Museum (Natural History) Historical Series* 2(7): 200–78(1963).

2. The great neurophysiologist Lord Adrian shared the 1932 Nobel Prize in Physiology or Medicine with Charles Sherrington for pioneering work on the electrical properties and functions of nerve cells. He was professor of physiology at the University of Cambridge 1937 to 1951, president of the Royal Society from 1950 to 1955, and master of Trinity College, Cambridge, from 1951 to 1965. In 1952 Alan Hodgkin and Andrew Huxley, researchers in the Physiology Laboratory at Cambridge, wrote a series of papers, now classics, presenting the results of a set of experiments in which they investigated the flow of (ionic) electric current through the surface membrane

of a nerve fiber of a squid. The papers culminated in a mathematical description of the behavior of the membrane based upon these experiments—the Hodgkin-Huxley model—which accounts for the conduction and excitation of the fiber. This model has been used as the basis for almost all other ionic current models since. For this work they were awarded the 1963 Nobel Prize in Physiology or Medicine. For the summary paper containing the model, see Alan L. Hodgkin and Andrew F. Huxley, "A Quantitative Description of Membrane Current and Its Application to Conduction and Excitation in Nerves," *Journal of Physiology* 117: 500–544(1952).

3. "Tripos" refers to the honors examination that was introduced at Cambridge in the eighteenth century. It was called the tripos after the three-legged stool used formerly at disputations. At first the test was primarily mathematical, but a classical tripos was instituted in 1824, and tripos in the natural sciences and the moral sciences were added in 1851.

4. Kenneth J. W. Craik, *The Nature of Explanation* (Cambridge: Cambridge University Press, 1943).

5. Horace B. Barlow, H. I. Kohn, and E. G. Walsh, "Visual Sensations Aroused by Magnetic Fields," *American Journal of Physiology* 148: 372–75(1947); William A. H. Rushton and Horace B. Barlow, "Single-Fibre Responses from an Intact Animal," *Nature* 152: 597(1943).

6. Horace B. Barlow, H. I. Kohn, and E. G. Walsh, "The Effect of Dark Adaptation and of Light upon the Electric Threshold of the Human Eye," *American Journal of Physiology* 148:376–381(1947).

7. W. H. Marshall and S. A. Talbot, "Recent Evidence for Neural Mechanisms in Vision Leading to a General Theory of Sensory Acuity," *Biological Symposia—Visual Mechanisms* 7: 117–64(1942).

8. S. Hecht, S. Shlaer, and Maurice Pirenne, "Energy, Quanta and Vision," *Journal of General Physiology* 25: 891–40(1941).

9. See Horace B. Barlow, "Retinal Noise and Absolute Threshold," *Journal of the Optical Society of America* 46: 634–39(1956).

10. Horace B. Barlow, "Summation and Inhibition in the Frog's Retina," *Journal of Physiology* 119: 69–88(1953). In this classic paper Barlow demonstrated a particular organization of inhibitory connections between retinal neurons (lateral connections between neighboring cells) and was able to provide accurate measures of retinal cell receptive fields: previous estimates were shown to be wrong as they were based on incorrect assumptions about functional network structure and did not take account of the inhibitory affect of surrounding cells. This paper gives the first suggestion that the retina acts as a filter passing on useful information; this is developed into the idea of certain types of cells acting as specialized "fly detectors"—an idea that was to become very influential.

11. See Konrad Lorenz, *King Solomon's Ring: New Light on Animal Ways* (London: Methuen, 1952), and Niko Tinbergen, "Derived Activities: Their Causation, Biological Significance, Origin, and Emancipation During Evolution," *Quarterly Review of Biology* 27(1): 1–32(1952).

12. Haldan K. Hartline, "The Receptive Fields of Optic Nerve Fibres," *American Journal of Physiology* 130: 690–99(1940).

13. Ragnar Granit (1900–1991), the great Finnish neurobiologist, shared the 1967 Nobel Prize in Physiology or Medicine with Haldan Hartline and George Wald for his work on vision. He also made many important contributions to the neurophysiology of motor systems.

14. The Ratio Club was a London-based dining club for the discussion of cybernetics and related issues; it is the subject of chapter 6 of this volume, by Philip Husbands and Owen Holland.

15. Philip M. Woodward, *Probability and Information Theory, with Applications to Radar* (London: Pergamon Press, 1953).

16. In this context sparse representation is the idea, later developed as "cardinal cells" in Barlow, "Single Units and Sensation: A Neuron Doctrine for Perceptual Psychology?" *Perception* 1: 371–94(1972), that stimulus features are represented by a few neurons within a large neuronal network. Thus, the representation of the stimulus feature is sparse within the population of neurons. More generally, a sparse representation is one that uses a small numbers of descriptors from a large set. The idea was further developed recently by various groups including D. J. Field and B. A. Olshausen, who coined the term "sparse representation"; see, for example Olshausen and Field, "Emergence of Simple-Cell Receptive Field Properties by Learning a Sparse Code for Natural Images," *Nature* 381: 607–9(1996).

17. William Rushton (1951). Conduction of the nervous impulse. In *Modern Trends in Neurology*, pp. 1–12, edited by Anthony Feiling (London: Butterworth, 1951).

18. This is a reference to Rushton's "Personal Record," a document the Royal Society asks its fellows to write. See also H. B. Barlow, "William Rushton, 8 December 1901–21 June 1980," *Biographical Memoirs of Fellows of the Royal Society* 32: 422–59 (December 1986).

19. Barlow, "Summation and Inhibition in the Frog's Retina." See note 10.

20. Jerry Lettvin, H. R. Maturana, Warren McCulloch, and William H. Pitts, "What the Frog's Eye Tells the Frog's Brain," *Proceedings of the IRE* 47: 1940–59(1959).

21. One way to compress a message, and thereby make its transmission more efficient, is to reduce the amount of redundancy in its coding. Barlow argued that the nervous system may be transforming "sensory messages" through a succession of

recoding operations that reduce redundancy and make the barrage of sensory information reaching it manageable.

22. Horace B. Barlow, "Sensory Mechanism, the Reduction of Redundancy, and Intelligence," in *Mechanisation of Thought Processes: Proceedings of a Symposium held at the National Physical Laboratory on 24–27 November 1958*, pp. 537–59, edited by Albert Uttley (London: Her Majesty's Stationery Office, 1959).

23. Horace B. Barlow, "The Coding of Sensory messages," in *Current Problems in Animal Behaviour*, pp. 330–60, edited by W. H. Thorpe and O. L. Zangwill (Cambridge: Cambridge University Press, 1961).

24. Horace B. Barlow, "Possible Principles Underlying the Transformations of Sensory Messages," in *Sensory Communication*, pp. 217–34, edited by W. A. Rosenblith (Cambridge, Mass.: MIT Press, 1961). See also chapter 19 of this volume, "An Interview with Jack Cowan," for further discussion of this meeting.

25. As more neurophysiological data became available, the notion of redundancy reduction became difficult to sustain. There are vastly more neurons concerned with vision in the human cortex than there are ganglion cells in the retinas, suggesting an expansion in redundancy rather than a reduction. Barlow now argues for the principle of redundancy exploitation in the nervous system. In relation to distributed neural "representations," learning is more efficient with increased redundancy as this reduces "overlap" between distributed patterns of activity. Learning exploits redundancy. For a more detailed discussion see A. R. Gardner-Medwin and Horace B. Barlow, "The Limits of Counting Accuracy in Distributed Neural Representations," *Neural Computation* 13(3): 477–504(2001).

26. Barlow, "Single Units and Sensation." See note 16.

27. A synaptic bouton is a small protuberance at the presynaptic nerve terminal that buds from the tip of an axon.

28. See Horace B. Barlow and B. Reeves, "The Versatility and Absolute Efficiency of Detecting Mirror Symmetry in Random Dot Displays," *Vision Research* 19: 783–93(1979), and Horace B. Barlow and S. P. Tripathy, "Correspondence Noise and Signal Pooling in the Detection of Coherent Visual Motion," *Journal of Neuroscience* 17: 7954–66(1997).

Figure 19.1
Jack Cowan. Image courtesy of Jack Cowan.

19   An Interview with Jack Cowan

Jack Cowan was born in Leeds, England, in 1933. Educated at Edinburgh University, Imperial College, and MIT, he is one of the pioneers of continuous approaches to neural networks and brain modeling. He has made many important contributions to machine learning, neural networks, and computational neuroscience. In 1967 he took over from Nicolas Rashevsky as chair of the Committee on Mathematical Biology at the University of Chicago, where he has remained ever since; he is currently professor in the Mathematics Department.

This is an edited transcript of an interview conducted on the November 6, 2006.

Philip Husbands: Can you start by saying a little about your family background, in particular any influences that might have steered you towards a career in science.

Jack Cowan: My grandparents emigrated from Poland and Lithuania at the turn of the last century; I think they left after the 1908 pogroms and they ended up in England, on my mother's side, and Scotland on my father's. My mother's parents had a clothing business in Leeds and my father's family sold fruit in Edinburgh. My father became a baker. My mother was clever and did get a scholarship to go to university but she had to decline because of the family finances. So I was the first member of my family to go to university.

PH: Did you get much influence from school?

JC: Yes, in that I went to a good school. My parents were very encouraging from an early age—my mother claims that I started reading when I was very young and that I was bossing the other kids in kindergarten! Anyway, we moved to Edinburgh from Leeds when I was six years old and I went to a local school there for about three years, but my parents could see that I had some aptitude so they got me into George Heriot's School, a very good private school. I got bursaries all the way through and ended up the top boy in the school—I was Dux of the school—and got a scholarship to Edinburgh University.

PH: What year did you go to university?

JC: I was an undergraduate from 1951 to 1955, studying physics. I remember when I was about fourteen we had the traditional argument between Jewish parents and their son—they wanted me to become a doctor or a dentist or lawyer or something like that and I kept telling them, "No way, I'm going to be a scientist." So I decided early on that I wanted to do science and I can't say there were any particular outside influences on this decision; it seemed to come from within.

PH: How were your undergraduate days?

JC: Well from being top boy at Heriot's my undergraduate career was a disaster. I found the physics faculty and the lectures at that time really boring. I didn't do well at all.

But after that I was rescued by a man called J. B. Smith, who was head of a section at Ferranti Labs in Edinburgh where I'd applied for a job. He had also been the school Dux at Heriot's—a decade or so before me—so I guess he took a chance and hired me. I was in the instrument and fire control section. I was there for three years from 1955, although in the middle of that I was sent to Imperial College for a year to work with Arthur Porter, one of the pioneers of computing in Britain. I also got to know Dennis Gabor, whom I hit it off with. As well as being the inventor of holography, he had a lot of interest in cybernetics, machine learning and things like that. He worked on adaptive filters and introduced the idea of using gradient descent to solve for the coefficients of a filter that was learning by comparing the input with the output. I would say that Gabor was a huge influence on me.

PH: Was it going to Imperial that sparked the direction your work took, leading you into machine learning and neural networks?

JC: To a large extent. But before that what really got me started, and actually I think what impressed Smith, was that I had read Norbert Wiener's book on cybernetics. I picked it up in the library when I was an undergraduate and found it very, very interesting. Also while I was still an undergraduate I heard a lecture by Gabor on machine learning, which was very influential.

PH: What kind of work were you doing for Ferranti?

JC: The first project they gave me was to work out pursuit curves. Ferranti worked on the computer guidance systems for the British fighter planes of the time; there was a consortium of Ferranti, English Electric, and Fairey Aviation involved in the computers that controlled air-to-air missiles. So they had me work with a couple of other people on the mathematical problems of prediction of missile trajectories and things like that. So I learned quite a bit of useful mathematical stuff doing that.

A year or two before I started at Ferranti, Smith and a colleague, Davidson, had built a machine that solved logic problems by trial and error. I got it working again and they arranged for me to take this machine down to London to the Electrical Engineering Department at Imperial College to demonstrate it to Porter and Gabor. That developed my interest in automata theory and machine learning. Anyway, Ferranti arranged for me to spend a year at Imperial doing a postgraduate diploma in electrical engineering. Porter wanted me to stay to complete a Ph.D., but I only had the year. I had started to play around with many-valued logics to try and solve logic problems in a better way than simple trial and error as embodied in Smith's machine. It was this that got Gabor interested in me and he became my mentor.

I met a lot of interesting people during that year: Wilfred Taylor, who was at University College and developed one of the very first learning machines, Raymond Beurle, from Nottingham University, who had written a very beautiful paper on the mathematics of large-scale brain activity and many others.[1] So I met all these guys, which was very inspiring, and in 1956, Ferranti also sent me to one of the earliest international meetings on cybernetics, in Belgium, where I met Grey Walter, with his turtles, and Ross Ashby, and that is where I first met Albert Uttley, who was working on conditional probability approaches to learning, among other things.[2] I also remember a very interesting lecture at Ferranti given by Donald MacKay. So by the time I was in my early twenties I'd already met most of the leading people in Britain working in the area that interested me. And there was a lot of very good work going on in Britain. As well as these interactions I came across a number of papers that would prove influential later—for instance, John Pringle's paper on the parallels between learning and evolution, which really set the foundation for competitive learning, and Turing's work on the chemical basis for morphogenesis, which would inspire my work a couple of decades on.[3]

A little later, essentially through Porter and Gabor, I ended up with a fellowship from the British Tabulating Machine Company to go to MIT. They ran a special scheme to send graduate researchers from Britain to MIT. This attracted me, so I applied and got it.

PH: When did you start at MIT?

JC: I arrived at MIT in the fall of 1958 as a graduate student. I joined the Communications Biophysics group run by Walter Rosenblith. I was in that group for about eighteen months, then I moved to the Warren McCulloch, Walter Pitts, and Jerry Lettvin group.

PH: How did that move come about?

JC: Well, my interests were a bit more theoretical than what was going on in the Communications Biophysics group—they were mainly interested in auditory psychophysics, which I didn't find as interesting as the more theoretical aspects of cybernetics. I had been working on many-valued logics at Imperial and through reading von Neumann's paper in Claude Shannon and John McCarthy's *Automata Studies* collection had got very interested in the problem of reliable computation using unreliable elements.[4] So I started to work on applying many-valued logic to that problem. That kind of thing didn't really fit in the Rosenblith group.

In 1959, while I was still in his group, Rosenblith organized a very interesting meeting at MIT on sensory communication.[5] That was a great meeting for a graduate student like me to attend, there were all kinds of very interesting people there (I've got the proceedings here): Fred Attneave, Horace Barlow, Colin Cherry, Peter Elias, J. C. R. Licklider, Donald Mackay, Werner Reichardt, Willie Rushton, Pat Wall, to name a few! It was an amazing meeting. The stand-out talks for me were Horace Barlow's, "Possible Principles Underlying the Transformations of Sensory Messages," where he talked about the possible role of redundancy reduction in the nervous system, and Werner Reichardt's "Autocorrelation: A Principle for the Evaluation of Sensory Information by the CNS," in which he presented an early version of the famous Reichardt motion-detector model.[6] That was also where I first heard about the Lettvin, Maturana, Pitts, and McCulloch work on the frog's visual system, which was also extremely good, and that was what got me really interested in joining the McCulloch and Pitts group.[7] McCulloch was also interested in the reliability problem, so I joined.

PH: What was MIT like at that period?

JC: In those days MIT was absolutely fantastic. I got to know a huge range of people; I consider myself to have been very lucky to be there at that time. I remember the first day I got there I was taken to lunch by Peter Elias and David Huffman, Huffman of Huffman coding and Elias who was one of the big shots in information theory, and they said to me, "You know, graduate school at MIT is not like in England. It's like a factory with an assembly line and you get on and it goes at a certain rate and if you fall off—too bad!" They warned me that it was very hard going and rigorous. They were right!

But it was an amazing place. As well as great names from cybernetics and information theory—Wiener, McCulloch, Pitts, Shannon—Noam Chomsky was down the hall, Schutzenberger was there working with him on formal linguistic theorems, Roman Jakobson was around. Some of the classes were incredible—for instance, being taught by the great pioneers of information theory.

PH: Who were the major influences on you from that time?

JC: McCulloch, Pitts, Wiener, and Shannon. I was very lucky that Shannon arrived at MIT from Bell Labs the year I got there. So I took courses on information theory with Bob Fano, Peter Elias, and Claude Shannon, I had the benefit of a set of lectures from Norbert Wiener, and I interacted all the time with Warren McCulloch and also to quite an extent with Walter Pitts.

PH: So Pitts was still active in the lab?

JC: He was still sort of functional. In fact I was one of the last students to really talk to him at length about his interests and work. He and Wiener probably had the biggest influence on me because it was through talking with them—separately, because by then Wiener had fallen out with McCulloch and Pitts—that I decided to start working on trying to develop differential equations to describe neural network dynamics and to try to do statistical mechanics on neural networks. Pitts directly encouraged me to look at continuous approaches to neural networks.

PH: I seem to remember that you have an unfinished thesis by Pitts...

JC: Well, I don't have a thesis but what I have is a fragment of an unpublished manuscript which I copied. He gave it to me for a while and let me copy it. So I hand copied it, imitating his writing, and then gave it back to him. Jerry Wiesner, who was then head of RLE, the Research Lab of Electronics, to which we belonged, actually offered money to anyone who could get Pitts to write something up and publish it so that they could give him a degree. But unfortunately this thing was only a fragment; he never finished it.

PH: It was on the beginnings of a statistical mechanics treatment of neural networks wasn't it?

JC: Yes. It was the beginnings of Walter's attempt to do something, but unfortunately it didn't go very far. But remember that when he did that, in the late fifties, this was long before any of the statistical mechanics techniques needed for solving the problem had been developed.

PH: Did you interact with Oliver Selfridge?

JC: I had some very nice talks with Oliver, who was working on the Pandemonium research at that time.[8] But he had also done some very nice earlier work with Wiener and Pitts on the origins of spirals in neural models, with possible applications to cardiac problems.[9] In fact, some of the stuff I work on now is closely related to what they were doing. Marvin Minsky also got involved with that work. There is a very nice study by them on reverberators and spirals.

PH: So when did your period at MIT end?

JC: 1962. So I was there for four years. During that period I recruited Shmuel Winograd, who went on to become a major figure at IBM, to the group. I was working on the reliability stuff with McCulloch, and Shmuel and I got interested in the capacity of computing devices. We developed a theory of how to design optimal reliable network configurations of computing elements. We came up with one of the earliest designs for a parallel distributed computing architecture. This work got us known and we wrote a monograph on it.[10]

PH: Would you say it was during this period that your interests started to move more towards biology?

JC: Yes. It was definitely at MIT, through the influence of McCulloch and others, that I moved from thinking about automata towards starting to think about the nervous system. So it was a defining period in that sense.

PH: At about that time approaches to machine intelligence began to diverge to some extent. Minsky and McCarthy and others were very active in exploring and promoting new directions in what they called artificial intelligence, and cybernetics was starting to wane. So things were at a cusp. What are your memories of the expectations people had?

JC: Well, there was always this tremendous hype about artificial intelligence around Marvin and McCarthy and Allen Newell and Herb Simon and so on. I remember Herb Simon coming to give a talk and it was the same message we got from Marvin; if we had bigger and faster computers we would be able to solve the problems of machine translation and AI and all kinds of stuff. But they set up the AI Lab and were instrumental in the development of lots of useful technology.

   Through McCulloch I got to know Marvin Minsky very well and in fact I recruited Seymour Papert to join our group, but by the time he arrived I'd gone back to England so he ended up working with Marvin.

PH: So what was the reaction in the McCulloch group to all the hype surrounding AI?

JC: Great skepticism.

PH: Do you remember what your own personal views were at the time on what was likely to be achieved and on what the important problems were?

JC: Well I was still in the middle of learning as much as I could and trying to think out what direction I should take. I had a strong bent towards applying the methods of theoretical physics and I was getting more and more interested in the nervous system and neural network models. As I mentioned earlier, Pitts and Wiener had influenced me to look in the direction of continuous approaches to neural networks, and I started to think that the statistical mechanics of neural networks was a very important problem. I remember sitting in the office I shared with McCulloch and having the idea that there is an analogy between the Lotka-Volterra dynamics of predator-prey interactions in populations and excitatory and inhibitory neuron interactions in neural networks, and that set me going for the rest of my career.

PH: How was the transition back to England in 1962?

JC: Well, in 1962 I was at a meeting in Chicago when I was approached by two gentlemen from the Office of Naval Research who asked me if I would like grant support. I said, "Well, yes!" and so they gave me my own personal grant that I was able to take back to England with me. I had to go back to Britain for at least a year because that was part of the terms for the fellowship I had that funded me at MIT.

So I went back to Imperial as an academic visitor. Meanwhile I got a master's degree at MIT, but neither Shmuel Winograd nor I decided to brave the doctoral program there, on the advice of Claude Shannon. After Claude had written his first famous paper, on the application of Boolean algebra to switching networks, he took the doctoral qualifying exam in electrical engineering and failed; I think he failed the heavy-current electrical engineering part. So he went to the Math Department and did his Ph.D. there. So we took his advice and Shmuel got his doctorate from NYU and I returned to Imperial without a Ph.D.

PH: How did your work develop at Imperial?

JC: So I went back to the Electrical Engineering Department at Imperial and got involved in a number of things. I started doing a bit of teaching, labs on numerical methods and computing and things like that, and I started supervising students, even though I was really technically still a student myself! I worked on the monograph on reliable computing from unreliable elements with Winograd, which got published by MIT Press after the Royal Society rejected it![11] We made the link between von Neumann's work and Shannon's work on the noisy-channel coding theorem and introduced the parallel distributed architecture thirty years before its time. After we finished that I turned to the problem of neural network dynamics, and by about 1964 I had the beginnings of a way to do the mathematics of neural networks using systems of nonlinear differential equations. I did a version of it that led to a statistical mechanics, but it wasn't quite the right version; it was a special case, the antisymmetric case.[12] This came from the analogy with population dynamics, where an excitor neuron is coupled to an inhibitor that is coupled back to it, so the weights are antisymmetric. In this work I introduced the sigmoid nonlinearity into neural models.[13] There was another special case that I didn't follow up at the time but it was followed up fifteen or so years later by John Hopfield, the symmetric case.[14] Hopfield networks were the other special case of the network population that I introduced in about 1964. Anyway, when I was doing that work in the sixties I realized that there was clearly a relationship between what I had done and Raymond Beurle's work on a field theory of large-scale brain activity—a kind of continuum model.[15] So I spent quite a bit of time working on that and wrote it all up in a report for the Office of Naval Research.

PH: So who else in the UK were you interacting with at that time?

JC: Mainly Gabor, Uttley, and MacKay at that stage, and a little bit with Christopher Longuet-Higgins and David Willshaw, who were doing interesting neural-network research in Edinburgh—associative-memory work. I also used to interact a bit with Richard Gregory, whom I got on very well with.

PH: You went and worked with Uttley's group, didn't you?

JC: Yes. I spent four years at Imperial, '62 to '66, and then in '66 to '67 I split my time—about a day a week at Imperial and the rest at the National Physical Laboratory at Teddington. Albert Uttley had invited me to go out there to work in his Autonomics Division. I mainly worked with Anthony Robertson, who was a neurophysiologist working in that group.

PH: What did you think of Uttley's ideas at that time?

JC: Well, I always liked Uttley's ideas; I think he was undervalued. He had some very good ideas which were precursors to more modern work on machine learning. He had the right ideas—for instance, using a conditional probability approach[16]—he just didn't have a clean enough formulation. Of course this was long before people discovered the relationship between statistics and neural networks.

PH: What about the wider field of theoretical biology that was gaining strength in Britain at about this time?

JC: Yes, that was another group of very interesting people I was involved in. My link to that started back in Edinburgh when I was growing up. One of my friends was Pearl Goldberg, who got married to Brian Goodwin, the theoretical biologist. We met up in Boston when I was at MIT and through me they ended up staying with McCulloch for a while. Anyway, Brian had developed a statistical-mechanics approach to cell metabolism. I liked that a lot and I realized my Lotka-Volterra thoughts on neural networks could be done the same way. So Brian's work was a trigger to my first statistical-mechanics approach to neural networks. When I got back to London in '62 I'd meet up with Brian, who was at Edinburgh University, and through him I got to know Lewis Wolpert, the developmental biologist. And so we had a discussion group on theoretical biology, which Michael Fisher used to come to occasionally, and so that's when I really started to get into the wider field. Then Conrad Waddington, who was also in Edinburgh, organized the "Towards a Theoretical Biology" meetings,[17] and through Brian I got to go to those. That was quite an interesting collection of people. The mathematicians René Thom and Christopher Zeeman were there, and so were Ernst Mayr, John Maynard Smith, and Dick Lewontin, the evolutionary biologists, and Lewis Wolpert, Donald Michie, who at that time was still working in genetics, Christopher Longuet-Higgins, Brian, and me.

Now Lewontin was on the lookout for someone to take over from Rashevsky at the University of Chicago. Nicolas Rashevsky had set up the Committee on Mathematical Biology in the late 1930s, but by 1965 he had resigned and they were looking for a replacement.[18] They settled on either Brian Goodwin or me, and Brian wasn't interested, as he had not long before moved to Sussex University. So I went for a long walk with Lewontin and Ernst Mayr in the woods outside the Villa Serbelloni, where we were having the meeting, which overlooked Lake Como. I remember Ernst was amazing, pointing out every animal and insect and plant in the woods. Anyway, they talked me into thinking seriously about taking the job. At that time I wanted to go to Sussex to work with Brian. I had applied to the UK Science Research Council for a grant to work on the statistical mechanics of large-scale brain activity and told them that if I didn't get the funding I'd have to go to the U.S. And they didn't give me the funding. The referees, who included Donald Mackay, claimed it was too speculative. So I ended up taking the job and moving to Chicago.

I'd been appointed a professor and chairman of the Committee on Mathematical Biology at Chicago and I still didn't have a Ph.D. So I decided it really was time and I took a week out to write up some of my work into a

thesis, and I had a viva exam with Gabor as my internal examiner and Raymond Beurle as the external. The viva lasted two minutes and then we
drank some champagne! So I got my Ph.D. on the statistical mechanics of
neural networks, the first ever Ph.D. in that area.

I arrived in Chicago with my wife, who was seven months pregnant, the
day after a monster snowstorm in the winter of 1967.

PH: Had the intellectual climate changed much in the time you'd been
away? I'm wondering if the AI bandwagon had had a negative impact on
funding in the areas you were interested in?

JC: Yes and no. It didn't do anything to mathematical biology, but it did
damage the field of neural networks. When Minsky and Papert published
their attack on the perceptron in 1969, and they'd been giving talks on
that stuff for a while before, they made the claim that you couldn't solve
the perceptron training problem.[19] In retrospect I had invented the machinery necessary to solve the problem, and show that they were wrong,
in the mid-sixties—the sigmoid model I used in my Lotka-Volterra-like
network-dynamics model. But I didn't work on that aspect; I put it aside.
There were two major things that I should have done but didn't at
the time. One, as I've already mentioned, was to do the other case of the
Lotka-Volterra network, which is essentially what Hopfield did, and the
other was to use the sigmoid model to do perceptron training, which is
what David Rumelhart, Geoff Hinton, and Ronald Williams did in 1986.[20]
So I kick myself for not doing either.

PH: How did things pan out in the Committee on Mathematical Biology?

JC: Well, I was chairman for six years and I built it into a department of
theoretical biology. I recruited people like Stuart Kaufmann and Art Win-
free, who both went on to become very prominent, and various other people. It actually had quite a decent influence on theoretical biology in the
U.S. and elsewhere. But then we merged with the biophysics department,
because it was thought that small departments were not so viable, but that
proved to be a mistake. The merged department then got further merged to
become part of something that also accommodated genetics and molecular
biology and other branches of biology. So in 1980, or thereabouts, I moved
to the mathematics department and I've been there ever since.

PH: What was the main focus of your work from the late 1960s?

JC: So my idea of correcting and extending Beurle's work paid off and I
was very fortunate to recruit a very good postdoc, Hugh Wilson, to work
with me on that. So Wilson and I published a couple of papers, in '72 and
'73, which triggered a great deal of activity.[21] We basically gave the first

nontrivial and useful field theory, what we would now call a mean field theory, for looking at large-scale brain dynamics. But even then I knew that that work wasn't really the answer to the problem I'd set myself of doing statistical mechanics of neural networks. Even when, in the late seventies, Bart Ermentrout and I showed that you could apply modern mathematical techniques to calculate the various patterns that could form in networks of that kind, which turned out to be useful for various kinds of applications, it still wasn't really getting to grips with what might be going on in the nervous system.[22] So I made a start on trying to do that in 1979 and discovered the key to doing it in 1985 while working at Los Alamos with two physicists, Alan Lapedes and David Sharp, and got a first version going in about 1990. I worked on it a bit more with a student, Toru Ohira, but it is only in the last two or three years, with a really bright graduate student named Michael Buice, have we actually solved the problem. So now we are in possession of a field theory for large-scale brain activity, which is exactly the kind of object that Norbert Wiener and Walter Pitts were clearly pointing at nearly fifty years ago. We've solved the problem that was put to me by Pitts and Wiener all those years ago. We finished the first paper on this only last week [October 2006], so it will see the light of day in due course. It uses Wiener path integrals as well as all the machinery of modern statistical mechanics and field theory, and it's made exactly the right contact with physics that I was hoping for and it's relevant to data at every level of analysis. It's a great boon at my age to be in the middle of all this new stuff. It might be the Rosetta Stone that unlocks a lot of how large-scale brain activity works.

PH: That sounds very exciting; I look forward to reading more about it. Can we just go back a little in that trajectory and talk about your work in pattern formation in neural networks and how it links to Turing?

JC: Well, the bulk of that research goes back to 1979, when I was working with another extremely bright graduate student, Bart Ermentrout. I went to a conference that Hermann Haken organized in Germany in 1977 on what he called synergetics—a modern version of cybernetics, but stressing the role of excitation. While at that meeting I realized that Turing's 1952 work on the chemical basis of morphogenesis could be applied to neural networks.[23] I realized that the stuff I'd done with Hugh Wilson was an analogue of the reaction-diffusion networks that Turing had worked on. There was a very good talk at that meeting by an applied mathematician from the U.S. called David Sattinger showing how to apply the techniques of nonlinear analysis, bifurcation theory as it's called, in the presence of symmetry groups, to things like fluid convection. And I realized there was an analogue of that in the nervous system. When I got back I mentioned this to Bart and he immediately saw what I saw.

We realized that we could apply it to the problem of what is going on in the cortex when people see geometric patterns when they are hallucinating. This happens after taking hallucinogens, or through meditation, or sometimes in other conditions. The Chicago neuropsychologist Heinrich Klüver did a lot of field work in the sixties to classify these types of geometric hallucinations. He mainly experimented on himself, using peyote.[24] Anyway he discovered that there were only four classes of patterns; they were the same for everyone experiencing these kinds of hallucinations. So we produced a first treatment of why people see these patterns—tunnels, funnels, spirals, and honeycombs—in the visual field. Applying the Turing mechanism we showed what kind of neural architecture would spontaneously give rise to these patterns and showed that is was consistent with the neuroanatomy that had been discovered by Hubel and Weisel and others going back to Sholl.[25] In recent years we've followed that up, working with Paul Bressloff, Martin Golubitsky, and some of my students, and we now have more detailed explanations.[26] We have a series of papers that will come out in due course that extend the model to cover hallucinations involving color, depth, and motion. We've extended the analysis to look at why people see themselves falling down tunnels with light at the end and so forth. We believe this work tells us quite a lot about what the architecture of the relevant parts of the brain must be like to generate these things. I was at a computational neuroscience and vision conference recently and I discovered that some of the techniques we have introduced in this work may be very relevant to computational vision, and that there may be some deep links between the field equations Wilson and I introduced and problems in vision such as color matching. So this is a new direction I am going to collaborate in.

PH: I wonder what your views are on the correct level of abstraction for brain modeling. There is an awful lot more known today about some of the low-level biochemical details, but still the higher-level overall picture is rather obscure.

JC: It's a very interesting question. We now have at Chicago Stephen Smale, who is a great mathematician—a Fields Medalist for his work on the Poincaré conjecture many years ago and many other honors—who has got interested in machine learning and vision recently. He's starting to work with a number of people in these areas and he has a very abstract way of thinking, but a very powerful way. There is a group of mathematicians who work on differential geometry and topology who are getting very interested in what goes on in the nervous system. There are many different levels of mathematical abstraction that can be applied to brain modeling. I think there are going to be rich developments over the coming decades in this area and we may see some rather different styles of modeling emerge than have been used to date.

PH: Historically, one of problems faced by theoretical work in neuroscience is indifference, or sometimes hostility, from the majority of those working in neuroscience. Do you see that changing?

JC: Well this is something I've had to struggle with for nearly fifty years, but I think it is changing. Most of the new young people coming through have a different attitude. Many are much better educated than their equivalents were even twenty-five years ago. I think more and more biologists will become at least open to mathematics whilst remaining very good empirical scientists. But the fact that experimental tools and methods have become more precise means that there is a lot more data that cries out for mathematical approaches. So I think attitudes are changing.

PH: If you put yourself back at the start of your career, right back to Ferranti, and try and remember your general expectations then, are you surprised at how far machine intelligence has come, or hasn't come?

JC: Well, something that has always surprised me is how many times ideas in this field are rediscovered by the next generation. For example I recently heard a very nice lecture from Tommy Poggio, who has been in the game a good while himself, on early vision. He used a mathematical device that actually had been invented in the 1950s by Wilfred Taylor at University College. Tommy wasn't aware of that. A lot of the ideas and machinery that is current now has actually been sitting in the field for a very long time. It's just that we haven't always seen the implications or how to use them properly. But am I surprised at how difficult it has turned out to do real machine intelligence? No, not at all. I always thought it would be much harder than the people in strong AI claimed. Now back in about 1966, Frank Schmitt, who ran the neuroscience research program at MIT, organized one of the first meetings on sensory coding and Shannon was at that meeting. I remember Shannon said something very interesting during the meeting. He said that he thought that while initially strong AI might make some interesting progress, in the long run bottom-up work on neural networks would prove to be much more powerful. He was one of the few people at MIT in 1958 who responded positively to a lecture Frank Rosenblatt gave on the perceptron. Most were extremely negative in their response to it, and I have to say it was a pretty bad lecture, with too many wild claims, but not Shannon. He said, "There could be something in this." I consider him to be amazingly perceptive, much more so than most others in the field. Him and McCulloch, Wiener and Pitts.

PH: What do you think are the most interesting developments in machine learning at the moment?

JC: Well there is some very interesting work on data mining that the mathematician Raphy Coifman at Yale and others have been involved in. If you have a very large database from which you want to extract information, you can get a big advantage if you can map the data space onto some lower-dimensional manifold in a systematic way. What they found was that simple things like averaging operators, smoothing Laplacian operators, and things connected to diffusion, are immensely powerful for doing that. In a strange way it's connected to what underlies the solution to the Poincaré conjecture because that involves the smoothing of manifolds, and smoothing plays a key role in this work on data mining. I've recently proposed that the resting state of the brain is Brownian motion, which is also closely related to that kind of operator. So I think there is something going on in the nervous system and something going on to enable machine learning that may be related and which will prove to be very interesting.

PH: Finally, is there a particular piece of your work that you are most proud of?

JC: Well, I think that the work I'm doing now with Michael Buice, which we discussed earlier, and which is the culmination of many years' work, is what I'm going to end up being most proud of. Even though I'm in my anecdotage, as they say, I like to look forward, and what I'm doing now I find is most interesting to me.

Notes

1. Raymond L. Beurle, "Properties of a Mass of Cells Capable of Regenerating Pulses," *Philosophical Transactions of the Royal Society of London* (series B) 240: 55–94(1956).

2. *Proceedings of the First International Congress on Cybernetics* (Namur, Belgium, June 26–29, 1956; Paris: Gauthier-Villars, Paris, 1958).

3. John Pringle, "On the Parallel Between Learning and Evolution," *Behaviour* 3: 174–215(1951); Alan M. Turing, "The Chemical Basis of Morphogenesis," *Philosophical Transactions of the Royal Society of London* (series B) 237: 37–72(1952).

4. John von Neumann, "Probabilistic Logics and the Synthesis of Reliable Organisms from Unreliable Components," in *Automata Studies*, edited by Claude E. Shannon and John McCarthy (Princeton: Princeton University Press, 1956).

5. Walter A. Rosenblith, ed., *Sensory Communication* (Cambridge, Mass.: MIT Press, 1961).

6. Horace B. Barlow, "Possible Principles Underlying the Transformations of Sensory Messages," in Rosenblith, *Sensory Communication*; Werner Reichardt, "Autocorrelation, a Principle for the Evaluation of Sensory Information by the Central Nervous System," in Rosenblith, *Sensory Communication*.

7. Jerry Y. Lettvin, H. R. Maturana, Warren S. McCulloch, and Walter H. Pitts, "Two Remarks on the Visual System of the Frog," in Rosenblith, *Sensory Communication*; Jerry Y. Lettvin, H. R. Maturana, Warren S. McCulloch, and Walter Pitts, "What the Frog's Eye Tells the Frog's Brain," *Proceedings of the IRE* 47: 1940–59(1959).

8. Oliver G. Selfridge, "Pandemonium: A Paradigm for Learning," in *The Mechanisation of Thought Processes*, volume 10, *National Physical Laboratory Symposia*, edited by D. Blake and Albert Uttley (London: Her Majesty's Stationery Office, London, 1959).

9. Oliver G. Selfridge, "Some Notes on the Theory of Flutter," *Archivos del Instituto de Cardiologia de Mexico* 18: 177(1948).

10. Shmuel Winograd and Jack D. Cowan, *Reliable Computation in the Presence of Noise* (Cambridge, Mass.: MIT Press, 1963).

11. Ibid.

12. Jack D. Cowan, "Statistical Mechanics of Nervous Nets," in *Proceedings of 1967 NATO Conference on Neural Networks*, edited by E. R. Caianiello (Springer-Verlag, 1968).

13. The sigmoid function is a differentiable nonlinear "squashing" function widely used as the transfer function in nodes in artificial neural networks (to compute node output from input). It turns out that this kind of function is necessary for various multilayered learning methods to work, including the back-propagation method (see note 20); the original perceptron used linear transfer functions.

14. John Hopfield, "Neural Networks and Physical Systems with Emergent Collective Computational Abilities," *Proceedings of the National Academy of Sciences* 79: 2554–58(1982).

15. Beurle, "Properties of a Mass of Cells Capable of Regenerating Pulses." See note 1.

16. Albert Uttley, "Conditional Probability Machines and Conditioned Reflexes," in Shannon and McCarthy, *Automata Studies*.

17. C. H. Waddington, ed., *Towards a Theoretical Biology*, volume 1: *Prolegomena* (Edinburgh: Edinburgh University Press, 1968). Several volumes in this series were produced.

18. Nicolas Rashevsky, a Russian physicist who arrived in the United States after various scrapes and near escapes during the civil war in his home country, set up

the Committee on Mathematical Biology at the University of Chicago in the 1930s. An influential pioneer in mathematical biology, he mentored many important theoretical biologists. He set up and edited *The Bulletin of Mathematical Biophysics*, which, among other notable works, published the pioneering papers by McCulloch and Pitts on neural networks.

19. Marvin Minsky and S. A. Papert, *Perceptrons* (Cambridge, Mass.: MIT Press, 1969). The perceptron, invented in 1957 by Frank Rosenblatt at Cornell University, is a simple linear single-layer feedforward artificial neural network.

20. D. E. Rumelhart, G. E. Hinton, and R. J. Williams, "Learning Representations by Back-Propagating Errors," *Nature* 323: 533–36(1986). This method for learning in multilayer networks, which overcame the limitations of perceptrons, had been independently described previously by Paul J. Werbos in 1974 in his Ph.D. thesis, "Beyond Regression: New Tools for Prediction and Analysis in the Behavioral Sciences" (Harvard University, 1974), and a similar method had been proposed by Shun-ichi Amari in "Theory of Adaptive Pattern Classifiers," *IEEE Transactions in Electronic Computers* EC-16: 299–307(1967).

21. See H. R. Wilson and Jack D. Cowan, "Excitatory and Inhibitory Interactions in Localized Populations of Model Neurons," *Biophysical Journal* 12: 1–24(1972), and H. R. Wilson and Jack D. Cowan, "A Mathematical Theory of the Functional Dynamics of Cortical and Thalamic Nervous Tissue," *Kybernetik* 13: 55–80(1973).

22. G. Bart Ermentrout and Jack D. Cowan, "A Mathematical Theory of Visual Hallucination Patterns," *Biological Cybernetics* 34: 137–50(1979).

23. Turing, "Chemical Basis of Morphogenesis." See note 3.

24. Heinrich Klüver, *Mescal and the Mechanisms of Hallucination* (Chicago: University of Chicago Press, 1966).

25. D. A. Sholl, *The Organization of the Nervous System* (New York: McGraw-Hill, 1956).

26. Paul Bressloff, Jack Cowan, Martin Golubitsky, P. Thomas, and M. Wiener, "Geometric Visual Hallucinations, Euclidean Symmetry and the Functional Architecture of Striate Cortex," *Philosophical Transactions of the Royal Society of London* (series B) 356: 299–330(2001).

About the Contributors

Peter Asaro has a Ph.D. in the history, philosophy, and sociology of science from the University of Illinios at Urbana-Champaign. He is a researcher in the Center for Cultural Analysis, Rutgers University.

Horace Barlow is Professor in the Department of Physiology, Development and Neuroscience, University of Cambridge.

Andy Beckett is a writer and journalist for *The Guardian* newspaper, London.

Margaret Boden is Research Professor of Cognitive Science at the Centre for Research in Cognitive Science, University of Sussex.

Jon Bird is a Research Fellow in the Centre for Computational Neuroscience and Robotics, University of Sussex.

Paul Brown is an Anglo-Australian artist and writer who has been specializing in art and technology for almost forty years. Examples of his artwork and publications are available on his website at http://www.paul-brown .com. He is currently visiting professor and artist-in-residence at the Centre for Computational Neuroscience and Robotics, University of Sussex.

Seth Bullock is Senior Lecturer in the School of Electronics and Computer Science, University of Southampton.

Roberto Cordeschi is Professor of the Philosophy of Science on the Philosophy Faculty of the University of Rome La Sapienza.

Jack Cowan is a Professor in the Mathematics Department, University of Chicago.

Ezequiel Di Paolo is Reader in Evolutionary and Adaptive Systems, Department of Informatics, University of Sussex.

Hubert Dreyfus is Professor of Philosophy in the Graduate School, University of California, Berkeley.

Andrew Hodges is Lecturer in Mathematics at Wadham College, University of Oxford. He maintains the website www.turing.org.uk.

John Holland is Professor of Psychology and Professor of Electrical Engineering and Computer Science, University of Michigan.

Owen Holland is Professor in the Department of Computer Science, University of Essex.

Jana Horáková is Assistant Professor in Theater and Interactive Media Studies, Masaryk University, Brno, Czech Republic.

Philip Husbands is Professor of Computer Science and Artificial Intelligence, Department of Informatics, University of Sussex, and Codirector of the Sussex Centre for Computational Neuroscience and Robotics.

Jozef Kelemen is Professor of Computer Science at the Silesian University, Opava, Czech Republic.

John Maynard Smith (1920–2004) was one of the great evolutionary biologists of the twentieth century. He was a professor at the University of Sussex.

Donald Michie (1923–2007) was Professor Emeritus of Machine Intelligence at the University of Edinburgh. In 2001 he received the IJCAI (International Joint Conferences on Artificial Intelligence) Award for Research Excellence. He was an important part of the British World War II Code-Cracking team at Bletchley Park.

Oliver Selfridge is associated with the MIT Media Lab, in Cambridge, Massachusetts, and also works at BBN Technologies.

Michael Wheeler is Reader in Philosophy in the Department of Philosophy, University of Stirling, Scotland.

Index

AARON, 275

ACE (automatic computing engine), 67, 135

Ackoff, R., 225–227

Adaptive teaching machines, 194–199

Adrian, Lord Edgar, 6–7, 94, 100, 109, 129, 411, 412–414, 415–417, 418, 421, 423, 426

Agre, P., 334, 335, 338, 339, 340, 341, 358, 362, 362n1

Alcohol, effect on control and communication, 120

ALGOL-60, 64

A-life. *See* Artificial Life

Al Jaziri, 260

Allende, S., 213, 214, 215, 216, 217

Analytical Engine. *See* Babbage, C.

Aquinas, T., 102

Aristotle, 286

Ars Elecronica, 277

Art, mechanization of, 259–283

Artificial Intelligence (AI), 20, 31, 32, 33, 35, 36, 65, 75, 86, 88, 97, 99, 108, 150, 169, 172, 180, 204, 219, 20, 221, 228, 236, 237, 244, 245, 246, 253, 283, 302, 393, 403–406

biologically inspired, 13, 205, 405

convergence with philosophy, 331–332

founding of, 12–13, 219–220, 389–390, 399, 436–437

good old-fashioned (GOFAI), 13–14, 222, 334, 335, 336, 341, 343, 348, 353, 362

Heideggerian, AI, 13–14, 334–349, 357–362

symbolic AI as degenerating research program, 332–334

Artificial Life, 1, 4, 7, 41, 94, 97, 180, 245, 275, 283

Ascot, R., 269, 273

Ashby, W. Ross, 108, 112, 149–182, 199, 202, 223, 224, 225, 244, 245, 274, 307, 376, 391, 393, 433

brief biography of, 94

correspondence with K. Craik, 109–110

correspondence with W. Hick, 110–111

correspondence with A. Turing, 135

on cybernetics in psychology, 153, 155

and designs for intelligence, 162–163

on equilibrium in adaptation, 150, 154–162

on intelligence amplification, 150, 168–177

on models and simulations, 161–162

and the problem of the mechanical chess player, 131, 163–168

on mechanisms of adaptation, 152–162

philosophy of, 11, 149–182

and Ratio Club, 10, 91, 93, 100, 113, 116, 117, 118, 119–121, 122, 124, 125, 126, 128, 129, 131, 132, 133–135, 141